PEOPLE OF THE

To Glen, James and Mark

PEOPLE
OF THE GREAT OCEAN

ASPECTS OF HUMAN BIOLOGY OF THE EARLY PACIFIC

Philip Houghton

University of Otago

CAMBRIDGE
UNIVERSITY PRESS

CAMBRIDGE UNIVERSITY PRESS
Cambridge, New York, Melbourne, Madrid, Cape Town, Singapore, São Paulo, Delhi

Cambridge University Press
The Edinburgh Building, Cambridge CB2 8RU, UK

Published in the United States of America by Cambridge University Press, New York

www.cambridge.org
Information on this title: www.cambridge.org/9780521119306

First published 1996
This digitally printed version 2009

A catalogue record for this publication is available from the British Library

National Library of Australia Cataloguing in Publication data

Houghton, Philip, 1937–.
People of the great ocean: aspects of human biology of
the early Pacific.
Bibliography.
Includes index.
1. Man, Prehistoric – Oceania. 2. Pacific Islanders –
Anthropometry. 3. Physical anthropology. I. Title.
572.899

Library of Congress Cataloguing in Publication data

Houghton, Philip.
People of the great ocean: aspects of human biology of the early
Pacific/Philip Houghton.
       p.   cm.
Includes bibliographical references and index.
1. Physical anthropology – Oceania. 2. Anthropometry – Oceania
3. Human remains (Archaeology) – Oceania. 4. Human settlements –
Oceania. 5. Oceania – History. I. Title.
GN58.034H68 1996   95–32622
573'.0995–dc20

ISBN 978-0-521-47166-4 hardback
ISBN 978-0-521-11930-6 paperback

CONTENTS

FIGURES

TABLES

ACKNOWLEDGEMENTS

Some of the content of this book derives from interpretation of data from human skeletal remains from the Pacific past. For many Pacific peoples the remains of ancestors are of great cultural and spiritual significance, and of recent years there has been, in places, as cultures reassert themselves, a reaction against study of such remains. Today anyone studying these has a duty of justification, and one aim of this book is to provide an attempt at a justification. It is for the descendants of the first Pacific peoples, ultimate custodians of ancestral remains, to consider works such as this and make their decisions as to whether such study is justified. What I acknowledge here, with gratitude, is the approval – sometimes relaxed, sometimes cautious, sometimes uneasy, but always courteous – that has been given in various places and at various times.

I have a particular debt to Martin Kean for our discussions over several years. The analysis of head form in Chapter 4 is some evidence of these, but from any one of them we would come away with enough thoughts and ideas to fill a year, if only we had a free year to fill. Daniel Levy is responsible for the mathematical aspects of the simulations of exposure in the Pacific environment. In earlier years the late Basil de Lambert generously provided x-ray facilities, and his interest and expertise. To start mentioning others would be invidious because so many people have provided help in many ways. Studies by students in various disciplines, and ranging from short assignments to doctoral theses, have contributed greatly to the data used in Chapter 6. These studies are referenced as appropriate, but their contribution needs special mention. I am indebted to Robin Harvey for allowing the use of the individual data from his New Guinea research, and to Peter Parsons for drawing my attention to his papers. John Miles allowed me to look at and cite a chapter of his forthcoming book on infectious diseases in the pre-European Pacific. I am grateful to Robbie McPhee for his assistance with some of the figures. Blair Fitzharris, Richard Walter and Boyd Swinburn have kindly read and commented on some sections, but of course cannot be held responsible for the errors that remain.

Several authors and publishers have allowed use of copyright material, as referenced through the text, and I gratefully acknowledge the permission of the following for extracts from these works: Peter Bellwood, 1985, On Polynesians and Melanesians, *Journal of the Polynesian Society* 95: 131–134; Tsunehiko Hanihara, 1992, Dental Variation of the Polynesian Populations, *Journal of the Anthropological Society of Nippon* 100(3): 291–302; Daris Swindler, *A Racial Study of the West Nakanai*; Penguin Books, J. E. Gordon, *The New Science of Strong Materials*, and J. E. Gordon, *Structures*; University of Hawaii Press, D. L. Oliver, *Oceania*; Oxford University Press, A. V. S. Hill and S. W. Serjeantson, *The Colonization of the Pacific: A Genetic Trail*,

J. S. Friedlaender, *The Solomon Islands Project* and J. S. Weiner and J. Huizinga, *The Assessment of Population Affinities in Man*; Alan R. Liss, S. R. Johansson and S. Horowitz, and C. L. Brace and K. D. Hunt, *American Journal of Physical Anthropology*; The Royal Anthropological Society, R. A. Fisher, *Journal of the Royal Anthropological Institute*; Harvard University Press, R. Lewontin, in J. S. Friedlaender, *Patterns of Human Variation*; Academic Press, M. Pietrusewsky, *Journal of Human Evolution*; Kent State University Press, R. Feinberg, *Polynesian Seafaring and Navigation*; *Ocean Travel in Anutan Culture and Society*; Peabody Museum Press, E. Giles and J. Friedlaender, *The Measures of Man*.

INTRODUCTION

For Pacific people in the more distant past it was limitless, foreground and background to the world, and probably needed no specific name. Later, in some Polynesian languages, it became a variant of *Te Moana Nui a Kiwa* – the Great Ocean of Kiwa. The European name for this largest of oceans tends to evoke images of white beaches and palm trees, coral reefs and trade winds, sunlight and sea, images of tranquillity as deceptive as the name itself. The first explorers of the Pacific and their descendants had the less romantic job of surviving in this unique oceanic world, something that required both cultural and biological adaptation. It is the human biology of the first Pacific people and the underlying theme of adaptation that occupy this book. Adaptation suggests evolution, yet there is a tendency to assume *Homo sapiens* has been exempt from evolutionary influences, at least in the short term, which is sometimes taken to mean tens of thousands of years. Probably this is because from our beginnings we have been the technological animal and until recently an ever-increasing control of the environment has made it easy for us to assume some independence of it and to believe that it has not shaped us to the extent that it has other organisms. However we are not exempt, and the biology of any human group needs to be considered in the light of its environment and adaptation to it.

The first chapter therefore outlines the nature of the Pacific environment and our present understanding of the sequence of settlement. While this can only be a glance across a vast field of study it is enough to give a background to the human biology. In later chapters some of these environmental aspects are considered more closely.

Chapter 2 considers the physique of Pacific peoples in general terms of height and weight, drawing on several sources of evidence: the historical record, the skeletal record from the past, and the record of measurements on the living. From these a picture of physique in the past and its variation across the Pacific is drawn.

The next chapter examines this variation in physique against a model of adaptation to a changing environment. Such a model relates to long-established biogeographic principles and considers in some detail the physiology of the morphological adaptations, quantifying the interrelationship of form and function. Included in this analysis is a computer simulation of human survival at sea using data from Pacific meteorological records. Alternative explanations for variation in physique are examined, particularly variation in nutrition and in patterns of disease.

Chapter 4 considers the skeletal record from the past, particularly that from the wider reaches of the Pacific. The distinctive morphology is described and interpreted in functional terms, again relating to the adaptation required to meet a changing environment.

Various models and views on human settlement of the Pacific and the evidence supporting these views are discussed at some length in Chapter 5. Initial discussion is of a linguistic model for Pacific settlement, against which much biological and quasibiological evidence has been interpreted. Racial typologies, and the phenotypic evidence of cranial and dental data and dermatoglyphics are discussed, and then the genetic data of classical gene frequencies and DNA analyses. Using some of these data, possible models for Pacific settlement other than linguistic ones are then considered. Explanation for the human variety is here sought in terms of population variation as discussed, for example, by Mayr (1970) and Parsons (1991, 1994). Finally there is some further consideration of the results of the computer simulation of survival at sea and possible implications for Pacific settlement.

Chapter 6 is an examination of health and disease in the Pacific past as a mirror of the conditions of the environment and human success in coming to terms with it. General evidence of the state of health of groups in the past, including aspects of life span and population growth on small islands is first discussed, and then specific evidence of disease.

The book closes with a consideration in Chapter 7 of possible relationships of some health problems in contemporary Pacific peoples to the evolutionary influences of the past. There is a final statement viewing variation in Pacific human populations as an expression of principles established in wider fields of biology.

A comment is needed on the human evidence, which is not well balanced. Polynesia past and present is quite well represented. Here, explorers and missionaries, archaeologists and anthropologists, doctors and scientists, have provided a fair range of records. For Melanesia the record from all these sources is patchier, and the prehistoric record in particular is rather thin. As in terms of human presence this is the oldest and in some ways the most interesting region, the poverty of the record is unfortunate. Still, it grows, and the gaps are partly compensated for by some thorough recent surveys in human biology (Friedlaender 1975, 1987a, b, Harvey 1974, Hornabrook 1977a). For Micronesia the prehistoric record is rather scanty, and because of drastic population change within historic time more recent studies cannot atone. So the picture is patchy, and is further influenced by my particular interests and limitations. The bias is towards the wider Pacific, that which is defined later as Remote Oceania, for this is the Great Ocean proper, and here are seen the greatest demands on the human frame. Hence the subtitle, *aspects* of human biology.

THE PACIFIC WORLD

1

Stretching from the Americas to Asia and washing Antarctica in the south, the Pacific, including such arms as the Philippine, Coral and Tasman Seas, which are part of it in all but name, covers about one-quarter of the globe. The equator traverses the ocean, and about half of it lies within the tropics. On such an expanse of water the influences of the sun and the spin of the earth are unimpeded. Within a band straddling the equator the heated tropical air rises, creating a sultry region of calms or light winds, shifting a few degrees north or south with the seasons. This equatorial band, the doldrums, is typically some 250 kilometres wide, and east of 160° west longitude lies permanently north of the equator (Figure 1.1). The rising equatorial air flows north and south and descends near the margins of the tropics to be drawn towards the equator again as a continually circulating 'cell' of air – the classical Hadley cell of climatology (Figure 1.2). The Coriolis force created by the west-to-east rotation of the earth deflects wind flow to the left in the southern hemisphere and to the right in the northern hemisphere. Thus are created the north-east and south-east trade winds. These dominate the tropical Pacific environment – winds from the east are recorded for over 80% of the time for many islands – as well as much of its romantic image. However there are periods when the trades do drop away and for days on end westerly winds dominate. This has implications for exploration. The western Pacific is also a region of hurricanes, intense winds spiralling round a low pressure centre. In the south Pacific these may extend as far as 160° west longitude. For people at sea the risk is clear, but it would have been scarcely less for small groups establishing themselves on small islands. At higher latitudes, beyond the trades, is a pattern of westerly winds that is particularly marked in the unobstructed reaches of the south Pacific, the Roaring Forties.

In the western Pacific, beyond about 150° east longitude, the influence of the land masses overrides to some extent the trade wind pattern.

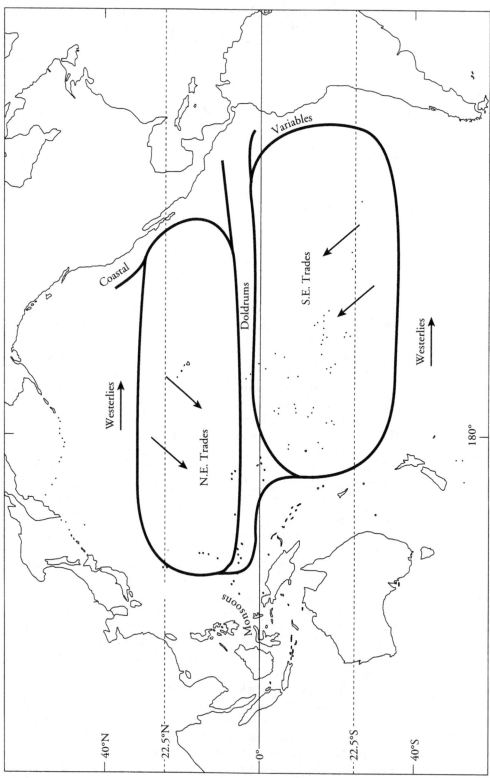

Figure 1.1 The general pattern of winds in the Pacific. The Coriolis force created by the rotation of the earth turns the strong equatorial flow into the south-east and north-east trade winds. Between 30° and 60° latitudes a westerly flow dominates.

Figure 1.2
Schematic figure of
the pattern of air
circulation above
the tropical Pacific.

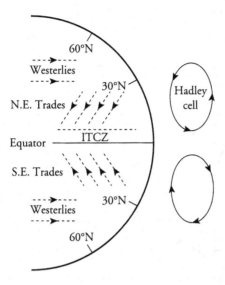

Here in the southern summer the north-west monsoon out of Asia reaches as far as the Solomon Islands.

The South-east Trade Wind and the North-west Monsoon carry on a continual struggle for mastery in these islands. However for two-thirds of the year the Trade prevails, viz., from April to November . . . the north-west and west winds set in about the end of November or the beginning of December, and prevail until the end of March . . . The period of the westerly winds in the Solomon Islands is also characterised by calms and variable winds. The exhilarating freshness of the Trades then gives place to the enervating influences of the Monsoon; and in consequence the period of westerly winds is the sickly season. (Guppy 1887: 362–363)

The periods of balance and change between these land and sea winds expand the belt of doldrums in the western Pacific, but the wind over any part of the ocean at any season has never been insignificant for human existence, and wind speed has been at least as influential as actual direction. Figure 1.3 shows – again, in the most general terms – mean wind speeds in the Pacific in the southern winter.

This pattern of winds – trades and westerlies – establishes a matching pattern of surface ocean currents (Figure 1.4). Between about 20° north and 10° south latitudes is a persistent east-to-west movement of water at about 25 km/day. Splitting this westerly current (it is a nautical quirk that winds are named by the direction from which they come, and currents by the direction in which they pass) a narrow band of water, the equatorial counter-current, passes in the opposite direction at a considerably faster clip of about 70 km/day. The main west-running streams diverge in the west. In the northern hemisphere the circulation of these waters is clockwise, north towards Japan, and then east to North America. In this course they mix with colder water coming through Bering Strait from the Arctic, and the consequent current down the west coast of North America gives rather chilly bathing as far south as the northern

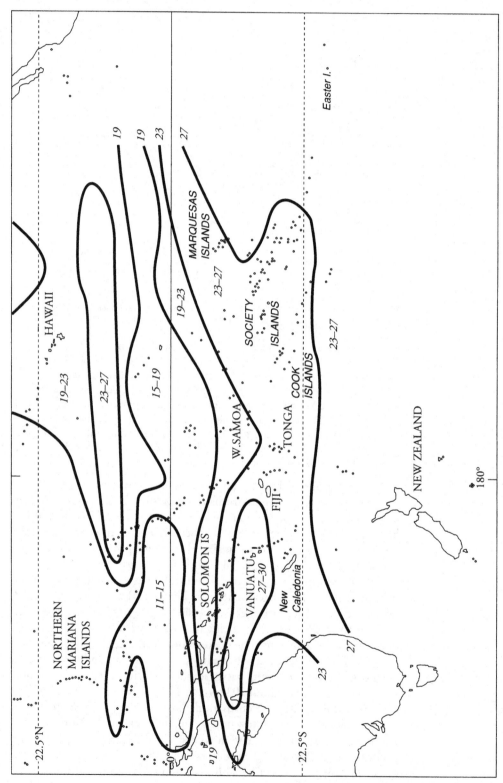

Figure 1.3 Mean wind speeds (km/hour) in the Pacific in the southern winter.

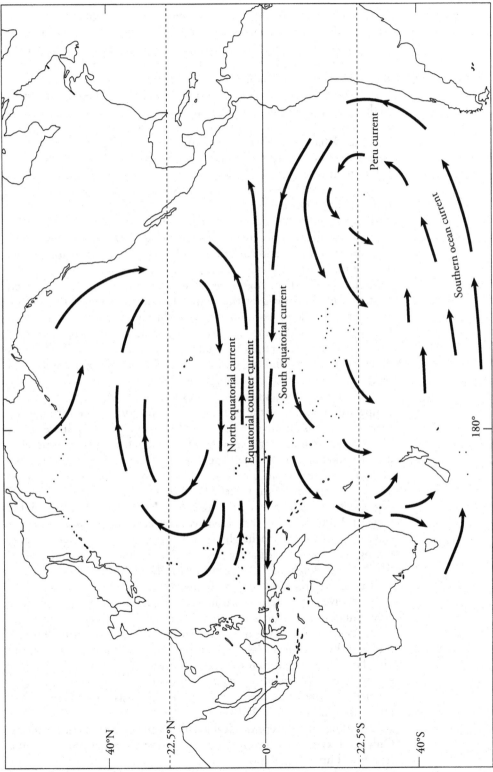

Figure 1.4 The general pattern of surface ocean currents in the Pacific.

Californian coast. In the southern hemisphere the equatorial stream turns south in an anti-clockwise course down the east coast of Australia to latitudes 30°–40° before turning eastward with the prevailing winds. New Zealand lies athwart this easterly current and thus, though well south, receives a reasonably warm wrapping of sub-tropical water.

In the far south the cold, food-rich Humboldt or Peru current rises to the surface to pass along the South American coast. The origins of the Humboldt are said to lie in a major deep-sea flow from the north Atlantic. Whatever the basis, the currents entering the tropical eastern Pacific from both north and south are rather cool, and this is reflected in the isotherms (Figure 1.5). With westward passage of the equatorial current the waters become warmer; all significant island groups of the tropical Pacific lie between the 21°C isotherms, and most groups west of 160° west longitude lie between the 27°C isotherms.

From time to time (the intervals being roughly four to seven years) this pattern of winds and currents is broken by the phenomenon known as *El Niño*. There are changes in atmospheric pressure, with a rise in the western Pacific and a fall in the east. Then the trade winds are weaker and westerlies may extend over the central Pacific for some of the season. Warmer water flows east and the cold Humboldt current pushes less firmly up the South American coast. Nowadays, the *El Niño* phenomenon is well studied, rather well understood in its pattern if not in its underlying causes, and somewhat predictable in its appearance and severity (Barnett *et al.* 1988, Ramage 1986).

In normal years, air temperatures near sea level closely match sea temperatures, and near the equator the temperature range through the year at sea level is minimal. Away from the equator the range increases. It is 5–6°C in Noumea and in Honolulu, enough to give a noticeably cooler winter season at these limits of the tropics. There is also temperature change with altitude, even in islands quite close to the equator. The Baining people of the Gazelle Peninsula of New Britain, within five degrees of the equator, live in 'rugged limestone country . . . at heights between 1300 and 3400 feet above sea level situated between mountain ridges . . . Malasait at a height of 1300 feet above sea-level is warm during the day and pleasantly cool at night, while Yalom and Komgi which are at a height of 3400 feet are relatively cool during the day as well as at night' (Kariks and Walsh 1968: 130).

Relatively is a useful word here. In the even 'cooler' highlands of New Guinea, at altitudes between 1300 and 2000 metres, the frost-tender sweet potato is the dominant crop, and heat stress is the human problem (Budd *et al.* 1974). What is perceived as cold is relative: in the Highlands Littlewood notes, 'As the temperature drops rapidly in the evening the people standing about in the open, particularly the children, characteristically cross their arms over their chests, shiver, and show other signs of chill' (1972: 14).

Through the eastern tropical Pacific altitude is not of consequence. Only the Hawaiian Islands reach up to the snowline, and this was not a region of human habitation.

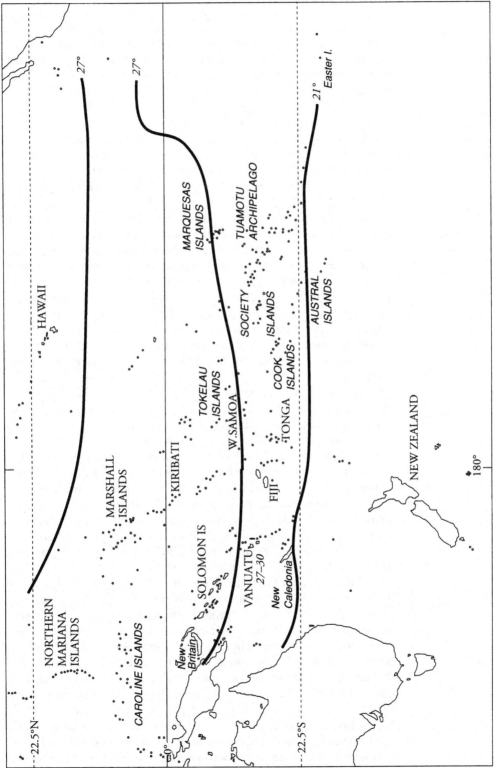

Figure 1.5 Water isotherms (°C) in the Pacific in the southern winter. The equator and the limits of the tropics are shown.

In this watery expanse, the land, in terms of area, is simply trivial. East of New Guinea a diminishing chain of intervisible islands ends at San Cristobal at the limit of the Solomons. Several of these islands are of fair size; New Britain, New Ireland and Bougainville all exceed 8000 square kilometres. Eastward, the islands are much smaller. In the Remote Pacific the thousand or so islets of the Marshall group have a combined land area of less than 180 square kilometres. The entire Tuamotu archipelago totals about 850 square kilometres. New Zealand, though of very large size by Pacific standards (260,000 square kilometres), is (one is reluctant to say) rather marginal to the unfolding of the human biological story; distinctive but peripheral.

Despite insignificance of size there is variety of island origin and structure in the Pacific, much of it a result of the vagaries of plate tectonics and continental drift. The Pacific plate occupies the greatest area, but in the west sinks below the Indian–Australian plate in a long and active boundary that passes from New Zealand north to Tonga, thence in an arc westward through Vanuatu, the Solomons and Island Melanesia, before turning north to the Marianas and Japan (Figure 1.6). The geochemical tumult created by the annihilation of the descending crust in the underlying magma is marked by ash eruptions that have in part formed the islands. Some episodes have been particularly violent. Major explosions, the last about 2000 years ago, created Taupo, the largest lake in New Zealand. In 1937 the craters of Rabaul in New Britain exploded and spread ash in notable sunsets around the globe (the 1994 eruptions were less violent). Mount Pinatuba in the Philippines did the same more recently. Lesser activity grumbles on in many of these boundary islands as an enduring sign of their origins.

On the plate boundary (the 'andesite line') and westward lie the relatively large islands of the Pacific, 'continental' in their geology. This is a mix of ancient metamorphic rocks, sediments, and volcanic rocks, markedly deformed by folding and faulting. The result is a dissected landscape, rather well designed to separate colonizing groups of *Homo sapiens* despite the relatively small land areas. This instability and activity along the western margin of the Pacific plate suggests young land, and even the largest, New Zealand, has looked something like its present shape only for the last three million years or so.

Though many of the islands east of the Pacific/Indian–Australian plate boundary are also fundamentally volcanic, their mode of formation has been different. Unlike the geologically complex islands on and west of the plate boundary, these intra-plate volcanic islands are of a more simple and stable, basaltic origin. Lava spills in a relatively peaceful way – at least compared with the activity of Rabaul or Pinatuba – from Kilauea on Hawaii. One theory is that these islands have been successively created as the Pacific plate passed, infinitesimally slowly, in a north-westerly direction, over 'hot spots' in the underlying mantle (Figure 1.6). This has created chains of islands, of which the most north-westerly in any chain is the oldest, and volcanically most quiescent or extinct (and often submerged), while the most south-easterly island is the youngest and

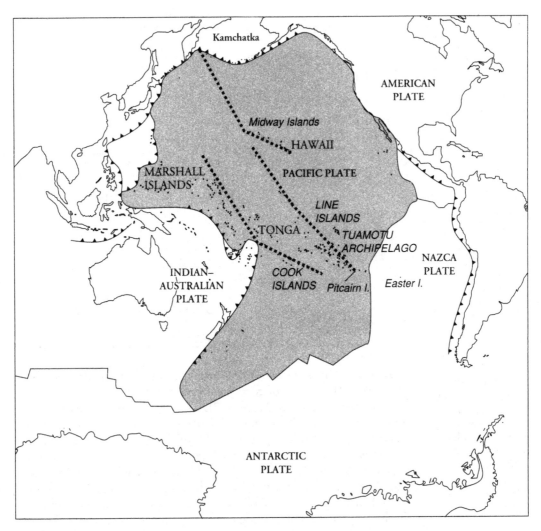

Figure 1.6
Tectonic plates and basaltic volcanic chains in the Pacific. The subsiding edges of plates are indicated by toothed lines. The three basaltic island chains within the Pacific plate are indicated by interrupted lines. Adapted from Stevens 1980.

volcanically most active. Three such chains make up a substantial number of the central Pacific islands. One is the Hawaiian chain, with active Kilauea on Hawaii in the south-east, perhaps only two or three million years old, extending in a series of increasing dormancy or extinction as a now-submarine chain of former volcanoes towards the north-west margin of the plate thousands of miles away, near the Kamchatka peninsula of Siberia. Similar patterns are traceable from Pitcairn Island (active within the last million years), through the Tuamotus and the Line Islands; and from the Austral to the Marshall Islands.

In this island-making the sea has played its part, and coral is as much part of the Pacific image as are the trade winds. At island margins, whether basaltic or andesitic, the coral polyp thrives within a fairly precise environmental range, as deep as 25 metres and in waters above about 19°C. The contribution of coral and its associated algae to the

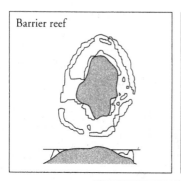

Barrier reef

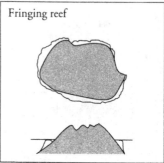

Fringing reef

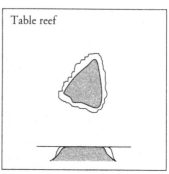

Table reef

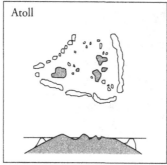

Atoll

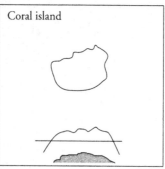

Coral island

Figure 1.7
Varieties of coral island. The inorganic land is stippled and the coral is unshaded. Sea level is indicated in the sections by the horizontal line. Adapted from Thomas 1963.

substance of any island today depends much on the history of the basic land mass. After a reef of coral has formed around an island there are several possible changes (Figure 1.7). The island core may start to disappear, as a result either of gradual descent into the depths, if the island lies towards the subsiding western edge of the Pacific plate, or of rising sea levels with climatic change. Sea levels have been lower in the past, as much as 100 metres lower in the late Pleistocene, around 16,000 years ago. The corals' response to rising sea level has been simply to grow and keep their heads above water, or at least somewhere close to the surface. In this way varieties of fringing reef and atoll have been created. Debris from the reef gradually builds up an island soil, but of a poor kind, for a long time lacking nutrients and humus.

Particularly to the west of the Pacific plate tectonic uplift of an island may have occurred. Uplifted coral will die but the skeletons remain as coralline limestone. Such uplift is obvious on the coasts of many islands. The Huon Peninsula of eastern New Guinea, for example, displays a flight of raised coral terraces that reflect changes over some 70,000 years. Guam shows an uplifted limestone coastline rising to more than 150 metres in parts, while Nauru is an example of a raised atoll.

With passage eastward across the Pacific the geology thus becomes simpler, the complex continental structure of the larger islands giving way to basalt, perhaps with coralline limestone, or coral alone. The biology also becomes simpler, at least in terms of species number. Five hundred and twenty species of bird are known from New Guinea, 54 from Fiji and 11 from the Marquesas, and similar clines are found for insects, reptiles, seed plants, and even reef fish. To some extent the decline in biota is spurious, as the land areas themselves are much less; but even allowing for this it is accepted that a genuine impoverishment does occur with movement east. Importantly, large-ish land animals, except those which have been domesticated, disappear – indeed most are left behind at Wallace's line, the deep-water biological boundary that bisects Indonesia. The implications of reduction in land size and biota for the food resources of neolithic human groups are clear – with passage eastward there is an increasing emphasis on the sea as a larder. It is better put thus than to say that land crops become unimportant, for that seems seldom to have been the case. In much of coastal Melanesia, even with the sea at hand, agriculture remains dominant. Thus:

The Trobriander is above all a cultivator . . . Fishing comes next in importance. In some villages situated on the lagoon it is the main source of sustenance and claims about half of their time and labour. But while fishing is prominent in some districts, agriculture is paramount in all. Were fishing made impossible . . . the population as a whole would find enough sustenance from agriculture. But when gardens fail in time of drought, famine inevitably sets in. (Malinowski 1935: 8)

Further east, on Treasury Island in the Bougainville Straits,

acres and acres of taro and banana plantations lie in the immediate vicinity of the village; and I passed through similarly cultivated tracts in the east and west islands of the district . . . When crossing the eastern part of the island of Morgusaia, I traversed for nearly a mile one continuous tract of cultivation. In the midst of the taro and banana plantations stood groves of the stately sago palm and clumps of the betel-nut palm . . . The diet of these islanders is essentially a vegetable one . . . Yams, sweet potatoes, two kinds of taro, cocoa-nuts, plantains, and sugar cane form the staple substances of their diet. (Guppy 1887: 81, 84)

Even on such a small and isolated island as Bellona, south of the main Solomon chain, the most important foods are yams, taro, bananas and coconut, and 'during a full year the average Bellona man is estimated to fish only about 45 times, using up to eight hours on each occasion' (Christiansen 1975: 30).

The major food plants of the Pacific fall into two groups; trees, semi-cultivated at best (at least before the copra industry), and root crops that usually require much care. Of the trees, the coconut palm (*Cocos nucifera*) tends to be concentrated where it has been planted by people. Sensitive to temperature, and not surviving below 25°C, it is very much a plant of the tropical strand – and part of the Pacific image. Its uses stretch far beyond food, for the various parts of the tree, fibre and leaf, were central to Oceanic technology. By contrast the sago palm (*Metroxylon* spp.), distributed from New Guinea as far into the Pacific as Samoa and the Caroline Islands, although a staple in parts of its western distribution is usually only a famine food in the east. Breadfruit (*Arctocarpus* spp.) and the banana are distributed throughout the Pacific, and various species of pandanus provide nuts for the diet.

Usually more important for food than these tree products are the root crops. Various types of taro (*Colocasia* spp.), swamp taro (*Cyrtosperma*) and yam (*Discoria* spp.) formed the staple of most diets. The sweet potato (*Ipomoea batatas*), which became an important food in some parts of the Pacific, seems to have been a late introduction, entering Polynesia from South America perhaps about A.D. 500. Work and skill were needed to encourage some of these root crops to thrive, particularly in the more difficult eastern reaches of the Pacific. From the archaeological record Kirch (1984) suggests that for the first few generations after settlement of smaller islands a higher proportion of food came from the sea, and only gradually did land crops start to dominate, presumably as propagation extended and soils improved.

Supplementing these plant resources, and probably always rather a luxury, were the few domesticated animals – pigs, chickens and dogs, of which only the last got through to New Zealand. Along with these was the rat, which, if not domesticated, seems to have been a pretty ubiquitous stowaway.

Supply of fresh water is sometimes a problem in this oceanic world. The equatorial region of convergence and ascent of the trade winds, while popularly known as the doldrums, goes under the more technical name of the Intertropical Convergence Zone (ITCZ). As the trade winds approach this region they become moister and warmer, and on rising create the only major rainforming mechanism over much of the tropical Pacific. Near the equator the ITCZ is always fairly close, so there are not likely to be dry periods. Away from the equator rainfall is concentrated in summer when the influence of the ITCZ is felt, and the further from the equator one progresses, the less the rainfall. On high islands there may be a dramatic difference between the wet, cloudy windward coasts and the relatively cloudless and dry leeward coasts. Such contrasts are clearly seen on, for example, the island of Hawaii, and on Viti Levu in Fiji.

On pure coral islands the sole source of fresh water is rain, which soaks quickly through the porous limestone leaving no significant ponds or streams. But the fresh water, being marginally less dense than the sea, accumulates in a sort of biconvex reservoir in the undersea stratum of the island, extending about a metre above sea level and fluctuating with the tides. The situation is not unique to such islands, being also found on continental fringes of similar structure, but the implications for biological survival are much greater on an island. As the fresh water gradually leaks into the surrounding ocean, constant replenishment by rain is necessary, and a prolonged dry season may be disastrous. The smaller the island the greater the leakage, and it seems that an island less than about one hundred metres in diameter is unable to sustain a fresh-water lens adequate for land life. The lens water is available naturally for deep-rooted plants, or, with cultivation, for shallow-rooted food crops. On an uplifted coral island the water table is too far down for most plants, and such places – for example Niue – have a characteristically meagre vegetation, and are really rather sterile places in biological terms.

Human settlement of the western fringes of the Pacific occurred a long time ago. Leaving aside the *Homo erectus* remains from Java and their debatable relationship to *Homo sapiens* – an argument outside the needs of this discussion – there is evidence for human settlement (but not accompanying human remains) on the Huon Peninsula of New Guinea 40,000 years ago. High on the uplifted coral terraces, and at a level dated to 45–54,000 years b.p. (before present), stone tools, including flakes, cores and waisted blades, have been recovered (Groube *et al.* 1986). Such uplift has occurred at the eastern end of most of the larger Melanesian islands, but unfortunately has tended to to be accompanied by a subsidence at the western ends. This, along with rise in sea level in the past 16,000 years, is likely to have drowned much evidence of early human existence. However the record grows. Inland sites in west New Britain have produced dates of 35,000 b.p. (Pavlides and Gosden 1994). A coastal midden site on the east coast of New Ireland has produced a date of 32,000 b.p., while a clutch of others from the same coast gives dates between 8000 and 20,000 years (Allen, Gosden and White 1989). The locations of these sites are shown in Figure 1.8.

These early sites tell of transportation of material by humans. The evidence for this comes from elemental analysis of obsidian flakes; any source of this black volcanic glass, useful for its cutting edges, has its own distinctive spectrum of trace elements, a reflection of the particular mix present in the formative eruption. Obsidian from Talasea on the northern coast of New Britain dominates in the early material; but after about 5000 b.p. obsidian appears from Lou in the Admiralty group of islands that form the northern perimeter of the Bismarck Sea. A trip between New Ireland and Lou involves a voyage of over 200 kilometres, and a large part of it out of sight of land.

East of these sites at the margin of the Bismarck Archipelago there is at present no evidence of human existence before about 3500 years ago, though the antiquity of settlement in north Bougainville suggests that earlier sites may yet be found further east. (On the other hand there is growing evidence for quite marked cooling of the waters of the tropical Pacific in the late Pleistocene (Anderson and Webb 1994) and thus a generally cooler climate at sea level, and this may have restrained the movement of people.) In the eastern Solomons the earliest of these much more recent sites are associated with the cultural complex known as Lapita. The name is taken from the pottery associated with the culture, which in turn derives from the village of Lapita in New Caledonia where the pottery was first identified in 1917. In its more florid form it has a distinctive dentate design. About half a century passed before, with the burgeoning of archaeological work in the west-central Pacific, the significance of its distribution was realized. The Lapita culture, on its pottery evidence, was spread in a broad arc across the west-central Pacific, from the Bismarck Archipelago to Samoa. This is not the place nor have I the competence to comment much further on Lapita. The pottery first appears about 3500 years ago, and, in the context of the broad limits of archaeological dating, spread very rapidly across its range of distribution. However the probability is that the western dates are

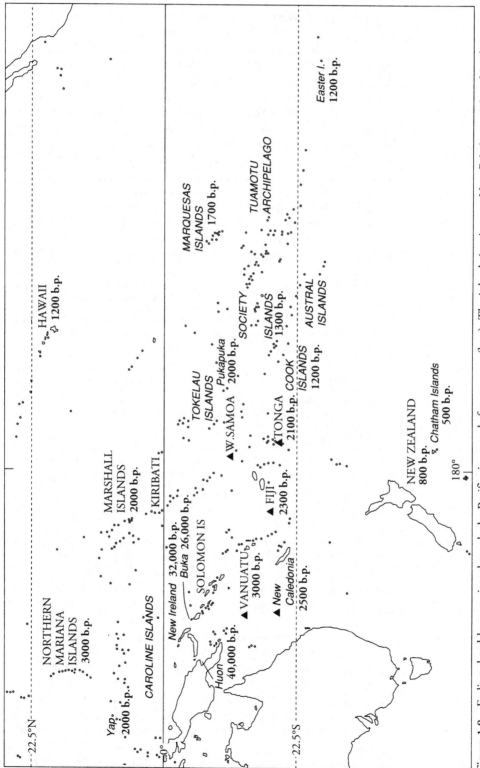

Figure 1.8 Earliest dated human sites through the Pacific, in years before present (b.p.). The inland sites in west New Britain, mentioned in the text and dated to 35,000 b.p., are not indicated. Places where Lapita or related pottery types are associated with earliest settlement are indicated by ▲. The eastern Polynesian dates are those suggested by Spriggs and Anderson 1993.

somewhat earlier and therefore likely to indicate the origins of the culture. Sites and dates are shown in Figure 1.8. The accumulating evidence for much earlier human movement in the Bismarck Sea region has tended, if not to diminish the significance of the more obvious Lapita culture, at least to require it to be set in a larger frame. Debate continues on the origins of the Lapita culture: whether it was a regional development, arising within what must already have been a long-established and sophisticated group of cultures, or whether it was introduced from the west, from or through Island South-East Asia. We will touch on these matters later. But in the east the carriers of the Lapita culture are equated with the immediate ancestors of the Polynesians.

The development of an oceanic technology of canoe construction and navigation method was clearly crucial to the settlement of the wider Pacific. There has been a lot of debate, some of it acrimonious, about the control these neolithic voyagers had over their wanderings. The idea that the settlement of the Pacific was a result largely of chance, of drift voyages ending further and further afield and happening by good fortune to carry the requirements for colonization, was largely dispelled by a computer simulation study by Levison, Ward and Webb (1973), which suggested that few islands of the wider Pacific could have been reached by drift voyages from the west. However the view that the settlement of the Pacific had the elements of a deliberate and competent search has had its opposition. One of the more fervent and thoughtful opponents in recent times was Andrew Sharp, who has probably been rather misread (or not read) and misunderstood. Sharp was particularly impressed by the inability of neolithic navigators (and many more recent, it must be said) accurately to determine longitude, or lateral displacement – at least in terms that the Western mind could comprehend. This led him to insist that

The essential feature of these voyages was that they were random, unnavigated ones. This does not mean that the voyagers lacked control of their vessels or were at the mercy of the winds and waves. When they came in sight of land they could make for it. This was exploration, although the actual sighting of land, like all discovery, came of necessity by chance. (Sharp 1963: 71)

The crucial word here is 'unnavigated'. It was the idea that neolithic voyagers could have been much in control over what they were about on the Great Ocean that seems to have been anathema to Sharp. Some other major contributors to our understanding of oceanic travel, themselves versed in the sea, have been more persuaded of an ability to explore in a controlled way and return home safely, of a 'search and return' strategy. Finney and his colleagues (Finney 1979, Finney *et al.* 1986, 1989) have been involved with the practicalities of sailing a large canoe, including its ability to push into eastern Polynesia by taking advantage of episodic lulls in the trade winds and the resulting periods of westerlies.

Lewis has emphasized signs whereby neolithic navigators could readily have inferred the presence of land, and has captured the perplexity of a recent Tikopian navigator when asked what he would do if he missed his home landfall: 'I know the way my island is . . . It is my island. It is where I follow the stars where to go – I cannot miss my island' (Lewis 1972: 232). The case for the ability to return to a known latitude and run along it to home has been lucidly set out by Irwin (1992). Of all these things, the only matter Sharp jibed at was the idea that pretty consistently the neolithic voyagers knew where they were in relation to home and were capable of finding their way back, whether or not new land was found. Ironically, contained within his own writing is something of the technique that Irwin deals with in detail: 'On west-east and east-west courses, the voyagers could gauge their broad direction at night by keeping their vessels in line with the east-west path of stars that passed overhead in that latitude' (Sharp 1963: 48).

This strategy of keeping to a latitude until the right longitude was reached was the same as that employed by the Dutch in their voyages from the Cape of Good Hope to Australia, thence north to the Indies. As Irwin (1992) points out, the error inherent in what is a rather crude estimate of latitude is rather neatly compensated for by the range over which the presence of land may be detected.

Scepticism has been expressed as to the ability of neolithic craft to hold together for any length of time in a seaway or when making to windward. Again from Sharp: 'Stone Age outrigger canoes . . . were tied together with vegetable fibres, and therefore were more vulnerable to stress than modern sailing craft, or than modern outrigger canoes shaped with iron tools and fastened together with bolts or nails' (1963: 56).

But the very matters that Sharp perceived as defects were in fact strengths. A modern multihull exposed to the racking strains of the sea has a tendency to fall apart at the rigid joints between the hulls. An engineer's view is:

In pure strength, apart from their flexibility, the lashings, sewings and bindings used by primitive peoples, and by seamen down to recent times, are more efficient than metal fastenings . . . Wood screws, beloved of amateur carpenters and boatbuilders, are the least efficient of all joints. (Gordon 1976: 154)

And,

the quality of being able to store strain energy and deflect elastically under load without breaking is called 'resilience', and is a very valuable characteristic in a structure . . . Since the invention of Fibreglass and other artificial composite materials we have been returning at times to the sort of fibrous nonmetallic structures which were developed by the Polynesians and the Eskimos. As a result we have become more aware of our own inadequacies in visualizing stress systems and, just possibly, more respectful of primitive technologies. (Gordon 1978: 90, 21)

Others, such as Heyerdahl (1952a), have felt that the wind patterns of the Pacific prohibited settlement from west to east, and that any approach must have been downwind, from the Americas. Irwin (1992) has discussed the risks of the downwind approach, which makes return home more difficult for a canoe and impossible for a raft. The Humboldt current generally makes it not feasible simply to push off for the wide

Pacific in a raft: the resources of the Peruvian navy were not available in prehistory. While the Humboldt does fluctuate, the Galapagos or Panama were more likely drift destinations for a raft leaving the South American coast. However all this is an old and well-worn controversy on which I do not wish to spend time. It is taken here that the Pacific was settled predominantly from the west by people who knew what they were about. That is what the archaeological and linguistic evidence says. The biological evidence is to be considered.

The conceptual view of their unique world that might have been held by different groups of Pacific islanders in the past is outside this discussion. But in practical terms, after settlement of a region people obviously maintained at least a knowledge of islands with which they continued to have voyaging contact, and, for a time, perhaps much more. The Society Islanders are said to have possessed, at the time of European contact, knowledge of every main group of the Polynesian Pacific with the exception of New Zealand, the Gambiers (surprising) and Hawaii. The Tongans, by contrast, seem to have had a wide knowledge of western Polynesia but to have known nothing, in geographic terms, of east Polynesia (Dening 1972) – but of course they had never been there. Island groups with which contact was lost presumably became more and more shadowy places in tradition, though, as with the Hawaiiki of the New Zealanders, often retaining spiritual significance (Orbell 1975).

The scientific, classifying Europeans changed all this. Dumont D'Urville, who was on a French expedition through the Pacific in the 1820s, seems to have introduced to general use the parcelling of the Pacific into Melanesia, Micronesia and Polynesia (Figure 1.9). These divisions have survived – I have been using them throughout this discussion – and whether for better or worse are now ineradicable concepts in any account of the Pacific world. On the useful side, the vast area of the Pacific does require some order and subdivision for the purposes of most discussion. On the negative side, the very basis of the names makes for confusion: Micronesia, region of small islands, and Polynesia, region of many islands, if interchangeable within the tropics, are at least basic geographic statements of a sort, though without very rational boundaries. Melanesia, Black Islands, derives from the skin colour of some of the people, and literally and genetically is a superficial and over-generalized statement. As Friedlaender aptly put it: 'People will inevitably persist in the naming of racial groups based on simple physical and even social attributes, but, at the very least they should be made aware how grossly simplified any such taxonomic system has to be, and the diversity which a name such as "Melanesian" masks' (1975: 215).

So, one great and enduring misconception arising out of this old Pacific trichotomy is that it is also some archaic and fundamental biological division in human terms. This cast of thought has bedevilled countless studies of Pacific human biology from D'Urville's time to the present day, where it still confuses some interpretations of analyses in molecular biology. The vast Pacific does need some ordering and, as

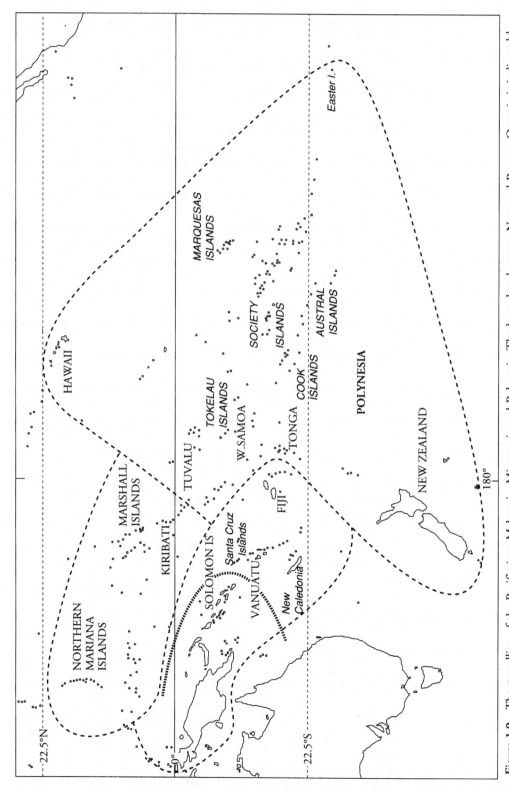

Figure 1.9 The parcelling of the Pacific into Melanesia, Micronesia and Polynesia. The boundary between Near and Remote Oceania is indicated by the dotted line between Vanuatu and the Solomon Islands.

labels, Melanesia, Micronesia and Polynesia – apart from being indelible – are of use. But the divisions were not founded on deep matters of human biology. Whether there *is* any biological coherence to any of them is something we will come to.

A more recent division of the Pacific, fundamentally geographic but more obviously useful in general biological terms, is Roger Green's concept of Near and Remote Oceania (Figure 1.9: Pawley and Green 1973, Green 1991). Near Oceania comprises the islands from New Guinea east to the end of the Solomons chain at San Cristobal. As already noted, these are intervisible islands and some are quite large. Beyond this limit any land is at least 350 kilometres distant, and this considerable gap marks the start of Remote Oceania – the vast, small-islanded expanse of the greater Pacific. Here, within the tropics, only Fiji and Hawaii are of respectable size. The relationship of this division to the attenuating plant and animal species, and to human resources, has been mentioned. The need for an increasingly sophisticated maritime competence is clear. The implications for the human organism have to be examined. The essential picture is of the immensity of the ocean and the insignificance of the land.

PHYSIQUE

2

In this chapter the physique of Pacific people is considered in basic terms. In essence this is a discussion of height and weight. Stature and mass are alternatives here, and I tend to use them indiscriminately. We are particularly interested in the physique as it was in prehistory, before the introduction of foreign genes, or any great change – in a broad sense, including diet and disease – of environment. There are several ways of gleaning this information. First there is the historical record. The journals of the early European seamen, accustomed to observe accurately, are valuable. At least for the English the keeping of such a record invariably was an order of their official instructions, and other agencies put in their plea. The first volume of the *Philosophical Transactions of the Royal Society* contained a set of *Directions for Seamen, bound for far voyages*, which advised and admonished to record. The importance of accurate drawing was emphasized, and most major expeditions of the eighteenth and nineteenth centuries took with them competent draughtsmen and artists. It is wise to focus on the original drawings of people, for by the time these had passed through the hands of two or three engravers on the way to publication the influence of Rousseau was usually strongly evident, all becoming plump and Arcadian.

Along with the nautical records are those of various scientists and medical men who travelled with the more sophisticated expeditions. Some of these, such as the record by H. L. Guppy, surgeon on H. M. S. *Lark* during her voyage through the Solomon Islands in the 1870s, are thoughtful and informative and still of use. On the New Guinea coast the remarkable Russian, Miklouho-Maclay (1975), left a fascinating and sympathetic record of cultures scarcely aware of intruders. Others, who were neither practical seamen nor of scientific bent, tend to be less useful chroniclers. Missionary records are usually infuriatingly reticent on physical detail, perhaps because some of it must have been disconcerting to those trying to concentrate on the unworldly. Always there are exceptions, such as the journal of the Reverend Thomas Williams from the Fiji of the early 1840s, which is a lode of information on the people and their culture.

A second and generally much later source of information on physique is actual measurements of living and relatively unadmixed groups. This record is patchy. Quite a number of anthropologically-inclined scientists found their way into various corners of the Pacific from early in the nineteenth century, and from these one might expect some good data. But like the missionaries they are often exasperatingly unhelpful, in this instance because of an obsession with that lodestone of early anthropology, the cranial index, the ratio of length to breadth of head. Innumerable patient Pacific islanders have been recorded variously as brachycephalic, dolicocephalic or mesocephalic – round-headed, long-headed, and something in between – to debatable biological end. This is a matter to be discussed in Chapter 4. However the meticulous Germans did a few anthropometric studies during their colonial reign in the western Pacific (Parkinson 1907, Friederici 1913, Schlaginhaufen 1929), and the Japanese did the same in Micronesia (Hiroko 1926, Hasebe 1928, Samejima 1938). In the 1920s and 30s the Bayard-Dominick Fund sponsored a series of expeditions out of the Bishop Museum in Hawaii to various islands, particularly in Polynesia, and from these there is a valuable record (Sullivan 1921, 1922, 1923, Shapiro 1930, Shapiro and Buck 1936). But for some regions, such as the Solomon Islands, detailed studies are as recent as the 1970s, and probably were done only just in time, before major cultural and environmental changes occurred. For some groups – New Zealand Maori women are an example – there are simply no data of substance.

The third source of information on the physique of people in the past is prehistoric human remains. Interpretation of these poses many a problem, but they are the unequivocal link and need scrutiny.

Clearly each of these sources has its strengths and weaknesses, but used judiciously one complements the others, so we can comment pretty confidently on the physique of many groups in the past. Here we will look at the historical record first.

The historical record

The chronology and distribution of this points up a contrast between European and Pacific seafaring. While prehistoric contact with the Americas must have occurred, it is now clear that the Pacific was originally settled out of Asia, from west to east. This, as already noted, is essentially upwind. By contrast, the Europeans in their Pacific journeyings far from home were obligatory downwind sailors. Their typical ship of the seventeenth and early eighteenth centuries could make little better that 90 degrees to the wind, and the exploration record reflects this. The Portuguese made a tedious approach to their Spice Islands via Madagascar and India. The Dutch supplanted them and learned to take the westerlies of the South Indian Ocean until they met the shore of Australia (sometimes too literally), then coasted north. The Spanish, constrained by the Pope's edict as to their half of the world, crept into the Pacific through the Straits of Magellan and then worked their way up the west coast of South America with something of a beam wind until the trades were reached. These they followed in a northerly track across the Pacific to the particular destination for the first couple of centuries, the Philippines and some of western Micronesia. Consequently, they encountered the occasional Polynesian island, much of western Micronesia, but, Mendana and Quiros apart, left scant record of Melanesia.

The first European crossing of the Pacific in 1521 was notably light on landfalls, with no people sighted between the South American coast and Guam. Here Pigafetta, Magellan's chronicler, noted the men to be 'as tall as we, and well built' (1906: 95). In 1568 Mendana reached Santa Isabel in the Solomons, where, though not too much was said about physical attributes, the impression comes through of a healthy, vigorous set of people: 'This island is densely populated by a fine and well-featured race of men . . . all the people are well grown and good looking' (Mendana 1901: 146, 289). In the last decade of the sixteenth century Quiros at the Marquesas wrote of a 'strong and healthy race, and indeed robust . . . about forty came on board, beside whom the Spaniards seemed of small stature' (1904: 16–17). In his final expedition, in 1606, still searching for *La Austrialia del Espiritu Santo*, Quiros captured in the Tuamotus 'a very strong man, muscular and of good stature. In spite of his being bound, four Spaniards could not manage him' (Munilla 1966: 164). At Rakahanga in the Cook Islands the natives were 'tall and well-built, tawny in colour' (Munilla 1966: 172). Quiros sailed on to note 'stout men' at Guam, and then south to encounter Taumako in the Duff group, north-eastern outliers of the Solomons, where the natives were 'tall as a rule, straight, vigorous, well-favoured, of a clear mulatto colour more or less'. At Espiritu Santo, in modern Vanuatu, people were 'of fine make and good colour . . . tawny men, very tall', and, advancing the cause for colonization, he asserted the climate to be 'very healthy . . . from the vigour and size of the natives' (Quiros 1904: 231, 241, 270).

In these historical descriptions we have to bear in mind the comparisons. The 'common size' of male Europeans certainly was smaller than their stature today. In the period 1836–1840 the average height of conscripts into the French army was 1650 mm or 5 feet 5 inches (Chamla 1964), while those of the army of the Netherlands has risen by nearly 140 mm or 5½ inches in the last century (Van Wieringen 1972). As these would have represented the healthy and robust section of the general male population it seems reasonable to say that in the early historic period of the Pacific a 'larger' European male would have been 1700 mm (5 feet 7 inches) or a little more. It is worth noting that 'stout', that recurring word in the descriptions of Pacific people, had none of the insinuation of fatness that crept in during the next century: 'strong; lusty' defined Dr Johnson (himself stout enough in the later sense), and it referred to muscularity. Even the term 'corpulent' tended to refer to large body size rather than specifically to fatness.

The Dutch gradually supplanted the Portuguese, and in 1642 Abel Tasman, in an expedition exceptional for its entrance into the Pacific by the westerlies south of Australia, thought the New Zealanders – on very limited observation – to be 'of ordinary height but rough in voice and bones, their colour between brown and yellow': or, as one of the seamen observed 'rough, uncivilized, strong, full of verve'. In Tonga the men were ' somewhat more than ordinary height', and the women 'comparatively quite as sturdy of body and limbs as are the men' (Tasman 1968: 121, 164). Passing west through Melanesia on his route to Batavia, he noted the people occasionally encountered off New Britain and New Ireland to be considerably blacker in skin colour than those further east.

Despite these incursions into Melanesia, the focus of European interest remained in the north-west, and late in the seventeenth century the Spanish control of the Philippines and the Marianas tightened. A few records exist of the native Chamorros of the Marianas before the devastation of colonization. In 1670 Mendoza recorded that 'in their bodies they do not resemble [Europeans] for they are as large as giants, and of such great strength, that it has actually happened that one of them, while standing on the ground, has laid hold of two Spaniards of good stature, seizing each of them by one foot with his hands, and lifting them thus as easily as if they were children' (Carano and Sanchez 1964: 17). 'The Marianos are in colour a somewhat lighter shade than' the Filipinos, larger in stature, more corpulent and robust than Europeans' (Garcia 1937: 21). After the close of the seventeenth century observations on the people of the Marianas become, at best, suspect as a record of the past because of the impact of Spanish rule. Between 1668 and 1710 the native Chamorro people were reduced in numbers from an estimated 50,000–100,000, to the 3439 counted in the first census (Cordy 1983). Thompson concluded that 'The physical type has been completely changed by extinction of the pure Chamorro through warfare, disease, and intermarriage with Spaniards, Mexicans, Filipinos, Chinese, Japanese and other foreigners' (1932: 7).

Not many other groups in the Pacific suffered the early severe fate of the Chamorros. In Polynesia the welcome to Europeans tended to be pleasantly warm rather than hot, but the locals retained control. The warlike Fijians made contact there a risky business, and further west in Melanesia this same tendency plus the barricades of endemic disease, particularly malaria, limited European incursions. For some of the less-frequented groups such as the Louisiade Archipelago of New Guinea, pristine descriptions continue into this century. And the scattered islands of eastern Micronesia were protected by their geographic insignificance.

In 1721 the Dutch expedition of Jacob von Roggeveen made the European discovery of Easter Island, finding people 'well-proportioned in limbs, having very sturdy and strong muscles . . . generally large in stature, and their natural colour is not black but pale yellow or sallow . . . Also these people have snow-white teeth, and are outstandingly strong'. In the Tuamotus the men were 'all strong and well-made fellows, in all respects similar to those of Paaslant' (Easter Island), as were the Samoans, 'lively fellows, fat and sleek, in colour brownish-red' (Roggeveen 1970: 97, 121, 152). Again at Easter Island, the Spaniard Francisco A. de Aguera y Infanzon in 1770 observed men 'very generally of large stature, very many exceeding 5' 11". Most of them attain 5' 6½", and there were two whom out of curiosity we measured, one of 6' 5" and the other of 6' 6½", all of their limbs being of proportionate dimensions' (1908: 99).

It was in the latter decades of the eighteenth century that the descriptive flood started, with the voyages of the meticulous chroniclers, who were allowed the luxury of more considered observation because of advances both in navigational theory (longitudes had become ascertainable even before Harrison's chronometer, though the observations and the mathematics were a bit daunting) and practical ship design (Cook chose for his vessel an east-coast collier that he knew could make 70 degrees to windward). In surveying this growing abundance of historical comment it is clearest to consider it island group by island group, while keeping in mind that these groupings are to some extent a product of European colonization. Islands now bracketed under one name may not have been much related in prehistory. Vanuatu comprises a dozen significant islands stretching over several degrees of latitude. The northern Cook Islands of Manihiki and Pukapuka are 400 nautical miles from the southern members of the group, which was not commuting distance in the past whatever the oceanic expertise. And on a larger island of Near Oceania, partly because of the dissected landscape, partly because of what someone has termed 'the rough quarantine of war', communities in adjoining valleys may have been strangers.

New Zealand, although on the periphery of the Pacific both geographically and in terms of prehistory, was visited early by a number of European expeditions and is rather well described. 'The Natives of this Country are a Strong, rawboned, well made, Active People, rather above than under the common size especially the Men' (Cook 1955: 278). At Tolaga Bay on the east coast of the North Island, Monkhouse, surgeon on the *Endeavour*, noted the people to be 'generally short: few of them measure 5' 8" – mostly stout made and seem capable of great activity'. He examined the man killed in the first and unfortunate encounter with the New Zealanders at Poverty Bay, a 'short but very stout bodied man – measured about 5f.3I' (Monkhouse 1955: 565–566). The same year, 1769, the French expedition of Jean-Francois-Marie de Surville reached New Zealand. Surville encountered canoes manned by '8 to 10 sturdy men, 5ft 7 inches to 5ft 11 inches tall' (Surville 1982: 63). (These are English measures: in the translation I have corrected for the old French foot, which is 0.788 inches longer than the English.) In 1772 another French expedition arrived, commanded by Marion du Fresne. There are several descriptions from Marion's officers.

These islanders are generally tall, well-built and with pleasant faces and regular features, seeming to be very agile and strong and vigorous-looking. We measured some of them, who seemed to be the tallest among them, and they were over six feet and well-proportioned. Their usual height, as far as I could see, is five foot nine to five foot ten inches. (Roux 1985: 139)

The Zealanders are commonly five feet ten to eleven inches tall. We saw several who were six feet tall. They are well built and well proportioned . . . their legs are well shaped and their feet the same. (Dez 1985: 317)

The comments on female physique are erratic.

The women are rather short and are not nearly as well-built as the men. (Montesson 1985: 245)
   Robust and strongly built, rarely less than five feet two or three inches tall. (Lesson 1971: 86)
   The women are small and ill-made. (Du Clesmeur 1914: 471)

In the Marquesas, at the far margin of eastern Polynesia, the inhabitants 'are without exception as fine a race of people as any in this Sea or perhaps any whatever . . . The Men in general are tall that is about Six feet high' (Cook 1961: 372–3). Lieutenant Clerke thought them

the most beautifull [sic] race of People I ever behold – of a great number of Men that fell under my inspection, I did not observe a single one either remarkably thin, or disagreeably Corpulent but they were all in fine Order and exquisitely proportion'd. We saw very few of their Women, but what were seen, were remarkably fair for the situation of their Country and very béautifull. (1961: 761)

William Wales, the astronomer, perhaps starry-eyed, concurred:

The Natives of these Islands taken Collectively, are undoubtedly the finest Race of People that I or perhaps any Person Else has ever seen. They are . . . almost without exception all fine tall stout-limbed, and well made People, neither lean enough for scare-crows, nor yet so fat as in the least to impede their Activity. (Wales 1961: 832)

At Tahiti, in the Society Islands, 'it is very common to see them measure six (Paris) feet and upwards in height. I never saw men better made and whole limbs more proportionate' (Bougainville 1772: 248). And 'With respect to their persons the men in general are tall, strong limb'd and well shaped, one of the tallest we saw measured Six feet 3 inches and a half, the superior women are in every respect as large as Europeans but the inferior sort are in general small' (Cook 1955: 123–4).

Cook visited Tonga in the course of both the second and third voyages. Astronomer Wales noted the men to be

in general strong and raw boned, pretty tall and well proportioned . . . of a very light copper Colour. The women also are tall, well formed, and have very regular and soft features, but are rather to[o] fat to be esteemd beauties any where but in Holland; and this fault seems general. (1961: 808–9)

Surgeon Anderson considered the people

[to] seldom exceed the common stature, though we have measur'd some above six feet, but [they] are very strong and well made especially their limbs. They are generally broad about the shoulders and though the muscular disposition of the men, which seems a consequence of much action, rather conveys a notion of strength than beauty, there are several to be seen that are really handsome. The women are not so much distinguish'd from the men by their features as their general form, which is for the most part destitute of that strong fleshy firmness that appears in the last . . . in general their bodys and limbs are well pro-portione'd. (1967: 925–6)

At his eponymous islands 'The Inhabitants . . . were thought by the gentlemen who were a shore, to be as fine a race of people as any they had seen in the Sea, and in general stouter and fleshier but I hardly think the difference so great as they immagined [sic]' (Cook 1967: 87). By the time Hawaii was reached, late in the third voyage, the descriptions had become rather casual – it had all been seen before. 'I have already observed that these people are of the same nation as the people of Otaheite and many others of the South sea islands, consequently they differ but little from them in their persons' (Cook 1967: 279). Or, as Ellis observed half a century later, 'The natives are in general rather above the middle stature, well formed, with fine muscular limbs' (1827: 32).

Although relatively large in the context of Remote Oceania, Fiji had little early contact. It was not a salubrious watering place for the needy seaman. In 1849 Captain John Erskine visited there in the course of an extensive voyage through the central Pacific: 'I never saw a people more prepossessing in appearance and manner; the men were in general of large stature and well-formed.' Of Thakambou, then at the height of his power: 'It was impossible not to admire the appearance of the chief: of large, almost gigantic size, his limbs were beautifully formed and proportioned' (Erskine 1967: 186). The missionary Thomas Williams, who lived in Fiji for several years around 1840, wrote of people who were:

generally above the middle height, well made and of great variety of figure. They exceed the white race in average stature, but are below the Tongans. Men above six feet are often seen, but rarely so tall as six feet six inches. Corpulent persons are not common, but large, powerful, muscular men abound . . . Most have broad chests and strong, sinewy arms, and the prevailing stoutness of limb and shortness of neck is at once conspicuous. (1858: 104)

In Samoa Erskine observed the men to be 'a remarkably fine-looking set of people, and among them were several above six feet high, with herculean proportions . . . One stout fellow . . . his arm measured above the elbow 15½", while that of one of our foc'sle men, probably the stoutest man on the ship was but 14"' (1967: 41).

Northward, eastern Micronesia escaped the early attention and devastation inflicted on the Marianas, and only the occasional descriptive comment emerged. In 1765, at Nukunau in Kiribas, Byron encountered 'clean limbed, tall well-proportioned Men, of a bright copper colour' (1964: 111–112) . The Marshall Islands were hardly known until the 1817 expedition of the Russian, Kotzebue, who found people 'tall and well made, of a dark brown complexion' (1839: 214). For the Caroline Islands, Hezel summarizes a contact record of 'sturdy, well-built people of medium stature' (1983: 53).

Further south, at the eastern fringe of Melanesia, the widespread and sometimes substantial islands of Vanuatu, favourably advertised by Quiros, evoked varied views – in fact from here on through Melanesia no single comment can sum up the human form on a large and scattered group of islands or of a single large island. At Malekula 'The people are slender and finely proportioned' (Forster 1982: 565); and at Erromanga 'They are a stout well-limbed race of people, better and stronger than those at Mallicollo' (*ibid:* 586). At Tanna the men 'are all well made, the greater part of them tall and stout, but none of them is corpulent and fat' (*ibid:* 622). At Big Bay on Espiritu Santo one Dr Corney recruited 'four men to go to Fiji with us for three years. They were all adults of about 20–24 years, tall, black and athletic young men, much above the average stature of the New Hebrides anywhere north of Eromango: and some other people of the locality appeared to me equally well built, and some 5' 10" or 5' 11" in height' (Quiros 1904: 274). The physique through Vanuatu was summed up thus:

As a rule the natives belonging to the southern portion of the New Hebrides, are stronger and better developed than those further north. In Tanna, they are finer, stouter and, I think, more brave than the inhabitants of any of the other islands. The Erromanga men are shorter; and in Santa Cruz they are all small and slight, but wiry and active. At Espiritu Santo, the men . . . were not so well formed as the Tannese. (Markham 1970: 238)

South-west, in New Caledonia, Cook found a 'strong robust active well made people . . . of nearly the same colour as the people of Tanna, but . . . a much stouter race, some who were seen measured Six feet four inches' (1961: 539). Pickersgill concurred: 'The people, both Men and Women, are tall, robust and well made, but large-limbed and heavy looking; the Men generally approaching 6 foot, some 6' 3 or 4 inches, the Women in proportion' (1961: 35).

Despite Markham's comment on the people of the scattered, small islands of the Santa Cruz group between northern Vanuatu and the larger islands of the Solomons, Carteret in 1767 there observed 'black woolly headed Negroes well beyond of the common Stature' (1965: 172). At adjacent Anuta, Shabel'skii observed 'men of considerable stature, with well-proportioned limbs and extremely regular facial features' (1990: 193). On the small island of Santa Catalina, off the eastern end of San Cristobal, last in the Solomons chain, Guppy, surgeon on H. M. S. *Lark*, felt that the natives were 'distinguished from all others in this part of the group, by their finer physique, lighter colour, and greater height' (1887: 103). He noted that 'A wide distinction exists between the inhabitants of the interior and those of the coast . . . The large islands . . . are but thinly populated in their interior by tribes of more puny physique . . . I was unable to make any measurements of these natives; but those I saw were usually of short stature' (*ibid:* 14, 120).

Later writers confirmed this distinction on the larger islands between coastal or 'saltwater' peoples and inland or 'bush' people. 'Relatively robust and more stocky constitutions are more common among coastal populations: hill people tend to be leaner and more gracile' (Ross 1973: 47), and the distinction persists through the Solomons chain, and on to Bougainville. On Buka, where the anthropologist Beatrice Blackwood lived for two years in the late 1920s, the coastal people such as the Halia, who seasonally made open-sea voyages north to the island of Nissan, were on the whole 'of medium stature. They are rarely fat, but both men and women frequently possess considerable muscular development' (Blackwood 1935: 208). Further west, the Tami and Siassi people, central to the trading network of the Vitiaz Strait region between New Guinea and New Britain, were noted for their 'Conspicuous muscularity' (Harding 1967: 128). The south-eastern outliers of New Guinea, the Trobriand and D'Entrecasteaux groups and the Louisiade Archipelago are inhabited largely by the Massim people, of Kula renown. Malinowski distinguished the southern Massim from the people further west and more related to mainland New Guinea by the 'extreme lightness of their skins, their sturdy, even lumpy stature' (1922: 36).

The historical evidence cited here is of course only a small fragment of that in existence. But it is, I think, a fair summary of the impressions of early observers, and from island to island the observations are generally consistent. Variation in physique within an island is sometimes noted, and has been used to support ideas of European genes in the Pacific, or of different native 'races'. Yet such variation was long ago put in context by, for example, the Forsters on Cook's second voyage or by the Russian, Bellingshausen:

Although many travellers assert the existence of various races amongst the inhabitants of Otahiti, I noticed nothing of the kind myself. The evident difference between the chiefs and the common people results from their different modes of life. Otahitians of high rank are of somewhat larger build, corpulent and olive-coloured, but the common people are more reddish. The Otahitian dignitaries lead a quiet, sedentary life; the common people are continually active, always without clothing, and are frequently all day long in the open, fishing on the coral reefs (Bellingshausen 1945: 282)

In Hawaii Ellis made the same point:

The chiefs in particular are tall and stout, and their personal appearance is so much superior to that of the common people, that some have imagined them a distinct race. This, however, is not the fact; the great care taken of them in childhood, and their better living, have probably occasioned the difference. (Ellis 1827: 32)

However this in itself is a point of interest which we will mention in the next chapter.

The larger picture that emerges from these many historical brush-strokes is of a singularly tall, muscular and well-proportioned people inhabiting the vast area of Remote Oceania ('the finest Race of People that I or perhaps any Person Else has ever seen'), while westward, as the islands enlarged, the people seemed somewhat diminished in size. Especially when close to the island-continent of New Guinea were they of particularly modest form. Contained within this picture is another consistent pattern for Near Oceania, of stockier coastal peoples contrasting with a more gracile inland form. And through the long island chain of Vanuatu, extending over some seven degrees of latitude or 600 km, there is a gradual change with movement south to a more robust physique.

The record of the living

Anthropometry, the measurement of living people, is the second source of information on the physique of the people of the Pacific. In this field some studies stand out. The Bayard-Dominick series of expeditions out of the Bishop Museum in Hawaii has been mentioned, and provides data from Tonga, Samoa, the Marquesas and the Society Islands, though with that unfortunate emphasis on head dimensions. Another major series of studies was the more recent Harvard Solomon Islands Project, carried out between 1966 and 1980 (Friedlaender 1987a). Admirable in concept and achievement, this survey of the human biology of several well-defined human groups within Island Melanesia will, along with the earlier studies in the region by Oliver (1955) and Friedlaender (1975), be sustaining analysis for many years. These biological data were captured just in time, before major environmental and political changes made such an exercise less relevant to the past, and less feasible. The International Biological Programme of the 1960s and 1970s studied many aspects of human biology in coastal and highland localities in Papua New Guinea (Harvey 1974, Budd *et al.* 1974, Hornabrook 1977a). Though becoming geographically marginal to the focus of this book, these studies make a valuable contribution to some later discussion. Other studies on a lesser scale help flesh out the picture. For example Peter Buck, *Te Rangi Hiroa*, measured the men of the Maori Battalion during their return by troopship from the First World War (Buck 1922–3), to provide our only data of substance on a New Zealand Maori group.

The earlier anthropometric studies, though often providing a superfluity of not-very-enlightening head measurements, generally did not measure body mass. This makes assessment of physique less satisfactory. Measures of body breadth, such as shoulder (biacromial) and pelvic (bi-iliac) diameters, can atone to some extent for the lack of mass data, but neither were these recorded much in the earlier studies. Breadths of limb bones are useful. In a very large study Frisancho and Flegel (1983) found a single biepicondylar (elbow) breadth measure to be a good predictor of general body frame size. Wrist and elbow breadths are highly correlated (0.85 and 0.78) with skeletal mass (Martin 1991).

In Tables 2.1 and 2.2 are anthropometric data from several Pacific groups; Figure 2.1 gives their geographic location. For two Fijian regions, the north-east coast of Viti Levu, and the Lau islands, there are comparative data which I have felt to be worth inclusion. Where possible I have used data from individuals in their third and fourth decades on the assumption that these cover the life span of most adults in prehistory. Unfortunately full summary statistics have not always been published. At this stage we need not say too much about this anthropometric evidence – it will be central to later discussion – beyond noting that it reinforces the impression from the historical record of increasing stature and a greater robustness of frame and body with passage from Near to Remote Oceania.

The skeletal record

Stature

The third line of evidence on the physique of Pacific people in the past comes from examination of prehistoric human remains and generally involves the determination of stature. A lot of work has been done on this since the Surgeon-Anatomist William Cheselden in 1712 communicated a paper to the Royal Society on 'The dimensions of some bones of extraordinary size which were dug up near St Albans' (Wells 1969: 455). From these Cheselden concluded that 'if all the parts bore a due proportion this man must have been eight feet high'. On the measurements he gives, his estimate may well be correct – the individual was possibly a pituitary giant – but the words 'if all the parts bore a due proportion' are pertinent. Nowadays the usual approach to the estimation of stature from human remains is by application of appropriate regression equations to the lengths of an individual's long (ie, limb) bones, a method based on the linear relationship between the two variables, stature and long bone length. A typical equation reads:

$$S = 1.880F + 81.306$$

where S = stature and F = length of femur,
both in centimetres.

Table 2.1 Male anthropometric data. Mass in kilograms, linear dimensions in millimetres. Sources: Kwaio, Baegu, Nasioi, Nagovisi, Aita, Ulawa, Ontong Java and Lau (Malaita), Friedlaender 1987a; Karkar, Harvey 1974; Baining, Kariks and Walsh 1968; Manus, Heath and Carter 1971; Tolai, Champness *et al.* 1983; West Nakanai, Swindler 1962; Ulithi, Lessa and Lay 1953; Fiji (coastal A), Hawley and Jansen 1971; Fiji (coastal B), Gabel 1958; Pukapuka and Tokelau, Prior *et al.* 1981; Tonga, Finau *et al.* 1983; Samoa, Baker *et al.* 1986; Maori (New Zealand), Buck 1922–23; Lau A (Fiji), Gabel 1958; Lau B (Fiji), Lourie 1972; Hawaii, Dunn and Tozer 1928.

Group		Mass	Stature	Biacromial breadth	Chest breadth	Bi-iliac breadth	Epicondylar breadth	Condylar breadth	Sitting height ratio
Karkar	mean	56.4	1610	362	244	263	65	89	52.2
	n	115	115	79	79	79	79	79	
	S.D.	5.13	57.3	14.1	11.7	14.6	3.60	4.00	
Kwaio	mean	57.7	1610	366	261	265	64	91	53.5
	n	46	46	46	46	46	46	46	46
	S.D.								
Nasioi	mean	57.7	1632	368	259	267	64	92	51.7
	n	28	28	28	28	28	28	28	28
	S.D.								
Baegu	mean	58.6	1620	362	258	267	66	88	52.0
	n	39	39	38	39	39	39	39	39
	S.D.								
Nagovisi	mean	58.6	1605	367	257	265	65	89	52.4
	n	29	29	29	29	29	29	29	29
	S.D.								
Baining	mean	60.1	1577						
	n	347	347						
	S.D.								
Manus	mean	60.2	1629	368		278			
	n	20		7		7			
	S.D.			25		14			
Tolai	mean	60.6	1636						
	n	20							
	S.D.								

Table 2.1 (continued)

Group		Mass	Stature	Biacromial breadth	Chest breadth	Bi-iliac breadth	Epicondylar breadth	Condylar breadth	Sitting height ratio
West Nakanai	mean	62.6	1642						51.2
	n	265	267						266
	S.D.	6.4	5.03						
Aita	mean	60.9	1596	396	263	267	66	90	53.0
	n	39	40	40	40	40	39	39	39
	S.D.								
Ulawa	mean	60.9	1629	373	271	262	67	88	51.6
	n	37	37	37	37	37	37	37	37
	S.D.								
Lau (Malaita)	mean	64.5	1640	386	279	273	69	91	52.9
	n	20	20	20	20	20	20	20	20
	S.D.								
Ulithi	mean	65.5	1635	373	269	284			54.0
	n	56	56	55	56	56			56
	S.D.	8.0	42	14.07	14.89	16.99			
Ontong Java	mean	67.7	1662	387	282	284	71	93	53.4
	n	75	75	75	75	75	75	75	75
	S.D.								
Fiji (coastal A)	mean	68.6	1735						
	n	154	154						
	S.D.	7.4	64						
Fiji (coastal B)	mean	72.0	1734	397	287	292			50.5
	n	210	210	210	210	210			210
	S.D.	22.8	58	7.6	7.8	5.9			
Pukapuka	mean	69.0	1688						
	n	27	27						
	(s.e.)	1.60	13.0						

Tokelau	mean	69.7	1674						
	n	26	26						
	S.D.								
Tonga (Foa)	mean	75.2	1713						
	n	198	198						
	(s.e.)	0.30	2.0						
Maori	mean	74.5	1706		279				53.8
	n	426	426		415				420
	S.D.								
Samoa	mean	75.9	1714						
	n	26	26						
	S.D.	15.1	55						
Lau A (Fiji)	mean	76.1	1708	402	285	294	70.5	95	54.0
	n	74	74	74	74	74	74	74	74
	(s.e.)	1.76		1.96	2.01	2.05	0.53	0.06	
Lau B (Fiji)	mean	75.6	1733	399	294	29.5			51.0
	n	73	130	120	120	120			120
	S.D.	19.30	60.0	6.10	7.20	4.10			
Hawaii	mean	77.3	1713						53.4
	n	60	60						60
	S.D.								

Table 2.2 Female anthropometric data. Mass in kilograms, linear dimensions in millimetres. Sources as for Table 2.1.

Group		Mass	Stature	Biacromial breadth	Chest breadth	Bi-iliac breadth	Epicondylar breadth	Condylar breadth	Sitting height ratio
Karkar	mean	47.0	1511	321	226	257	56.6	80.4	52.3
	n	128	128	113	113	113	113	113	
	S.D.	5.9	57	14.2	12.4	14.3	2.7	3.8	
Baining	mean	47.9	1477						
	n								
	S.D.								
Nasioi	mean	48.2	1523	325	237	261	57	84	
	n	24	24	24	24	23	24	24	
	S.D.								
Manus	mean	48.2	1510	322		254			
	n	38	38	10		10			
	S.D.			13		10			
Kwaio	mean	48.6	1497	319	242	267	57	85	
	n	49	49	49	49	49	49	49	
	S.D.								
Baegu	mean	49.1	1580	325	240	263	57	79	
	n	43	42	42	42	43	43	43	
	S.D.								
Nagovisi	mean	49.1	1513	324	236	260	56	82	
	n	37	37	37	37	37	37	37	
	S.D.								
Ulawa	mean	50.0	1510	322	238	261	58	78	
	n	51	51	51	51	51	51	50	
	S.D.								

Aita	mean	54.1	1498	332	248	266	58	84
	n	49	49	49	49	49	49	49
	S.D.							
Tolai	mean	55.1	1557					
	n							
	S.D.							
Lau	mean	55.9	1534	341	258	277	60	81
	n	38	38	38	38	38	38	38
	S.D.							
Ontong Java	mean	59.6	1560	349	262	285	63	83
	n	119	119	119	119	119	119	119
	S.D.							
Pukapuka	mean	60.7	1572					
	n	23	23					
	(s.e.)	1.3	11					
Fiji (coastal)	mean	61.5	1615					
	n	192	192					
	S.D.	8.4	52					
Hawaii	mean	69.4	1626					
	n							
	S.D.							
Samoa	mean	70.4	1592					
	n	49	49					
	S.D.	12.5	56					
Tokelau	mean	70.6	1610					
	n	41	41					
	(s.e.)	1.8	8					
Tonga (Foa)	mean	71.0	1618					
	n	171	172					
	(s.e.)	0.4	2					

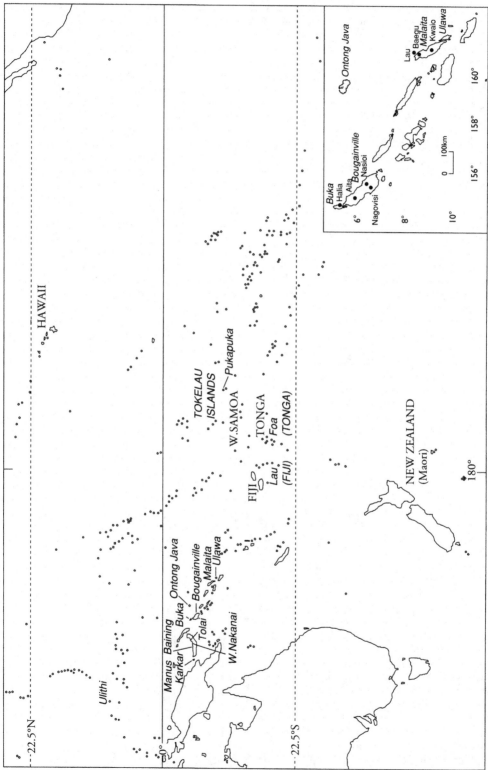

Figure 2.1 Locations of the groups providing the anthropometric data of Tables 2.1 and 2.2. The Solomon Islands and Bougainville are illustrated in greater detail in the inset.

This approach originates with Karl Pearson (1899; that is one of his equations), who also cautioned that such equations should not be extended from one group to another – his own study was based on data for 100 oldish men and women from southern France – as body proportions differ from group to group. Subsequent studies have confirmed both the usefulness of the general method and the differing body proportions of different human groups.

The most obvious variables in body proportions are the relative contributions of the lower limbs, and of trunk and head (= axial length), to total stature. Leonardo's famous drawing of the proportions of Man divides stature equally between lower limbs and body axis: a circle with its centre at pubis skirts the top of the head and the soles of the feet. This may have represented the ideal Florentine of the fifteenth century (though Leonardo was actually illustrating principles set down centuries earlier by the Roman architect Vitruvius; a quotation from Vitruvius is written in Leonardo's mirror-image script below the drawing) and probably still does approximate European body proportions today. But through the world proportions vary. Some people, such as the Australian Aborigines and many African groups, have particularly long legs, contributing up to 54% of stature, with the axial contribution being appropriately less. Other groups, including many Asian peoples, show an opposing tendency, with axial length contributing up to 54% of stature and the lower limbs being relatively short. A stature equation that assumes a particular bone length to contribute a set proportion to total stature should not be expected to cover the human race. (Some recent studies question this view, and later we will say a little more about these.)

In anthropometric studies, these major body proportions may be established with reasonable accuracy by measuring axial length: with the person seated, the measurement is taken from the chair platform to the top of the head. Expressed against total stature this is the 'sitting height ratio'. As an example, a sitting height ratio of 54 indicates that the person has relatively short legs. For any group, the female invariably has a slightly greater sitting height ratio than the male; that is, female limbs are relatively shorter. There are also differences of proportion within the limbs, such as the contribution of thigh (femur) and lower limb (tibia) to total leg length. The long legs of the Australian Aborigine have been said to show a particularly long tibial segment relative to femur (Abbie 1957), though as a general statement this has been questioned (Murphy and Wood 1983). However, if it is so, the opposite is true in New Zealand, where A. S. Thomson, a nineteenth-century surgeon, saw the Maori as

[having] longer bodies and longer arms with shorter legs than Englishmen of similar stature. The lengthening of the arms occurs in the forearms, and the shortening of the legs in the bones below the knee; the leg bones of New Zealanders are indeed an inch and a half shorter than these bones are in Englishmen. Their long bodies are produced by the size of the spinal bones and the cartilages between these bones. (1859: 69–70)

So body proportions matter in these studies, and ideally stature equations should be derived from a group genetically and phenotypically close to those they are being applied to. Many series of equations have been derived, the most comprehensive and rigorous probably being those of Trotter and Gleser (1952) for American Negroes and Whites (I use the terms of the original paper). Other groups studied include Mesoamericans (Genoves 1967), East Africans (Allbrook 1961), Chinese (Stevenson 1929) and Indians (Rosing 1983). Despite these and other options, an attempt at development of new stature equations was necessary for Pacific people because with other stature formulae prehistoric Polynesians were simply far below the historical and anthropometric record. Table 2.3 illustrates this, using the data for the Maori Battalion obtained by Peter Buck. The best result from other equations gave a mean stature almost two centimetres below the observed mean, and most estimates were much worse. Pearson's original European equations were almost seven centimetres below true. The disparity with the historical record is clear. Clearly, sensible interpretation of remains from the Pacific past required new and specific equations.

Table 2.3 Average stature in millimetres of full-blooded Maori of the Maori Battalion: calculated by applying formulae for other groups to the average leg (tibial) length of 363 mm, corrected to dry bone length. The actual average stature of the Maori group was 1706 mm.

Stature equation	Calculated stature of Maori	Difference from true stature	Source
Southern European	1637	−68	Pearson 1899
American White	1687	−19	Trotter 1970
American Negro	1645	−61	Trotter 1970
British	1673	−33	Allbrook 1961
Bantu	1557	−149	Allbrook 1961
Mesoamerican	1637	−68	Genoves 1967
North Chinese	1651	−55	Stevenson 1929
South Chinese	1646	−60	Wang *et al.* 1979
Japanese	1637	−69	Fuji 1960

It happens that in the Pacific the human skeletal material at present available for study is largely from Polynesia, with some from Micronesia, and very little indeed from Melanesia. That reduces the problem slightly, but only slightly. The difficulties and pitfalls in the way of establishing sensible stature equations, relevant to the past, for groups that now have much genetic admixture, as well as major changes in health, diet and environment, would be difficult to exaggerate. Ironically, the only group for which it has been attempted within Polynesia is the New Zealand Maori (Houghton, Leach and Sutton 1975), now one of the most admixed of all.

As a starting point (and hopeful that they might be useful for Polynesia generally), we used Buck's data, for his 424 men of the First World War were assuredly more full-blooded than any modern Maori group, and the influence of half a century and more of environmental change was avoided. As well as stature, leg length, and thigh length of the Maori Battalion, we used the maximum lengths of long bones from 98 individuals (44 males and 54 females) from New Zealand's prehistoric era. Variation in measuring technique offers plenty of scope for confusion, so that used by Trotter and Gleser (1952) in their monumental study of stature estimation in Americans was strictly followed. The steps towards derivation of these stature equations are less than enthralling when detailed, and the reader is referred to the original paper for the subtler points. The general procedure went thus:

1. Buck's data were used to derive linear regression equations linking height with living thigh length and with lower leg length for male Maoris.

2. The equations were adjusted to compensate for the difference between anthropometric measurement in the living and the equivalent measurement on the dry bones. Thickness of skin, subcutaneous tissue and cartilage, as well as the effect of drying on the bone itself, have to be allowed for here. A 5 mm reduction from living leg length to dry tibia length was allowed, and a 7 mm reduction from living thigh length to dry femoral length (Ingalls 1927). These corrections are, if anything, on the conservative side, tending to slight under-estimation of height in the final calculation. At this stage also, regression equations had to be developed to link oblique trochanteric length (the skeletal equivalent of thigh length in the living) with total femoral length; the correlation between these two skeletal measurements is very high as one is largely a component of the other. Of course there are difficulties in transferring from living segmental lengths to dry bone lengths, but here the men were thin after war service and bony landmarks would have been relatively obvious.

3. At this point the matter of sexual dimorphism arose. An analysis of a wide range of stature equations for other racial groups revealed most strongly that sex-based increments in regression coefficients were practically a constant; moreover, the slight variations which occurred caused fluctuations in stature estimates that were far less than the standard errors of the estimates of the equations. It was therefore feasible to estimate with some confidence the female equations linking height with both femur and tibia from the corresponding male equations.

4. Buck's original measurements were all made on the right side of the body. Stature equations have generally ignored asymmetry, although it is universal, and probably related to handedness (some would challenge this). Most people have right upper limb bones 2–4 mm longer than the left, while in the lower limb the left side tends to be slightly longer, which may arise from the tendency to

brace oneself on the left leg when working with the right hand. These minor differences ultimately derive from the slightly increased blood flow to the favoured limb over a long time, particularly during the active growth period. A growing limb that is the site of a chronic infection, with its accompanying increase in blood flow, shows the same tendency to overgrowth. The average bilateral asymmetry was assessed on the 98 individuals and an appropriate slight adjustment made for the left side.

5. It was now possible to derive the multiple regression equations linking the three variables of stature and femur and tibia lengths, and to extend the equations to cover the upper limb bones.

Several of these equations are given in Table 2.4. Because they are an actual component of stature, leg bones give more consistent results with a lower standard error than do arm bones. In Table 2.5 stature estimates for several Pacific groups are presented; these are derived from skeletal data gleaned from various studies and using the Polynesian equations.

As data from them will recur, profiles of the skeletal populations used in Table 2.5 are now given. Figure 2.2 shows their location, with their time in prehistory if an estimate is available. The abbreviation, b. p., denotes years before present, and is the uncalibrated radiocarbon date for the site of the population. Of course any radiocarbon date is

---

Table 2.4 Polynesian stature equations for single long bones, and a few other examples. Units are millimetres. S = stature, G = sex (10 = male, 20 = female), F = femur, T = tibia, f = fibula, H = humerus, R = radius, U = ulna. Sides are subscripted thus: F_L = left femur.

Single long bones, right side	Standard error of estimate (mm)
1 $S = 2.137\ F - 5.184\ G + 830.7$	15.7
2 $S = 2.210\ T - 5.247\ G + 978.6$	4.7
3 $S = 2.103\ f - 6.056\ G + 1045.0$	14.4
4 $S = 1.782\ H - 7.339\ G + 1226.4$	28.4
5 $S = 2.475\ R - 5.642\ G + 1160.7$	20.7
6 $S = 2.257\ U - 6.738\ G + 1182.7$	24.5

Single long bones, left side	
7 $S = 2.176\ F - 4.528\ G + 796.8$	20.0
8 $S = 2.077\ T - 5.602\ G + 1029.6$	7.1
9 $S = 2.164\ f - 5.721\ G + 1023.4$	18.0
10 $S = 2.520\ H - 4.440\ G + 963.1$	21.5
11 $S = 2.500\ R - 5.415\ G + 1154.8$	20.5
12 $S = 2.009\ U - 6.911\ G + 1257.6$	26.5

Equations using combinations of bones	
13 $S = 1.369\ F_R + 1.136\ f_L - 4.325\ G + 761.4$	10.8
14 $S = 3.013\ T_R - 0.240\ U_R - 0.820\ f_L - 5.742\ G + 1047.6$	0.9
15 $S = 3.546\ T_L - 1.032\ f_L - 1.058\ R_L - 6.874\ G + 1139.9$	1.0

Table 2.5   Mean maximum lengths of some long bones and stature estimates for several of the Pacific prehistoric groups discussed in the text. Dimensions in millimetres. H = humerus, R = radius, F = femur, T = tibia. The stature estimates are from the Polynesian formulae (Houghton *et al.* 1975) and where individual data are unavailable stature has been calculated from the mean length of the femur.

			H	R	F	T	Stature
Marianas	*male*	*mean*	309.8	242.8	448.9	372.0	1730
		S.D.	16.82	14.08	11.70	20.46	
		n	6	5	9	4	30
	female	*mean*	303.0	220.0	415.7	342.5	1632
		S.D.	16.97		7.02	2.12	
		n	2	1	3	2	5
Easter Island	*male*	*mean*	320.5	246.3	444.3	372.9	1728
		S.D.	12.78	6.04	23.00	15.51	49.15
		n	10	8	14	8	14
	female	*mean*	287.6	221.8	411.0	340.3	1600
		S.D.	9.55	7.25	6.90	13.24	16.00
		n	7	8	6	7	8
New Zealand	*male*	*mean*	324.4	254.0	447.9	367.0	1736
		S.D.	10.70	9.55	18.51	17.96	39.89
		n	33	33	81	78	124
	female	*mean*	294.1	226.7	416.6	336.2	1615.6
		S.D.	11.31	12.30	16.24	14.7	44.76
		n	37	33	41	61	98
Taumako	*male*	*mean*	326.4	271.3	454.8	379.6	1750
		S.D.	16.18	7.46	18.83	15.08	38.55
		n	26	4	33	23	40
	female	*mean*	304		428	350	1633
		S.D.	6.78		15.61	15.27	28.00
		n	8	0	11	5	20
Watom	*male*	*mean*	340.0	285.0	477.0		1784
		S.D.					
		n	2	1	1		3
	female	*mean*		247.0			1659
		S.D.					
		n		1			1
Tonga (To-At-1/2)	*male*	*mean*	326.0	249.6	466.1	383.9	1766.0
		S.D.	13.67	9.35	21.50	19.56	46.14
		n	11	8	8	8	11
	female	*mean*	315.4	219.0	433.3	378.0	1653.2
		S.D.	18.28	12.77	21.46	15.38	41.06
		n	5	3	3	4	5
Tonga (To-At-36)	*male*	*mean*	321.6	242.5	439.7	359.8	1722
		S.D.	17.17	12.85	18.53	9.99	31.77
		n	5	6	7	6	10
	female	*mean*	307.5	233.0	418.3	347.0	1624
		S.D.	17.68	7.07	26.31		35.98
		n	2	2	3	1	9

(continued)

Table 2.5 *(continued)*

			H	R	F	T	Stature
Marquesas	*male*	*mean*	328.0	249.0	450.0	369.0	1741
		S.D.	12.00	10.30	17.30	18.10	
		n	8	6	7	9	
	female	*mean*	284.0	218.0	391.0	332.0	1563
		S.D.	8.00	14.20	17.50	21.00	
		n	3	4	4	3	
Hawaii	*male*	*mean*	318.0	249.0	440.0	362.0	1719
(Mokapu)		*S.D.*	12.90	9.80	18.60	17.50	
		n	47	42	49	46	
	female	*mean*	292.0	223.0	410.0	335.0	1603
		S.D.	11.80	10.90	18.50	16.30	
		n	82	71	87	84	
Hawaii	*male*	*mean*	312.2	240.6	429.9	376.3	1698
(Keopu)		*S.D.*	21.30	12.81	20.21	10.39	
		n	13	15	7	12	
	female	*mean*	288.4	219.6	418.6	347.4	1622
		S.D.	17.82	16.29	21.34	16.07	
		n	24	23	26	23	
Sigatoka	*male*	*mean*	326.6	250.2	460.6	370.6	1740
		S.D.	9.13	10.79	16.98	9.27	
		n	6	5	8	8	
	female	*mean*	308.5	243.0	432.4	373.3	1644
		S.D.	15.01	19.97	16.38	21.96	
		n	8	3	5	3	
Cook Islands	*male*	*mean*	316.4	242.7	446.2	367.4	1732.4
		S.D.	16.05	16.32	17.85	17.39	38.15
		n	12	10	14	13	14
	female	*mean*	296.25	217	410.25	330	1603.7
		S.D.	15.96		13.67	8.71	29.21
		n	4	1	4	3	4

accompanied by an estimate of error and there are plenty of complexities within the method. However this is not the place for such subtleties. The purpose is simply to give some idea of where in time the group is located.

The New Zealand group is a country-wide series dating principally from the last 250 years of prehistory, that is, about A.D. 1500–1750. This may seem rather a wide spread in time and space to contain a coherent group, but the evidence is that the New Zealand gene pool in prehistory was very limited and that there may have been only one or two founding groups, all from the eastern Pacific. An overview of this material has been presented (Houghton 1980a). On present evidence, settlement time of New Zealand is rather recent, no greater than 1000 years ago and possibly only about 800 years (Anderson 1991), and the

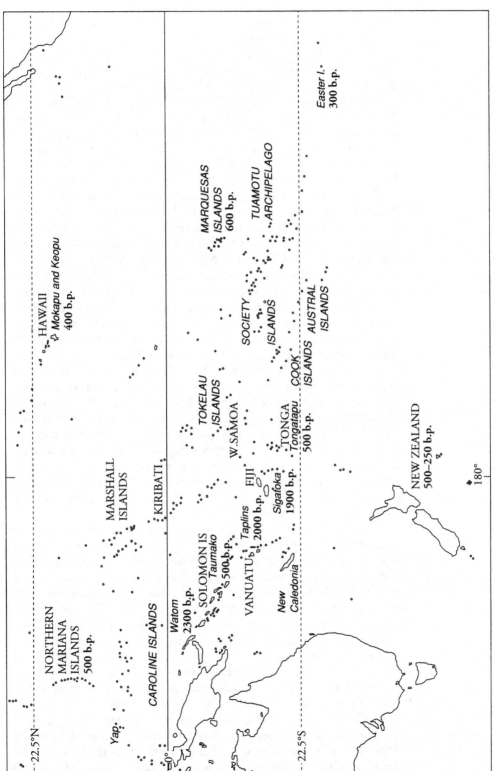

Figure 2.2 Locations of the skeletal groups discussed in the text and providing data for Tables 2.5 and 2.6. The figures within the frames are an indication of the era of each group in years before present (b.p.).

individuals of the skeletal series may be taken as descendants of the original Polynesian settlers.

The Sigatoka population derives from a burial area exposed by erosion of the sand dunes near the mouth of the Sigatoka River in southern Viti Levu, Fiji. Since at least the 1940s the area, with its abundance of pottery fragments scattered along the foot of the dunes, has been recognized as one of ancient habitation. The sudden erosion of part of the dunes prompted a salvage excavation under the auspices of the Fiji Museum, and this was largely carried out in 1988 under the supervision of Simon Best. This important series of about 50 individuals in a lamentable state of preservation was saved from complete disintegration by the conservation skills of Dilys Johns. It is dated to about 1900 b.p. A detailed analysis of the human remains has been made (Visser 1995).

The Marianas data are from remains excavated at various sites on Saipan and Tinian (Roy 1989): on Saipan, the Latte House site in the north (720 +/− 100 b.p.); the Marianas High School site; the San Antonio site on the south-west coast (360 +/− 100 b.p.); and the Hafa Dai Beach site (possibly historic). And on Tinian, the Latte House site. As so often in the Marianas, the remains were initially exposed during construction works and their placing in history or prehistory is sometimes uncertain. They are often fragmentary, and such matters as the defining of individuals, and sexing, are often insecure.

The 196 individuals from the from Taumako in the Duff group of the north-eastern Solomons, excavated by Janet Davidson and Foss Leach in 1978, I believe have been dated to between A.D. 1530 and 1690. No archaeological report on this important series has ever been published. There is a report (Houghton 1980b) on the human remains, intended as an appendix to the archaeological publication, and behind this again is a lot of unpublished data. Despite meticulous excavation the remains are much eroded and fragmentary. Some masterly skull reconstructions have been carried out by Dr Kazumichi Katayama and colleagues.

The minimal Watom series is from the island of that name in north-eastern New Britain. This significant Lapita site (SAC, Reber-Rakival) was first excavated by Jim Specht in the 1960s (Specht 1968), and the excavation extended in the 1980s (Green, Anson and Specht 1989). The dating is 2100–2500 b.p. Various reports on these remains have appeared (Houghton 1989, Pietrusewsky 1989, Turner 1989a).

The first Hawaiian series is from Mokapu on Oahu. Exposed initially by erosion and then by excavation over a lengthy period, this late prehistoric series of more than 1000 individuals has been examined by Snow (1974). The second Hawaiian series is fron Site 50-Ha-D8–30, Keopu, North Kona, Hawaii, and contained remains of 355 people. The site is dated to the A.D. 1400–1800 era (Han 1986).

The first Tongan series, from Atu on Tongatapu (Site To-At-1: actually two adjacent mounds) was excavated by Janet Davidson in 1964. Dating seems rather vague but apparently is in the 800–400 b.p. range. A report on the (often very fragmentary) remains of some 99 people has been published (Pietrusewsky 1969) and I have also used here some of my own unpublished observations. The second Tongan series from Site To-At-36, also on Tongatapu, was excavated by D. Spenneman, and contained some 40 people. This series has been studied by Van Dijk (1993). No dates seem to be available but it is probably late prehistoric and more or less of the same era as To-At-1.

The Marquesan data, the series from the Hane dune site on Uahuka, was excavated during 1964 and 1965 by Dr Yashi Sinoto of the Bishop Museum. A report on the remains of 42 individuals has been published (Pietrusewsky 1976). Sinoto's dates for level IV of the site, from which most of the burials come, lie between A.D. 1110 +/– 110 and A.D. 1635 +/– 90 years.

The Easter Island series of 33 people was excavated under the direction of C. S. Smith and W. Mulloy during 1955 and 1956 as part of the programme of the Norwegian Archaeological Expedition to Polynesia. A detailed report on the human remains has been published (Murrill 1965). Most of the remains are from the Late period (A.D. 1680–1868) and none is earlier than the Middle period (A.D. 1000–1680).

The data of Table 2.5 have been hard to collate. In some series the remains have been much mixed when exposed, and sexing of individuals must have been done simply on the basis of general size, which will distort the summary statistics. Sometimes it is not clear whether the mean value for a bone takes in both right and left from some individuals. The minimal numbers of some important series such as Watom are evident. And so on. In general where sides have been defined I have used data from the right side. However even if the secureness of some of the data is in some doubt, for comparative purposes it is unlikely that any of the mean values are significantly astray. An exception to this statement is the first Tongan series, where it will be observed that in the lower limb the female means exceed the male. This curiosity arises from the massive stature of one purported female (Pietrusewsky 1969) and the small number in the series. In my own notes for this group, I see a comment that this individual might well be male. On the other hand there is clear evidence from more recent times of women of majestic proportions amongst the Tongans. Whichever the case, this individual does distort the figures.

I am sure there are other good data, particularly in the French literature, but I have not been able to locate them. The paucity of material from Near Oceania is striking. However this does mean that there the problem as to the preferred stature equations has hardly yet arisen. Through Melanesia axial height is relatively large, prompting the (very tentative) suggestion that the Polynesian (Maori) equations might work for the people of the western Pacific.

The old concept of a fairly consistent ratio of femoral length to stature as being ubiquitously useful in the estimation of adult stature has been re-examined by Feldesman, Kleckner and Lundy (1990). Some of their results are remarkably accurate, which is a bit depressing for those who have spent a lot of time developing appropriate regression equations for particular groups. However one must remain uneasy about the idea, because body proportions really do differ between groups. Feldesman *et al.* (1990) give a range of femur/stature ratios from 25.45 for three unsexed Eskimo to 27.81 for 577 US black males. The suggested 'universal' (my term) ratio of 26.74 gives a stature of 166.4 cm for 130 prehistoric New Zealand males, and 156.2 cm for 111 females. This is not in harmony with the evidence from other sources and the results from the regression equations. It gives the Maori Battalion a mean stature of 161 cm against the reality of 170.6 cm. It is likely that while the femur/stature ratio may be useful for a quick estimate of stature, it still needs to be determined for a particular group, and of course in archaeological studies the femur may not be present. The same principle presumably could be developed for other bones. Incidentally, in our study the tibia emerged as the most accurate indicator of stature.

In the prehistoric New Zealand material the ratio of femoral length to stature is 25.9. To now use this ratio to calculate stature is somewhat circular because the stature equation itself derived partly from femoral length. However not entirely, for tibial length was also used, and the exercise is not without interest. Use of this ratio gives a mean stature of 170.2, compared with the regression figure of 173.4 cm.

Skeletal frame and robusticity

So far, this interpretation of physique from the skeletal record has been of stature, which is where most studies stop. Yet, just as with the living, a statement on physique that deals only with stature is a very two-dimensional and incomplete thing. Measures of skeletal breadth, indicative of frame size, are necessary to build up a fuller picture of body form.

For total body breadth only the width of the articulated pelvis can readily be determined from the skeleton. Theoretically, biacromial or shoulder breadth could be estimated by articulating clavicles with sternum, but getting the orientation correct would be tricky. However lengths and ratios derived from the clavicle may be informative. The useful breadth dimensions of limb bones are those taken near joints. Bones are expanded at their ends primarily to reduce the disrupting forces on the articular cartilage. This has good compressive strength but a rather low tensile strength, and does not particularly like the shearing forces of everyday movement. Expanding the bone ends reduces the force/unit area on the cartilage. The size of a joint surface should be proportional to the forces acting across it and correlate with an individual's mass. Measurements on the living have shown such breadths to be good indicators of frame size (Frisancho and Flegel 1983, Martin 1991), and in Table 2.6 several of these breadth dimensions are compared across Pacific groups. (These are bony dimensions on skeletal series: breadths on the living are in Tables 2.1 and 2.2.) As with the previous table, the data in Table 2.6 are dominated by Remote Oceania, so a trend across the Pacific cannot be discerned. But they do suggest that in the small-islanded expanse of the Pacific, people were of substantial frame.

Indices of robusticity figure in osteometric studies, though the term 'robusticity' seems rather imprecise and worthy of more rigorous scrutiny – not something being undertaken here. Anyway, these robusticity indices express some breadth dimension as a proportion of the length of a long bone. They need to be interpreted cautiously, being an expression only of the relative breadth of a bone, containing no statement as to overall size. For this, absolute dimensions are necessary. The breadth measures usually chosen for robusticity indices, diameters at midshaft, are not the best. The case for bone ends has just been made and, as well, shaft diameters are small compared to the overall length of the bone; thus slight variation in measurement here may significantly alter an index. On the femur the *linea aspera* is a particularly variable structure. The population from Nebira, Papua New Guinea (Bulmer 1979, Pietrusewsky 1976) illustrates the problems of interpretation. These people I see as being of slight build, with little sexual dimorphism and with skeletal evidence of anaemia which may relate to endemic malaria: yet they have a femoral robusticity index above 13, greater than that of most people of Remote Oceania. I have not included such traditional robusticity indices in the table, and leave an explanation of their significance to those who publish them.

Body mass

To – literally – round out the picture of physique in the past, a knowledge of body mass would be useful. This is a problem that has been scrutinized more by workers in the field of hominoid evolution (Jungers 1988, McHenry 1988, Ruff 1988) than those studying recent *Homo sapiens*. An assessment of mass offers the possibility of going beyond a dry numeric record, and of advancing the analysis into the realms of what could be called palaeophysiology. Bone dimensions that have been used in the estimation of body mass include skull length (Steudel 1981), tooth size (Gingerich, Smith and Rosenberg 1982), femoral length cubed (Lovejoy, Heiple and Burstein 1973) and vertebral joint size (McHenry, 1975). McHenry provides a formula that allows application to individual *Homo sapiens*, but it also has practical disadvantages in that the vertebrae employed, twelfth thoracic and fifth lumbar, are in a region prone to deformation and degeneration, so that their normal dimensions often cannot be determined. However the theory is sound. Many studies attest to the strongly positive relationship between body mass and joint dimensions (Frisancho and Flegel 1983, Martin 1991, Ruff 1988). Within *Homo sapiens* this particular relationship should be general, being gravity dependent, and not differ between groups because of different genetic backgrounds, or even between the sexes. In this it differs

Table 2.6 Skeletal breadth dimensions and total body mass estimates for several of the Pacific prehistoric groups discussed in the text. Mass in kilograms, dimensions in millimetres.

Group		Clavicular length	Bi-iliac breadth	Epicondylar (humeral) breadth	Condylar (femoral) breadth	Wrist breadth	Ankle breadth	Estimated body mass
Maori	*male* mean	148.5	260.2	60.2	82.1	50.5	68.9	75.6
	S.D.	7.70	11.98	3.10	4.00	2.42	2.92	4.73
	n	31	9	34	33	15	15	15
	female mean	130.6	259.3	56.1	73.7	45.3	63.7	57.5
	S.D.	7.97	12.77	3.61	3.05	1.68	2.54	4.98
	n	14	14	18	18	18	18	18
Hawaii (Mokapu)	*male* mean	147.9	253.3					
	S.D.	9.30	12.80					
	n	67	52					
	female mean	133.3	248.7					
	S.D.	7.80	13.70					
	n	105	86					
Easter Island	*male* mean	138.7	265	58.4	78.7			
	S.D.	4.19		1.90	3.73			
	n	6	1	10	12			
	female mean	125.4	255	51.8	69.6			
	S.D.	3.74		2.39	3.62			
	n	7	2	9	8			
Tonga (To-At-1/2)	*male* mean	141.6						
	S.D.	6.92						
	n	10						
	female mean	137.0						
	S.D.	4.97						
	n	4						

Marquesas (Hane)	male	mean	151.0					
		S.D.	7.60					
		n	10					
	female	mean	127.0					
		S.D.	5.50					
		n	3					
Watom	male	mean	137.0	61.0				
		S.D.						
		n	1					
	female	mean			75.0			
		S.D.						
		n			1			
Taumako	male	mean	144.0	62.5	82.4	50.0	68.0	78.8
		S.D.	7.48	4.42	2.96			
		n	7	24	9	2	2	2
	female	mean	129.3	55.0				
		S.D.	5.03	3.08				
		n	3	9		2	2	2
Sigatoka	male	mean		60.1	75.8	45.8	64.6	70.0
		S.D.						
		n		10	7	8	8	
	female	mean		53.1	73.7	56		55.0
		S.D.						
		n		8	5	2	8	
Cook Islands	male	mean		60.0	80.4			
		S.D.		5.24	3.89			
		n		9	8			
	female	mean		56.3	74.3			
		S.D.		4.57	4.03			
		n		4	4			

from stature, though stature has to be considered in estimation of an individual's mass.

Despite the desultory interest in determining body mass from the skeleton there is actually a respectable lineage of studies of the problem. In 1921 Matiegka used limb joint dimensions to estimate skeletal mass, an analysis extended by Trotter (1954). Subsequently Behnke, in a notable series of papers (Behnke 1959, Wilmore and Behnke 1968, 1969), developed regression formulae for lean body mass derived from various skeletal dimensions. (Lean body mass is essentially the body without adipose tissue, the fat under the skin; not without fat altogether, for various lipids are integral to many tissues, notably the nervous system.) For exploratory purposes, the original formula, derived by Behnke (1959) from radiological data on a group of US navy personnel on whom detailed body composition studies had been carried out, is useful. This formula is:

lean body mass (LBM) = 0.123 (D^2. S)

where D = sum of joint diameters (Figure 2.3)

and S = stature

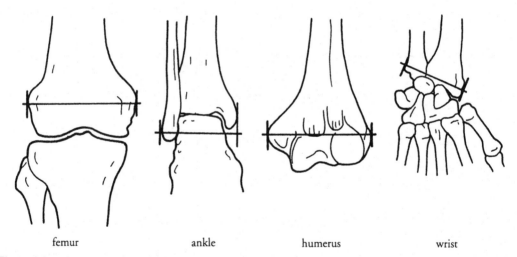

femur   ankle   humerus   wrist

Figure 2.3
Skeletal limb and joint diameters used in the mass equation.

Both stature and sum of diameters are expressed in decimetres in the formula. The squaring of the linear joint dimensions recognizes, albeit crudely, that the significant joint dimension is an area. In Behnke's original study the two sides of the equation had a correlation coefficient of 0.81.

The approach does not seem to have been pursued, probably because of growing awareness of the problems of radiation. Newer techniques, such as magnetic resonance imaging or ultrasound, should allow a fresh look at the problem of determining body mass from the skeleton. In the meantime the Behnke equation gives a starting point, for effectively he measured on radiographs what may be determined from dry bones.

We will look below at an attempt to assess how the Behnke approach works for a prehistoric series. First though it is useful to assess the accuracy of the joint/stature combination in a living group for whom the necessary anthropometric dimensions are available. Dr Robin Harvey has generously provided his anthropometric records of the people studied on Karkar Island, and at Lufa in the Highlands, during the International Biological Programme in Papua New Guinea in the late 1960s (Harvey 1974). Three joint diameters (wrist, elbow and knee) were available as well as stature and mass. In this analysis I first used data from 98 adults of both sexes from Lufa.

Mass was regressed against the square of the sum of the three joint diameters, times stature:

ie, mass = diameters (wrist + elbow + knee)^2 . stature.

In this calculation both the sum of the joint diameters and stature are expressed in decimetres. The F ratio for the regression of these normally distributed data was highly significant, and the coefficient of determination (r^2 value) was 0.6. The derived regression equation was:

mass = 19.5 + 0.5 (D^2. S)

where D = sum of joint diamters, and S = stature

In a confirmatory analysis this equation was then tested against 36 adults (20 male, 16 female) of the Karkar Island series. The actual mean mass of these adults was 52.4 kg (S.D. 7.3) and the calculated mass was 50.1 kg (S.D. 5.7). The coefficient of determination (r^2 value) was 0.7. A paired t-test on the predicted and actual values showed a highly significant relationship (p < 0.002).

These results suggest that the joint-area/stature combination is useful in the assessment of mass. In then exploring the determination of body mass from prehistoric skeletal material from the Pacific we returned to the Behnke equation and worked through the following steps. (Be warned, this is an unsophisticated analysis.) Prehistoric New Zealand Polynesians were studied initially. Four joint-related dimensions were measured on each side of each individual; humeral biepicondylar width; wrist width; femoral bicondylar width; ankle width (Figure 2.3). In this series no significant differences were apparent between right and left joint dimensions, despite the slightly different right and left long bone lengths resulting from handedness, and it seems justifiable to replace a missing dimension with its opposite, or simply to double right or left side measurements.

For each individual the sum of the eight dimensions was multiplied by 1.02 to correct for shrinkage of dry bone. Ingalls (1927) found shrinkage of 1.5–1.8% in the femoral bicondylar widths of his series, in which the bone had been dried for several months. In view of some two hundred years burial and then many decades in museum boxes for most of the present series, it was felt that an allowance of 2% shrinkage between the fresh and dry state was reasonable. Thus the formula becomes:

lean body mass = 0.123 (cD^2. S)

where cD = sum of diameters corrected for shrinkage.

The stature of each individual was calculated from tibial length using our Polynesian formula.

The mean total body mass of the individuals in the Behnke series was 78.3 kg, 23.5% greater than the lean body mass. (In terms of total mass, 19% is not accounted for by lean body mass.)

Therefore:

$$\text{total mass} = 1.235 \ (0.123 \ cD^2. \ S)$$
$$(\text{ie, Behnke LBM} + 23.5\%)$$

The original equation estimated lean body mass, which had been derived by other means (density studies) on the recent American group. The residual 19% component of total body mass was fat. It is most unlikely that prehistoric males in general reached this value of fat contribution to weight, and a value in the lower part of the range for westernized young adult males is probable; Damon and Goldman (1964) give a range of 12–18% body fat for athletes. Specifically regarding the Polynesian, the historical record of earliest contact does not suggest the general existence of obesity. (The emphasis is on the word 'general'; we shall come later to the record of obesity in the Pacific.) Skinfold thicknesses on relatively unwesternized Polynesian males (Prior et al. 1981, Finau et al. 1983) suggest a body fat contribution of about 12%. The same sources suggest body fat content of about 16% in young relatively unwesternized Polynesian females compared with values of 16–39% in young westernized females (Young et al. 1962). However all these considerations of fat are here slightly beside the point – they are taken up again in the next chapter – as joint dimensions relate to total mass regardless of body composition. They just mean that in prehistoric groups the missing 19% will be distributed between fat and lean mass.

Total body mass was thus estimated using the modified formula:

$$\text{total body mass} = 1.235 \ (0.123 \ cD^2. \ S)$$

Mass estimates for these prehistoric New Zealanders were 57.5 kg for 17 females and 75.6 kg for 15 males.

It is useful now to set these results alongside Buck's data on the Maori Battalion. New Zealand probably was settled by a rather small group from eastern Polynesia, with a restricted gene pool, and both the men of the Maori Battalion and the people of the prehistoric series are of that lineage. The mean mass obtained for the prehistoric group with this modified formula is 0.4 kg greater than the mean mass for the Maori Battalion. The estimated stature is 1.7 cm greater for the prehistoric group. Regarding the Maori Battalion, Buck (who was their Medical Officer) commented:

It cannot be argued that, as soldiers, the men were above the average in weight and physique . . . Towards the close of the war, the pick of the race had passed through civilised warfare with its huge toll from disease, wounds and violent death so that anyone who could pass the medical tests was sent away to keep up the strength of the Maori Battalion in the field. Consequently the returning Battalion contained men of 9 stone odd in weight and five feet in height, so that the measurements give a fair average of the race with, if anything, a tendency to the low side. As regards weight, it must also be remembered that the men had been in constant active physical training and had not had time during the voyage to regain the superfluous flesh of civilian life . . . Therefore our figures are, if anything, on the light side. (1922–23: 43)

In view of these comments the correspondence between the prehistoric and historic data seems very good. We have, unfortunately, no early data on female Maori mass. However Behnke makes the point that in young adults the female/male mass ratio is about ¾ or 75%. The suggestions for prehistoric Maori mass of 57.5 and 75.6 kg respectively arrive encouragingly at a ratio of 76%.

Stature does play a part in this calculation of living body mass and it would be best to use stature formulae derived from some group close to the one being studied. Yet in the New Zealand series the use of the White American formulae of Trotter and Gleser (1952), which do not take into account the short distal segment of the Polynesian lower limb, does not lead to a substantial change in the mass estimate. The male value becomes 74.5 kg, a reduction of 1.2 kg, and the female value becomes 57.2 kg, a reduction of 0.9 kg. This rather minor influence of stature on the result is significant. It might be argued that the claims for tall stature in the prehistoric Pacific simply result from the use of equations derived from a living group – the Maori Battalion. This is so, though it ignores the other evidence. However the body breadths dominate in the mass estimates, which are therefore rather robust (that woolly word again) against variation in stature.

These osteometric data of lengths and breadths, estimates of frame size and of body mass, are clearly dominated by the groups from Remote Oceania. Even the large series of large people from Taumako on the fringes of Near Oceania is rather late in the prehistoric sequence, and may have affinities with Remote Oceania: Taumako is one of the Polynesian outliers. This brings up the point that these various sources are not always giving evidence of temporal change in physique during the initial west to east settlement of the Pacific, for this aspect is sometimes clouded in prehistoric time by a subsequent return of people from east to west. But overall, these data do tell of large people in the small-islanded Pacific.

In this chapter, using a range of sources and approaches, we have been looking to establish the human physical form through the Pacific in prehistory. The findings from the historical record are supported by the anthropometric and, within the limits of its sources, the skeletal evidence; with passage into Remote Oceania people get bigger, much bigger. Why this might be so is next examined.

PEOPLE AND ENVIRONMENT

3

We have looked at the Pacific environment and the physique of Pacific people. This chapter considers the shaping of the people by this environment.

Biogeographical rules

In terms of morphology, one view of the relationship between any warm-blooded species and its environment is summed up by Damon: 'Climate . . . does indeed seem to be the major regulatory factor for . . . body size and proportion' (1977: 221). In this statement are subsumed the classical biological rules of Bergmann and of Allen. In 1847 Carl Bergmann, a German physiologist, published 'The relationship of the conservation of heat in animals to their size' ('a very original paper' noted D'Arcy Thompson), in which he examined the problem of heat loss and heat production. Aware that surface area varies as the square of a body's linear dimensions, whereas mass varies as the cube, Bergmann reasoned that in the same environment a small animal has to produce more heat per unit of mass than does a large animal, in order to keep pace with surface loss: 'this extra heat production means more energy spent, more food consumed, more work done' (Thompson 1942: 34). Thus we come to such comparisons as that a man may eat one-fiftieth of his body weight in a day whereas in the same time a mouse must eat half its weight, and a warm-blooded animal much smaller than a mouse is an impossibility. The principle Bergmann arrived at explains why, within a warm-blooded polytypic species, those living in cold regions tend to have greater body mass than those living in warm regions. In various expressions this has become known as Bergmann's Law or Rule. Some mammals and birds with a wide geographic distribution illustrate this rule well: Arctic hares and foxes and bears are heavier creatures than their tropical cousins. The largest penguins are in the Antarctic and the smallest in the tropics. And so on.

In 1877 J. A. Allen, an American zoologist, formulated a related rule, that animals living in cold climates tend to have shorter and stockier extremities than those living in hot climates. The Arctic fox has shorter ears and muzzle and tail than the desert fox. This makes sense, for the temperature drop along even a stocky limb may be 20 degrees. For *Homo sapiens* too, Allen's Rule predicts that limbs will be relatively shorter and stouter in the colder parts of the species range. The Inuit and the Suomi have legs contributing only 47–48% to their stature, while, as earlier noted, some long-limbed Negro groups (those who now dominate Olympic sprints and hurdles) and Australian Aborigines have legs contributing 53–54% to their stature.

Allied to Allen's Rule is the suggestion of a relationship between climate and the shape of the human head. This, on first hearing, might seem rather unlikely, but there is no doubt that the head can be an important source of heat loss: at 4°C, some 50% of the body's resting heat output can be lost through the uncovered head (Froese and Burton 1957). Working from the cranial index, the ratio of head length to head breadth, Beals (1974) suggested that a brachycephalic head – round, with small surface area relative to volume – was a cold climate adaptation, and a dolicocephalic head – long, with a relatively large surface area – a warm climate feature. Beal's data are compelling (though he did not have any Pacific people in his series), but a more refined assessment of surface area than just the cranial index seems desirable.

The validity of Bergmann's Rule has been much debated, and facts such as the largest land animals nowadays being in the tropics, or examples of rather small creatures in the Arctic, are proposed as refutation. However nothing in biology is simple, and in climatic adaptation there is no need for an obsession with morphology. Adaptation may be morphological, physiological, biochemical or behavioural, or more probably a subtle combination of some or all of these. A species may hibernate or burrow or huddle to deal with climatic pressure. Amongst *Homo sapiens* a very few groups, such as the Australian Aborigine, have a particular ability to constrict superficial blood vessels in order to reduce heat loss (Scholander *et al.* 1958, Hammel 1959). This adaptation, along with the long thin Aboriginal frame, is ideal for a very-hot-by-day/dry-cold-by-night environment – and much handier than carrying around fat or furs for night-time insulation. Such a physiological adaptation eases the problem of morphological adaptation to a varying environment. However, while keeping in mind these options for an evolving animal, it does seem that the more the so-called exceptions to Bergmann's Rule are scrutinized, the more fundamentally robust it appears. That is, morphological variation is frequently the adaptation that adjusts a species to differing environments.

For *Homo sapiens*, a viewpoint in disagreement is given by Shephard:

Such hypotheses of natural selection imply a substantial evolutionary pressure from some adverse feature of the environment, with the prospect of relief from this pressure through adoption of a particular body form or physique . . . In practice, it is difficult to assign a numerical value to the importance of such a contribution to human adaptability. Successful colonization of a given habitat depends upon the amount of physical work needed for survival, population pressures and available technology . . . In some, if not all, environments a diversity of physical challenges or the importance of intellect relative to brute strength may well have precluded the emergence of an advantageous body form . . . It is possible to demonstrate substantial differences of genetic markers when superficially similar peoples are compared from one habitat to another . . . but the overall physique and body composition seems remarkably similar not only within a given ethnic group, but also from one indigenous population to another. Most of the small documented differences of body build could easily have arisen through differences in nutrition and physical activity rather than through genetic factors. (1991: 168)

Later we can assess how Shephard's comments stand up.

It was a surprisingly long time before *Homo sapiens* was scrutinized for fit to Bergmann's Rule. In human terms, this anticipates that larger and more muscular people should be found in colder climates, when a

Table 3.1   Stature/mass ratios for Pacific and other groups. Sources of Pacific data as for Table 2.1. Data marked * from Bogin 1988, Alacahuf from Elsner 1963, remainder from Schreider 1975.

Male				Female			
Pacific groups	*stature/ mass*	*Other groups*	*stature/ mass*	*Pacific groups*	*stature/ mass*	*Other groups*	*stature/ mass*
Karkar	2.85	Bushmen	3.86	Karkar	3.21	Vietnam	3.42
Kwaio	2.85	Burma	3.24	Nasioi	3.16	Andaman Is.	3.25
Baegu	2.82	Kikuya	3.17	Manus	3.13	Japan*	3.04
Nasioi	2.81	Korea	2.90	Nagovisi	3.08	Italy	3.01
Nagovisi	2.79	Japan*	2.65	Baegu	3.07	Netherlands*	2.87
Aita	2.77	England	2.58	Kwaio	3.08	US*	2.85
Ulawa	2.72	Netherlands*	2.57	Baining	3.08	Cambodia	2.83
Manus	2.71	Eskimo	2.56	Ulawa	3.02	France	2.83
Ontong Java	2.56	US*	2.47	Tolai	2.83	China (Canton)	2.74
Lau (Malaita)	2.52	Alacahuf	2.45	Aita	2.77	Eskimo	2.71
Ulithi	2.49	Finland	2.44	Lau	2.74	North Russia	2.70
Fiji coastal	2.41	Turkestan	2.34	Fiji coastal	2.61	Somalia	2.64
Pukapuka	2.39			Ontong Java	2.62		
Tonga (Foa)	2.28			Samoa	2.47		
Maori	2.27			Pukapuka	2.39		
Samoa	2.26			Hawaii	2.34		
Lau (Fiji)	2.24			Tonga (Foa)	2.28		
Tokelau	2.24			Tokelau	2.22		
Hawaii	2.22						

large muscle mass can produce more body heat, and a relatively smaller surface area will lessen heat loss. Smaller-bodied, or at least more linear people should be found in hotter climates, where endogenous heat production is less necessary and a relatively larger surface area allows for more efficient cooling. It turns out that Bergmann's Rule does seem to hold up rather well for humans. For example Roberts (1953) found a highly significant association between mean body weight and mean environmental temperature for a global series of indigenous peoples. One expression of the rule is the stature/weight ratio, values for which range from under 2.6 for individuals from cold climates such as Finland and England, through to values above 3.2 for warmer regions such as Vietnam and Burma (Table 3.1). Where apparent exceptions to the rule are found, they generally involve the problem of how to define body size. In a study of sub-Saharan African groups Hiernaux and Froment found that 'The two main indicators of body size (stature and weight) vary in contradiction with Bergmann's rule ... in sub-Saharan Africa stature tends to be taller where heat is more extreme' (1976: 765). However Roberts' study correctly emphasized that mass and surface area were the significant parameters. The relevance of Bergmann's Rule to humans received further support from Schreider (1975), who published exten-

Table 3.2 Mass/surface area ratios. Sources of data as for Table 3.1.

Male				Female			
Pacific groups	*mass/ surface*	*Other groups*	*mass/ surface*	*Pacific groups*	*mass/ surface*	*Other groups*	*mass/ surface*
Karkar	35.56	South China	30.90	Karkar	33.54	North Vietnam	32.83
Kwaio	35.62	Nigeria	35.56	Nasioi	33.81	Japan*	34.43
Baegu	35.82	Aust. Aborigine	35.63	Manus	34.03	Netherlands*	35.20
Nasioi	35.84	North China	36.02	Nagovisi	34.34	US*	35.49
Aita	36.00	Egypt	36.59	Baegu	34.42	France	35.83
Nagovisi	36.10	Japan*	36.81	Kwaio	34.42	Cambodia	36.32
Ulawa	36.53	Netherlands*	37.08	Baining	34.46	Eskimo	37.12
Manus	36.60	Italy	37.15	Ulawa	34.76	Russia	37.26
Tolai	36.63	US*	38.11	Tolai	35.94		
Halia	36.96	Finland	38.23	Aita	36.57		
Baining	37.44	Alacahuf	38.74	Lau	36.64		
Ontong Java	37.71	Eskimo	39.07	Fiji coastal	37.41		
Lau (Malaita)	38.15	Germany	39.14	Ontong Java	37.53		
Ulithi	38.37			Samoa	38.78		
Fiji coast	38.81			Pukapuka	39.50		
Pukapuka	39.09			Hawaii	39.78		
Tonga (Foa)	40.11			Tonga (Foa)	40.44		
Maori	40.23			Tokelau	41.09		
Samoa	40.30						
Lau (Fiji)	40.47						
Tokelau	40.47						
Hawaii	40.75						

sive lists for the more pertinent body mass/surface area ratio for worldwide groups. The values ranged from under 30 for hot-climate groups to as high as 39 for cold climates (Table 3.2). Schreider concluded that his results disagreed with the assumption the characteristics of the human form are roughly the same where climates are similar:

Our figures show that low values for the weight/surface ratio may be due to various anatomical variations; to the extreme elongation of the limbs with a notable slimming of the trunk, as in certain tall African populations, or to global reduction of all body measurements, of which the various pygmoid groups constitute an extreme example. (1975: 529)

In fact, as Hiernaux and Froment (1976) point out, savannah people tend to be rather tall, for there sweating is a most effective mode of heat loss, whereas forest people, living in the same mean temperature, are small and slight, for the still, humid environment in which they live requires absolutely minimal body size to minimize body heat production and maximize relative skin area.

Bergmann at sea

If morphology is strongly influenced by environment, then, in light of the climatic data presented in Chapter 1, showing a pretty consistent tropical pattern right across the central Pacific, there seems little reason to suppose that the appearance of people should change significantly. (The major matters of nutrition and disease we will come to.) Yet, as the first European visitors noted so clearly, human physique does change across the Pacific. Compared with the peoples of Near Oceania, and particularly its western fringes, those of Remote Oceania were notably large at the time of European contact. The Remote Oceanic values of between 2.21 and 2.34 for the stature/mass ratio (Table 3.1) are below those of any high-latitude/cold-climate group. For the physiologically more informative mass/surface area ratio there is an increase with movement away from Island Melanesia (Table 3.3). The male mean for all Remote Oceania is 39.6. A solitary high-latitude/cold-climate group lies (just) within the Polynesian range – a series of German males has a value of 39.04. (Behind this displacement of Pukapuka in the northern Cooks lies a story. Early last century a tidal wave swept away all but 14 of the inhabitants of the atoll. The chiefly lineages were lost and the few survivors were of less than the usual Polynesian physique: Shapiro 1943.) No other group enters the range of the Remote Oceanic females, who have a mean ratio of 39.4. Likewise, lower limbs are relatively short in Remote Oceania, contributing some 47–48% to stature (though it is interesting that this seems to apply also through Near Oceania). Heads in Remote Oceania are usually round, with a mean cephalic index of 82

Table 3.3 Cranial indices in living Pacific groups. Sources as for Table 3.1, remainder from Shapiro 1933.

Group	Index
Ulawa	72.3
Loyalty Is	72.5
Halia	72.8
Nasioi	72.9
Ontong Java	74.1
Erromango	74.7
Kwaio	75.0
Lau (Malaita)	75.0
Nagovisi	75.3
Baegu	75.4
Santa Cruz	76.5
New Caledonia	76.5
Maori	77.7
Tanna	78.9
Marquesas	79.4
Aita	80.6
Tonga	81.1
Samoa	81.3
Fiji	81.8
Baining	83.7
Hawaii	84.0
Society Is	85.0

or 83, which is in the purported cold-extreme part of the distribution (Table 3.3).

In terms of human morphology this is the Pacific paradox. In apparent suspension of Bergmann's Rule, some of the largest and most muscular people on earth – probably, in the past, *the* largest and most muscular – are found across a region that lies firmly within the tropics. This is not a matter that has excited much comment. It has been ascribed to their living on 'salubrious islands', but such an explanation owes more to the tourist than the scientific industry; in Chapter 6 discussion of some patterns of disease shows all was not paradise. Rather, resolution of the paradox lies in exploration of the influence of the real oceanic environment, which is not obvious from the bland climatic data.

Firth described on the small island of Tikopia, in latitude 12°S, the period of the trade winds 'that blow steadily from the east between April and September. At this time the sky is frequently overcast for several days at a time and the weather is often wet and even chilly' (1963: 30). This is a land-based comment. But it is particularly for people at sea in small craft that the tropics can prove cold, the more so if it is wet, dark, or the sky clouded. Even on a modern yacht life can be chilly. In the past, whatever the merits of the voyaging canoes of Oceania, such as the Fijian *drua* or the great Polynesian double canoes, there was no possibility of staying dry. A recurring wetness, ranging from dampness to outright soaking, would have been inevitable, particularly when working to windward. No one who has spent time in the open sea in a small boat will have any argument with this, and anyway there is an abundance of anecdotal evidence.

In 1977 Finney sailed in *Hokule'a*, a reconstruction of a Polynesian double canoe, from Hawaii to Tahiti:

May 4 . . . It is cold and wet. We might be in the tropics, but the 15-knot trade wind blowing across the deck plus the nearly constant spray from the head seas makes it downright cold, especially at night when there is no warming sun. I still feel chilly, even when wearing a jogging suit under my foul weather gear . . . Even when wearing foul weather gear as pyjamas we could not shut out the drips, dribbles and spurts of water that found their way into ears, down necks and into pants. (Finney 1979: 106)

On *Kon-Tiki*, in latitude 9°S, Heyerdahl 'stumbled about the deck bent double, naked and frozen' (1952b: 156). And through Bligh's journal of his small boat voyage to the west following the *Bounty* mutiny, in the latitudes of Tonga, Fiji and northern Vanuatu, runs the same theme, again and again the words 'miserably wet and cold'. May 5 1789: 'Among the hardships we were to undergo, that of being constantly wet was not the least: the night was very cold, and at daylight our limbs were so benumbed, that we could scarce find the use of them.' May 7: 'Heavy rain came on at four o'clock . . . being extremely wet and having no dry things to shift or cover us, we experienced cold and shiverings scarce to be conceived.' On other days: 'Our situation was miserable, always wet, and suffering extreme cold in the night . . . We suffered extreme cold and everyone dreaded the approach of night . . . At noon it was almost calm, no sun to be seen, and some of us shivering with cold' (Bligh 1961: 152–163).

O'Connell described events in a lifeboat following shipwreck in the Carolines in 1826:

Even in a latitude which must have been within fifteen degrees of the equator, a night passed without sleep or food, in an open boat, washed by the continual breakings of the sea over it, chilled our whole frames . . . Broiling heat succeeded the chills of the night . . . through the night the wet chills, and the same heat and calm upon the next day. After two days and three nights of exposure the daughter died about ten o'clock on the third day . . . The mother, in her weak state . . . in a few hours followed her daughter. (1972: 104)

Gladwin described a trip on a Caroline sailing canoe in latitude 8°N: 'Just about that time a rainsquall hit, the wind shifted, and we spent the next two hours tantalizingly in sight of Pulawat, chilled to the marrow, trying to claw our way into the pass against the wind and the rush of the outgoing current' (1970: 58).

In 1972 Feinberg took part in a voyage from the small island of Anuta, east of the Santa Cruz Islands, to the isolated rock of Patutaka, in latitude 12°S. The distance was only about 55 kilometres, but perhaps this will be the last record of a voyage taken under neolithic conditions, without any significant technical contribution from another culture.

As we paddled through the rain showers, which also packed what must have been close to gale-force winds, the cold became intense. But except for Pu Toke removing his shirt to give to Nau Rongovaru, and Nau Rongovaru's crouching in the 'hold' for some slight protection from the wind as she guarded the fire stick, my companions seemed oblivious to the temperature and went about their activities as if it were 80 degrees. I felt somewhat bashful about putting on my plastic raincoat, but I finally conceded to my better judgement. It made the difference while paddling between comfort and discomfort, and while sailing between discomfort and what would otherwise have been almost intolerable cold. (Feinberg 1988: 138)

Even modern survival technology is not absolute protection: this fact is evident from a New Zealand newspaper report of a pilot who had to ditch his plane at sea. He had been delivering light aircraft throughout the Pacific for about fifteen years, and during that time had never had reason to suspect just how cold the Pacific could be at night: 'My biggest problem was keeping warm. I had a small space blanket type thing . . . I wrapped myself in the raft's canopy to maintain just a little bit of body heat' (Otago *Daily Times*: February 1990).

Heat balance calculations

The reality is that the persistent wetness and wind at sea require a fresh perspective on the tropical environment. For the early Pacific explorers and colonizers with their neolithic technology, life was not like the tourist brochures or the images of Gauguin. The oceanic environment is potentially and frequently very cold, whether for voyaging, or for the more mundane but routine activities of coastal fishing in a small-island existence, and perhaps occasionally even for life ashore. The suggestion is that the people of Remote Oceania are big because of their particular and uniquely cold environment. The task is to quantify the advantages of a large and muscular body in these wet-cold conditions.

Stable body temperature requires an equilibrium between the heat produced by the body and the heat lost to the environment. Both factors need to be considered, and we will start with heat loss. Loss of heat from the body may be expressed in kilojoules (kJ), and for any set of environmental conditions varies with the surface area of the body. (The estimates of body surface area used here derive from the study by DuBois and DuBois (1915), which has stood the test of time.) For people with neolithic technology, exposed at sea, heat loss for different physiques may be derived from studies such as that of Siple (1968), which established a loss from the naked body of 1356 kJ/m²/hour with a 23 kph wind and air temperature of 21°C. These are climatic figures compatible with the Pacific environment. This heat-loss value is approaching the subjective 'cool' level but takes no account of evaporative heat loss. The thermal conductivity of water is some 25 times greater than that of air, and even in water at 24°C, fit young adult subjects cannot tolerate more than 12 hours immersion (Hayward and Keatinge 1981). Of course canoe travellers are not usually immersed, but a person in wet clothing, especially in the presence of wind, is in much the same situation as a person in water at a slightly higher temperature (Pugh 1966a). That is, when this factor of dampness or wetness is added, the rate of heat loss in air readily doubles (Pugh 1967, Maclean and Emslie-Smith 1977, Kaufmann and Bothe 1986) to enter the subjective 'very cool' to 'very cold' range, which is in harmony with the written accounts. The combined insulative value of air and permeable clothing in these wet conditions is only about 0.3 Clo. (The Clo is a quaint but reputable unit of insulation, one Clo being defined as the insulation required to keep a seated subject comfortable at an air temperature of 21°C in an air movement of 0.1 m/s: such insulation is provided by an ordinary suit. There is another delightful unit called the tog – not capitalized, apparently – which equals 6.45 Clo: Clark and Edholm 1985: 204.) A heat-loss figure of 2000 kJ/m²/hour derived from a study of nude men at temperature 14.5°C, 91% humidity and 16 kph wind (Iampietro et al. 1958) approximates typical wet-cold conditions of the 'tropical' Pacific; if anything it is on the conservative side, under-estimating the possible effective coldness. In these conditions a smaller person with a surface area of about 1.6 m² has a heat loss from the body of 3100–3200 kJ/hour, while a larger person with surface area of about 1.8 m² has a loss of 3500–3700 kJ/hour (Tables 3.4 and 3.5).

Studies such as these allow the construction of a wind-chill graph. Figure 3.1 is an example. From such graphs hourly heat loss per square metre of body surface may be determined for any combination of air temperature and wind speed, and for wet or dry conditions. The graph shows effectively the importance of quite low wind speeds in body cooling, and the significant – or alarming – influence of wetness.

We turn now to the other side of the equation, heat production by the body. While at the population level Basal Metabolic Rate (BMR: Resting Metabolic Rate or RMR is a synonym), the basic energy turnover of the body at rest, may be inversely correlated with environmental temperature (Roberts 1978), any such differences are inconsequential under exposure conditions. Heat production above basal level is related directly to activity of skeletal muscle. At basal level, skeletal muscle contributes only 20% of body heat production, but any increase above basal level comes almost entirely from this tissue. A level of metabolic activity some five times basal is achievable by shivering, which corresponds to heat production of about 23 kJ/kg body weight/hour. Heat balance studies generally use oxygen consumption as an indicator of heat production; a litre of oxygen (at a mean respiratory quotient of 4.825) provides 21 kJ of energy. In a 75 kg person a five times basal level of metabolic activity requires an oxygen consumption of about 1.4 litres/minute, equating with heat production of some 1750 kJ/hour. In a 56 kg body the consumption is about 1.1 litres/minute, or 1300 kJ/hour.

Strictly speaking, these are rather (some would say *very*) optimistic

Figure 3.1
Wind-chill graph.
The figures within
the graph are the
rate of heat loss
from the body in
kJ/m²/hour. The
lower set of figures
is for dry conditions
and the upper set for
wet conditions. The
subjective impression
for each rate of loss
is given. Adapted
from Clark and
Edholm 1985.

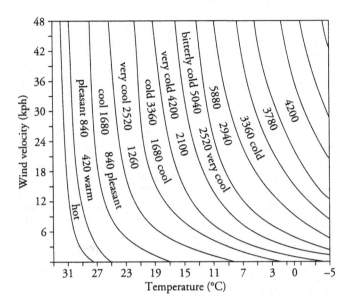

figures for heat production by shivering. The sustainable level of heat production from shivering may be only about twice basal level, that is about 700 kJ/hour in a big person. However the figure of five times basal level seems reasonable if an amount of active, deliberate activity is allowed for in attempts to keep warm. (Strictly speaking again, deliberate activity suppresses shivering. One activity which is quite strenuous and for which there would have been plenty of incentive – and this is an utterly serious statement – is bailing. An old maritime adage has it that the best bilge pump is a frightened person with a bucket.) The maximum heat production achievable by shivering (= 5 times basal) for the various Pacific groups is set out in Tables 3.4 and 3.5.

These heat loss and heat gain figures suggest that for relatively inactive, shivering and exposed people in wet-cold conditions at sea, even in the tropics, smaller individuals are able to produce only 42–43% of body heat lost, with an hourly deficit of 1800–1900 kJ. Larger individuals can supply 47–50% of body heat loss with an hourly deficit of 1750–1950 kJ. The larger people are better off, but for anyone such exposure is simply unsustainable for more than an hour or so. For example, a drop of core temperature from 37.5°C to 36.4° in 25 minutes has been noted in an inactive subject in wet-cold conditions of 5°C and a 14 kph wind (Pugh 1967).

However these figures can be related to the realities of survival at sea. During wet-cold exposure conditions, whether or not progress under sail was possible, physical activity for most crew members would be restricted; they would be huddled down out of the wind and protected by whatever (damp) coverings were available. The wind-chill graph (Figure 3.1) shows the reduction in heat loss achieved by lowering wind exposure to near zero. An optimistic estimate of the extent of this reduction is detailed for each group in Tables 3.4 and 3.5. It can be seen

Table 3.4 Male heat balance data at 14.5°C with shelter from wind.

Group	Mass (kg)	Stature (cm)	Surface area (sq m)	Heat prod'n kJ/hour	Sheltered heat loss kJ/hour	Total hourly deficit (kJ)	Rate of change of body temp. °C/hour	Body temp. after 8 hrs °C
Karkar	56.40	161.0	1.59	1297	1586	−289	−1.46	25.3
Kwaio	56.36	160.5	1.58	1296	1582	−286	−1.45	25.4
Baegu	57.27	161.3	1.60	1317	1599	−282	−1.40	25.8
Nasioi	57.73	162.2	1.61	1328	1611	−283	−1.40	25.8
Aita	59.55	165.2	1.65	1370	1654	−284	−1.36	26.1
Nagovisi	57.27	159.6	1.59	1317	1587	−269	−1.34	26.3
Ulawa	60.00	162.9	1.64	1380	1642	−262	−1.25	27.0
Manus	60.20	162.9	1.64	1385	1645	−260	−1.23	27.1
Tolai	60.60	163.6	1.65	1394	1655	−261	−1.23	27.2
Halia	63.60	167.9	1.72	1463	1721	−258	−1.16	27.7
Baining	60.10	157.7	1.61	1382	1605	−223	−1.06	28.5
Ontong Java	64.09	164.3	1.70	1474	1700	−226	−1.01	29.0
Lau (Malaita)	64.55	162.6	1.69	1485	1692	−207	−0.92	29.7
Ulithi	65.65	163.5	1.71	1510	1711	−201	−0.87	30.0
Fiji coastal	71.40	172.0	1.84	1642	1840	−197	−0.79	30.7
Pukapuka	70.60	168.8	1.81	1624	1806	−182	−0.74	31.1
Tonga (Foa)	75.20	171.3	1.87	1730	1875	−145	−0.55	32.6
Maori	75.20	170.6	1.87	1730	1869	−140	−0.53	32.8
Samoa	75.90	171.4	1.88	1746	1883	−137	−0.52	32.9
Lau (Fiji)	76.10	170.8	1.88	1750	1880	−130	−0.49	33.1
Tokelau	76.10	170.8	1.88	1750	1880	−130	−0.49	33.1
Hawaii	77.30	171.3	1.90	1778	1897	−119	−0.44	33.5

that the heat deficit of the large muscular person is now kept low, hour by hour, by shivering, supplemented perhaps by some degree of voluntary activity. Smaller individuals however are unable to keep pace with the heat loss, even in the sheltered situation, and would range in condition from chilled to hypothermic or worse, depending on body size and the time the wet wind-chill conditions persisted.

For a person thus sheltered, the changes in mean body temperature in one hour resulting from the calculated deficits in Tables 3.4 and 3.5 are derived from the heat balance equation (Clark and Edholm 1985: 5):

$$S = M.Cp.Td/Ad$$

where S = heat storage (ie, heat balance, +ve or −ve, in kJ/m^2 of body surface)

M = body mass (kg)

Cp = specific heat of body tissues

Td = rate of change of body temperature (°C/hour)

Ad = skin surface area (m^2)

Td, the loss (or gain) of body temperature/hour, is the required figure.

Table 3.5 Female heat balance data at 14.5°C with shelter from wind.

Group	Mass (kg)	Stature (cm)	Surface area (sq m)	Heat prod'n kJ/hour	Sheltered heat loss kJ/hour	Total hourly deficit (kJ)	Rate of change of body temp. °C/hour	Body temp. after 8 hrs °C
Karkar	47.04	151.1	1.40	1082	1402	−321	−1.95	21.4
Nasioi	48.18	152.3	1.42	1108	1425	−317	−1.88	22.0
Manus	48.20	151.0	1.42	1109	1416	−308	−1.82	22.4
Nagovisi	49.09	151.3	1.43	1129	1429	−300	−1.75	23.0
Baegu	49.09	150.8	1.43	1129	1426	−297	−1.73	23.2
Kwaio	48.64	149.7	1.41	1119	1413	−294	−1.73	23.2
Baining	47.90	147.7	1.39	1102	1390	−288	−1.72	23.2
Ulawa	50.00	151.0	1.44	1150	1439	−289	−1.65	23.8
Tolai	55.10	155.7	1.53	1267	1533	−266	−1.38	26.0
Aita	54.09	149.8	1.48	1244	1479	−235	−1.24	27.1
Lau (Malaita)	55.91	153.4	1.53	1286	1526	−240	−1.23	27.2
Fiji coastal	62.00	161.8	1.66	1426	1657	−231	−1.07	28.5
Ontong Java	59.55	156.0	1.59	1370	1587	−217	−1.04	28.7
Samoa	63.90	157.7	1.65	1470	1648	−178	−0.80	30.6
Pukapuka	66.50	158.7	1.68	1530	1684	−154	−0.66	31.7
Hawaii	69.42	162.6	1.75	1597	1745	−148	−0.61	32.1
Tonga (Foa)	71.00	161.8	1.76	1633	1756	−123	−0.49	33.1
Tokelau	71.80	159.7	1.75	1651	1747	−96	−0.38	33.9

Rearrangement of the equation gives:

$$Td = S.Ad/M.Cp$$

S requires:

a. body mass (to give heat production: taken here as 5 times basal = 23kJ/kg/hour)

b. environmental data (to give heat loss): ie, temperature, wind speed, wet or dry conditions

The wind-chill equation (Siple and Passel 1945, Toner and McArdle 1988: 392) may be used to estimate heat loss:

$$Kc = 1.16 (10v^{0.5} + 10.45 − v) (33-Ta)$$

where Kc = wind-chill (Watts/m^2)

v = wind velocity (metres/second)

Ta = air temperature (°C)

However in practice, heat loss can be determined with sufficient accuracy from the wind-chill graph.

Ad requires mass and stature, which are applied to the DuBois and DuBois (1915) equation for surface area:

$$Ad = (0.00718Mass^{0.425}) (Stature^{0.725})$$

Cp is 3.5

The mean body temperature is defined (Burton and Edholm 1955) as

$$T = (0.66Tc)+(0.33Ts)$$

where Ts is the mean skin temperature and Tc is the mean temperature of the deeper, vital organs and tissues, particularly of trunk and head;

generally termed the 'core' temperature. It is the maintenance of this deeper core temperature that is crucial to survival. A mean body temperature drop of 1.1°C/hour results in a core temperature of about 32°C after about eight hours. At this level the body's ability to shiver disappears, and a further drop in core temperature is likely to be abrupt and catastrophic. The smallest individuals could only avoid this temperature deficit by increasing oxygen intake by 250 ml/minute, 23% over the intake of 1.1 litres, which is already five times basal. Such increase would have to arise from continuing and vigorous deliberate muscular activity, and in reality would not be sustainable. (Another reality is that most of the people in Tables 3.4 and 3.5 would be dead after eight hours exposure in the prescribed conditions.)

During moderate exercise, body oxygen consumption is about 25ml/kg body weight/minute, which equates with body heat production of 30 kJ/kg/hour. A 75 kg body will then produce about 2250 kJ/hour in body heat, and a 56 kg body 1660 kJ/hour. This level of moderate exercise is a reasonable estimate for much of canoe activity, whether voyaging or local fishing. Voyaging canoes were sailing craft, but could be paddled. The physiology of physically fit canoe paddlers of Polynesian ancestry has been investigated. In what must have been an unusually entertaining experiment, Horvath and Finney (1976) determined a mean oxygen uptake of 2.2 litres/minute for mean speeds of a little over 5.5 kph sustained for eight hours of paddling. This is equivalent to an hourly heat output of some 2600 kJ, which is pretty warming stuff. During periods of sailing the level of activity would be less, but still above basal levels. As the wind dropped below about 8 kph and some paddling became possible, heat loss from the body would rapidly reduce, heat production increase, and for the larger body any deficit in body heat would readily be restored. What is clear is that in this environment seemingly minor changes in wind exposure may substantially influence heat balance and as Pugh observed in studies of exposure, 'apparently trivial differences in metabolism or total insulation can have disastrous consequences' (1966b: 1281).

Any set of conditions, of air temperature, wind speed and wetness, may be used in estimating heat balance. At 26°C and wind speed of 15 kph, which sound rather pleasant conditions, the 'bush' phenotypes of Island Melanesia are only just maintaining body temperature if they are dry. A minor deterioration in conditions, clouding of the sky and rain, or increase in wind speed with spray coming aboard, plunges this physique into heat imbalance. Later in the chapter the effects of some realistic variations in the weather are explored.

Despite the advantages of a large body, heat imbalance must have occurred at times for any individual, and the fluctuation between effective high and low temperatures in this oceanic environment needs emphasis. Unfortunately we do not seem to have the ability possessed by species such as the camel to accumulate a significant heat credit against cold hours to come. But in the tropical Pacific there is always the chance of a bitter night being followed by a very warm day, or of the wind dropping and vigorous exercise becoming possible, with recovery of heat balance. This contrast came through earlier in the extracts from the journals of Bligh and O'Connell. Biggs has translated the account of two men adrift in a small boat between Futuna and northern Fiji:

The things we suffered from were the heat of the sun, and thirst . . . Then it was dark and, alas, we suffered from the cold of the night, because our clothes were wet. For myself I was unable to sleep by day or night for the whole voyage . . . The different things that we suffered from during the eight days and seven nights that we drifted were . . . the heat of the sun and the cold of the nights. (1974: 362–3)

Lability of effective temperature thus is a distinctive feature of the tropical oceanic climate. Whatever temperatures might be reached by day it may reasonably be claimed that the environment of Remote Oceania is effectively the coldest to which *Homo sapiens* has adapted, and that at the time of Western contact the people displayed (and frequently still display) the supreme cold-climate body form. The cold could have occurred by day or night, for neolithic technology provided no effective insulation against the pervading wind and wet at sea. By contrast, such technology may be highly effective in dry cold regions. The Inuit are generally regarded as the archetypal cold-climate people, yet under their furs they bask in tropical warmth. A double layer of caribou skin offers nearly three times the insulation of current Arctic military clothing (Shephard 1991). Newman and Munro point out that it is not realistic 'to think of the Eskimo crouched motionless over a seal hole in the bitter, all-pervasive arctic cold', and cite Stefansson's (1946) account of inland temperatures of nearly 100°F, day and night (for of course there is no night in summer in the Inuit environment). 'Before white man's influence spread over the North American Arctic, the typical Eskimo house in the afternoon and evening resembled a sweat bath rather than a warm room . . . There were streams of perspiration running down our bodies constantly, and the children were occupied in carrying round dippers of ice water from which we drank great quantities' (1955: 14).

If the view that the neolithic Pacific was effectively a cold place contradicts the common perception, it is only because the typical Pacific dweller or visitor nowadays is simply not exposed under the conditions of neolithic times, in the face of which selection and evolution occurred. As Templeton and Rothman comment: 'The worst conditions an organism faces play the dominant role in evolution and not average conditions' (1974: 411).

Fat

It has been suggested that the natural insulation provided by body fat had survival value in oceanic colonisation. (Fat is widely distributed through the body, being integral, for example, to the nervous system. Strictly, subcutaneous fat should be termed 'adipose tissue'. However I cannot keep writing this ponderous term, so in the present context 'fat' should be taken as synonymous with subcutaneous fat or adipose tissue.) Impressed by the obesity of some westernized Polynesians, Baker (1984: 213) has suggested in an anecdotal way that selection for fatness occurred as the Polynesians evolved within the Pacific. While clear that he is 'not suggesting that the Polynesians would have died of hypothermia on the voyages', the fat is seen as providing both insulation against cold and a store of calories to cope with the 'metabolic heat production induced by the cold'. Beaven (1977) has commented along similar lines. (Most discussions of the modern Polynesian tendency to run to fat refer back to Neel's (1962) attractive though biochemically hazy epithet of the thrifty genotype: of which more in Chapter 7.) Certainly, many studies have shown the value of a good layer of subcutaneous fat in conserving body heat, particularly in cold-water immersion.

However the subjects of these studies, such as long-distance swimmers, have been westerners with invariably a higher body fat content than non-westernized groups, and they usually have been, as Pugh and Edholm put it, 'fat, and many of them grossly fat' (1955: 768). A lot of fat does seem to be needed to provide worthwhile insulation. Burton and Edholm (1955) estimate that a thickness of 2.5 cm is required to provide one Clo of insulation. Rennie *et al.* (1962) are more optimistic, estimating one centimetre of fat to provide one Clo of insulation. Another study (Daniels and Baker 1961) reported only slight difference in conductance (heat transferred to body surface/m²/°C difference between internal and surface temperature) between thin and fat men under cold conditions, though under warm conditions the difference in conductance was substantial – which is just when interference with heat loss is not wanted. Whatever its insulation value, such fat advantage 'is probably to be regarded as little more that a fortunate benefit provided by a food store' (Keatinge 1969: 20) and such levels of food storage are probably only generally attainable in the modern Western environment. There is no evidence that indigenous people active in cold environments have a general tendency to accumulate significant quantities of subcutaneous fat. In fact the opposite is the case. Studies of groups such as the Alacahuf Indians of Tierra del Fuego, Alaskan and Canadian Inuit, and Quechua and Arctic Indians, show skinfold thicknesses much less than in modern Western populations (Elsner 1963, Ducros and Ducros 1979). If fat had primarily an insulative function this would not be the case. The Hae Ngo of Korea, women professional divers who until about 20 years ago wore only light cotton clothing, in those days had a mean fat thickness marginally less than non-diving Korean women (Rennie *et al.* 1962). A group exposed over several weeks to a cold, wet, sub-Antarctic climate recorded a loss of subcutaneous fat (Budd 1965). And so on. And even for westernized subjects, a study by Hayward and Keatinge determined that 'the most notable general finding about people's ability to stabilize body temperature in cold water was that although it depended greatly on the internal insulations achieved, which in turn were closely related to their subcutaneous fat thicknesses, it also depended to about the same degree on the size of their metabolic responses to cold' (1981: 233). One may note also that muscle itself is a pretty good insulator.

In advocating the role of fat as insulator, reference is usually made to sea mammals. These may possess massive amounts of blubber, and at first sight insulation may seem its clear and necessary function. Davenport gives a strong caution against this assumption (1992: 98). He sees fat primarily as an energy reserve for large mammals (whether polar or tropical, land or sea) and often for fuelling long migration periods or seasonal shortages of food. It seems probable that the larger whale species have more of a problem in keeping their temperature down than up. Fat also provides a nitrogen sink during periods of prolonged deep diving, and material for buoyancy and streamlining (subtleties of this last function may be very important). 'Obviously none of these functions of fat exclude an insulative function [but] the proposition that peripheral fat is an insulative material must always be treated with great caution'.

While there is no doubt that Polynesians have a strong tendency to obesity in westernized conditions, the view taken here is that this is a secondary and indirect consequence of selection for muscle mass as an evolutionary adaptation to cold. The matter is discussed further in Chapter 7. Of course it is unwise to be too dismissive of the usefulness of fat. If, in Remote Oceania, we are seeing what is effectively, if intermittently, perhaps the coldest of global climates, and people displaying the supreme cold-climate body form, then it seems reasonable that every possible morphological advantage should be utilized. Baker's comment might seem to make survival sense, the fat being not only insulation but a reserve for muscle metabolism and heat production, and also endogenous water. However voyaging canoes were big enough to carry ample food for whatever voyage might be contemplated and more, and I do not think food and water would often have been a survival constraint.

The historical record has something to say about fat, though it should be remembered that this record is generally of settled and mature societies. In the Marianas in 1683, according to Sanvitores the people were 'larger in stature, more corpulent and robust than Europeans . . . They are so fat they appear swollen' (Garcia 1937: 21). But at Easter Island the people were 'totally free from corpulency' (Clerke 1961: 760), and in Fiji Williams in the mid-nineteenth century observed 'Corpulent persons are not common, but large, powerful, muscular men abound' (1858: 104). At the Marquesas, 'of a great number of Men that fell under my inspection, I did not observe a single one either remarkably thin, or disagreeably Corpulent' (Clerke 1961: 761). In New Zealand, 'some are remarkable for their large bones and muscles, but few that I have seen are corpulent' (Anderson 1967: 809). In Tahiti 'we over took a fat Chief whose Body was so unweildy (sic) that he could not move fast enough' (Pickersgill 1961: 773). In Tonga, of a chief: 'If weight in body could give weight in rank or power Poulaho was certainly the most eminent man in that respect we had seen, for though not very tall he was of a monstrous size with fat which rendered him very unwieldy and almost shapeless' (Anderson 1967: 880). At Hawaii in 1819 (almost 50 years after first European contact), Freycinet noted that 'Among the higher ranks . . . the obesity of the great majority is most remarkable, especially among the women, who attain, even while still quite young, a really monstrous weight' (1978: 53).

In interpreting these comments care must be exercised, because in the past the term 'corpulent' does seem often to be used in the sense of large body size, signifying muscle as much as fat, and fatness is mentioned as a separate and specific feature. Even the *Oxford English Dictionary* is not totally clear as to the time of divergence of the meanings. I think a fair summary of all this is that in Polynesia the people who are unequivocally fat in the historical record were predominantly of the aristocracy, in what were at the time of European contact mature and sophisticated – even affluent – societies, in firm control of their environment and its resources, and with a clear social hierarchy. Neither the written nor the extensive artistic record bears evidence of a general tendency towards fatness. Even in the favoured Pacific, neolithic life would have continued to require agility and speed of action for many activities. There are also major disadvantages in carrying a cloak of fat in the heat. At rest the temperature gradient from deep to superficial tissue amounts to about 1.7°C/cm subcutaneous fat thickness (Hatfield and Pugh 1951) and if the metabolic rate is increased fivefold by moderate exercise, the temperature gradient rises to 8.5°C/cm of fat.

Measurement of fat by skinfold thickness is a relatively recent anthropometric innovation, and the oceanic data are limited and inevitably are from groups with some degree of westernization. Even so, the skinfold thicknesses in the least acculturated groups from Remote Oceania are, particularly for males, in the lower part of the range for modern Western groups. They are similar to those to be found in young active Western adults (Tables 3.6 and 3.7: there is further discussion later). The astonishing fat cloak carried by some modern Polynesians is confined to the most westernized, such as Hawaiians and Californian Samoans (Baker *et al.* 1986), those to whom cold nights at sea are but an atavistic memory. Of course fatness need not be equated with physical weakness. Active fat people may be very strong indeed (Polynesians threaten to dominate the world of Japanese Sumo wrestling) for with every movement they are doing that extra bit of weight training.

Up to now we have been considering fat as insulation for the trunk, for protection of the crucial body core of heat. There is also the matter of limb insulation and function. The anthropometric data show the Remote Oceanic limb (if that is an acceptable phrase) to be relatively short and thick. Studies of limb cooling (Mitchell *et al.* 1970, Eberhart 1985) as well as basic physics suggest that the internal temperature of a limb is strongly influenced by its diameter. As the often rapid onset of physical impairment in wet-cold conditions has been shown to be a consequence of local muscle cooling of the extremities, and not to a developing general hypothermia (Vanggaard 1975), the desirability of a thick limb in exposure conditions seems clear. This suggests an advantage in having a limb wrapped in fat when in the cold at sea: but again we

Table 3.6 Male limb fat and muscle data. Sources of basic data: Tokelau, Wessen 1992; Samoa, Bindon and Baker 1985; Maori, Buck 1922–3; others from Page *et al.* 1977. Sample sizes in parentheses, dimensions in millimetres and sq. mm.

Group	Skin fold		Limb circumference		Upper arm muscle diameter	Upper arm muscle circum.	Total upper arm area	Muscle area (+ bone)	Fat area
	triceps	*subscapular*	*upper arm*	*calf*					
Lau (Fiji) (74)	7.8	13.3	309	389	88.0	284.5	7598	6441	1157
Tokelau (875)	12.6	12.7	327		87.4	287.4	8509	6573	1936
S.D.	5.45	6.25	37.7						
Samoa (86)	11.5	16.7	315		85.0	278.9	7896	6188	1708
S.D.	7	10.1	38						
Maori (426)			292	379					
Aita (81)	5.6	10.5	272	344	79.2	254.0	5887	5150	737
(s.e.)	0.14	0.31	3.7						
Nasioi (59)	5.4	8.3	267	335	77.8	250.0	5673	4975	698
(s.e.)	0.22	0.27	2.5						
Nagovisi (109)	6.9	9.9	276	329	78.7	254.3	6062	5147	915
(s.e.)	0.18	0.3	2.02						
Lau (Malaita) (77)	6.2	9.3	295	346	85.7	275.5	6925	6041	884
(s.e.)	0.38	0.46	2.59						
Baegu (126)	5.2	8.3	264	338	77.2	247.7	5546	4881	665
(s.e.)	0.1	0.14	1.95						
Kwaio (127)	5.4	8.8	260	338	75.6	243.0	5379	4700	679
(s.e.)	0.1	0.2	1.82						
Ulawa (106)	7.3	10.3	276	340	78.2	253.1	6062	5096	966
(s.e.)	0.2	0.32	183						
Ontong Java (145)	7	9.4	284	353	81.1	262.0	6418	5463	956
(s.e.)	0.31	0.33	1.97						
USA (3091)	13	15	307		80.5	266.2	7500	5637	1863

Table 3.7 Female limb fat and muscle data. Sources as for Table 3.6. Sample sizes in parentheses, dimensions in Millimetres and sq. mm.

Group	Skin fold		Limb circumference		Upper arm muscle diameter	Upper arm muscle circum.	Total upper arm area	Muscle area (+ bone)	Fat area
	triceps	*subscapular*	*upper arm*	*calf*					
Samoa (127) S.D.	29.2 14.00	31.7 14.90	322 55		63.7	230.3	8251	4219	4032
Aita (88) (s.e.)	10.5 0.36	13.2 0.57	260 2.22	344	68.8	227.0	5379	4101	1279
Nasioi (59) (s.e.)	7.9 0.39	9.8 0.49	233 2.47	335	63.7	208.2	4320	3449	871
Nagovisi (101) (s.e.)	10.1 0.34	12.6 0.53	242 2.21	329	63.6	210.3	4660	3518	1142
Lau (Malaita) (101) (s.e.)	10.9 0.52	12.7 0.78	263 2.96	346	69.3	228.8	5504	4164	1340
Baegu (111) (s.e.)	8.4 0.22	10.2 0.32	228 1.77	338	61.4	201.6	4137	3234	902
Kwaio (114) (s.e.)	7.9 0.34	11.8 0.61	228 2.08	338	62.1	203.2	4137	3285	852
Ulawa (83) (s.e.)	11.9 0.56	17.9 1.05	248 2.74	340	63.2	210.6	4894	3530	1365
Ontong Java (197) (s.e.)	17.3 0.43	21.4 0.63	282 2.29	353	66.8	227.6	6328	4124	2204
USA (3581)	22	18	284		61.2	214.9	6418	3674	2744

stumble into the reality that much of the time in Remote Oceania things are pretty hot and such fat insulation would be a major disadvantage. Substantial differences in deep muscle temperature in limbs of fat and thin indviduals (for example 4.5°C difference in forearm between individuals whose body fat differs by 30%: Petrovsky and Lind 1975) have been shown, as well as differences in temperature at which the most efficient isometric muscle function occurs. Isometric contraction may be defined as that effecting no movement about a joint, and shivering falls into this category. This temperature of greatest efficiency, at 27°C, is surprisingly low, and above and below this body temperature isometric endurance falls away markedly (Clark, Hellon and Lind 1958). In water immersion studies Rennie *et al.* (1962) estimated their American subjects to be utilizing a non-fatty insulative shell composed of (relatively) unperfused muscle of thickness 1.8 cm. Their more muscular Korean subjects, both male and female, with better cold resistance, utilized a muscle insulation of 3.2 cm thickness. Temperature gradients have been studied in lower limb muscles in rather severe cold-wet conditions, with findings, after 40 minutes at 5°C, of temperatures of 27°C at 2 cm depth in both leg (extensor compartment) and thigh (quadriceps), rising at 4 cm to 31°C in the leg and 37°C in the thigh (Pugh 1966a). The insulative effect of muscle is at least half that of fat and a thick muscular limb has a better prospect of remaining warm and functional, at least in its deeper part, when circulation and metabolic function would have become severely restricted in the thin limb, and at the same time avoiding the disadvantage of fat in warmer times. In survival conditions efficient limb function may be vital.

It seems likely that there are subtleties in fat distribution between groups. Maling *et al.* (1994) derived a 'central body fat index', calculated from peripheral and central skinfold thicknesses (biceps + triceps/subscapular + suprailiac), for healthy young adults of a sample of contemporary New Zealand Maori and European. The European central body fat index of 0.88 was significantly different from the Maori (0.66), and indicated that the Maori propensity is to put fat on the trunk rather than on the limbs.

In an interesting study, Bell, Tikuisis and Jacobs (1992) investigated the relative contribution of different muscles to shivering. They estimated that 71% of heat production originated from trunk muscles, 21% from thigh muscles, with the upper limb and distal lower limb contributing 8%. One suspects that a Polynesian morphology would provide a rather larger contribution from both upper and lower limbs. As a subjective observation, the limbs of young non-obese Polynesian adults today, such as seen in some rugby players, often are astonishingly massive.

Muscle size in the upper arm may be quantified if the mid-part of the arm is (reasonably) regarded as cylindrical: then, from circumference, and triceps skinfold thickness, an estimate may be made of the contributions of fat, and of muscle and bone, to the cross-section. Some comparative data are given in Tables 3.6 and 3.7. They are rather a rag-bag set in the sense that the data are incomplete for several groups and summary statistics are not always available. Recent North American data are provided for comparison. Two fat/muscle calculations are given in the table. One is Garn's (1956) 'muscle index', which estimates the diameter of the arm without subcutaneous fat by the equation

$$\text{muscle index} = (\text{upper arm circumference}/3.14) - (1.33 \text{ triceps skinfold})$$

Triceps skinfold is multiplied by 1.33 rather than 2 to compensate for the 25% compression of the skinfold in measurement.

The other calculation is by the equation

$$Cm = Ct - 3.142 \ T$$

where Cm is the muscle circumference in the cross-section, Ct is the total circumference, and T the thickness of the triceps fold. This approach derives from Jelliffe (1966), and the equation is given by Frisancho (1974).

Further calculations from the results of this equation separate the fat and the muscle (and bone) contributions to the upper arm cross-section. This is a useful approach. For males, the recent American figures for muscle size – from young adults in an optimal environment where the secular trend has probably plateaued – are exceeded in Near Oceania by the Malaitan Lau (and, nearly, the Ontong Javanese), and all the groups from Remote Oceania – the Fijian Lau, the Tokelauans, and the Samoans. On the more limited female data, the recent American figures for muscle size are exceeded by the Aita, the Malaitan Lau, the Ontong Javanese, and the Samoans. In another study a small series of 9 men and 6 women, simply classified as 'Polynesian' (Reid *et al.* 1990) gave a mean upper arm muscle area of 6100 mm^2 (s.e. 400) for males and 4500 mm^2 (s.e. 400) for females, compared with the values of 5637 and 3674 respectively for contemporary American males and females. (The members of this small Polynesian series, incidentally, were living a sedentary urban existence.) Overall these data do support the view that those who have been involved with the Oceanic environment are particularly muscular. It is a pity that comparable data are not available for the lower limb, for it is here that the Oceanic muscularity often is so particularly striking.

All in all, I think the balance of evidence comes down against fat being a primary and significant factor in cold resistance for the prehistoric settlers of Remote Oceania, and against obesity being a consistent feature of their physique.

Helen Leach has suggested to me that the common practice in the tropical Pacific of liberally smearing the body with coconut oil may help to provide some insulation against cold. Long-distance swimmers smear their bodies with substances such as lanolin, and it has been calculated that 2 kg of this grease would provide a mean body coating of about one millimetre thickness (Pugh and Edholm 1955). Lanolin is much thicker in consistency than coconut oil and would be retained better. If we accept the estimate of Rennie *et al.* (1962) that one cm of fat provides the equivalent of one Clo of insulation, then one millimetre provides the lesser equivalent. This does not seem too impressive, but combined with a water-shedding effect might make a perceptible difference.

Female survival

If body size is important, one might expect females to be at a disadvantage in wet-cold oceanic conditions, and that a constraint in the settlement of Remote Oceania may have been female survival, with its demographic implications. Although the data in Tables 3.4 and 3.5 suggest that females of Polynesian dimensions would fare almost as well at sea – or sometimes even slightly better – than males, caution is needed. The more favourable female data, though from relatively unwesternized groups, are fairly recent, and their skinfold thicknesses are close to recent Western female figures. From their anthropometric studies in Island Melanesia, Friedlaender and Rhoads noted:

The difference between men and women in the pattern of weight loss and gain with ageing is striking. While it is the women who lose fat and weight in largest amounts in the more traditional groups, it is the women, again, who put on the most fat in those populations subject to rapid acculturation.
. . . It is quite possible that women expend less energy in manual labour in modernizing groups than they do in traditional ones (1984: 300).

There is also the likelihood that the energy drains of pregnancy and lactation are more easily met, or avoided, with acculturation. Whatever the basis, the Polynesian female weights used here in the heat balance calculations may be rather optimistic for the conditions of prehistory. There is no doubt that children would have fared poorly, which raises questions about the pattern of growth and maturation of Polynesians. These are touched on in Chapter 7.

In Western society there does exist an impression (and it is largely correct) that women tolerate cold rather better than men. There is a tendency to misinterpret this situation, and to extrapolate it to the past. I have seen this misunderstanding worded thus: 'Even though women are smaller and often, according to simple equations, more susceptible to cold stress, they have been shown time and time again to be more successful in handling cold – the physiology of the system on a sex basis is more important than body build considerations' (this is an anonymous referee's report). However in unadorned *Homo sapiens*, considerations of body mass and surface area dominate body heat balance, and objective equations are preferable to banalities such as the 'physiology of the system'. Nothing in the 'physiology of the system' gives women an inherent advantage. Unless pregnant or lactating, they tend to have rather lower basal metabolic rates than men, and as Hayward and Keatinge determined: 'The tendency of women to have thicker subcutaneous fat entirely accounted for their tendency to be able to stabilize in colder water than men; in fact, among people of given fat thickness there was a clear tendency for women to be slightly less well able than men to stabilize in the cold' (1981: 233).

From their study of the Korean diving women, Rennie *et al.* concluded that 'women appear better adapted to endure cold exposure than men due to their thicker layer of subcutaneous fat . . . There was no evidence for vascular adaptation of the diving women' (1962: 964). The evidence is that women did not have this fat advantage in the past, and they may have been more susceptible than men to cold. The only glimpse of another possible adaptation to cold was the observation that the diving women had an elevated shivering threshold. (A similar response is seen in male Australian Aborigines, so it is not sex-specific.) Since shivering decreases tissue insulation and increases convective heat loss, its suppression is economical, provided that the duration of cold exposure is brief and is terminated before severe hypothermia results. Both these conditions were met by the living conditions of the Aborigines and the diving women, but cannot be guaranteed in the tropical Pacific: 'At noon it was almost calm, no sun to be seen, and some of us shivering with cold' (Bligh 1961: 163).

Pacific people

Against the heat balance data of Tables 3.4 and 3.5, it is instructive to look more closely at the local environments of the living groups discussed in the previous chapter. A comment on each group from Near Oceania follows.

Karkar Island, off the north coast of New Guinea, is 'high volcanic, lacking beaches and fringing reefs along much of the . . . coastline', and the people are 'inland cultivators, mediocre seamen', with generally only small canoes (Harding 1967: 23–24). 'Although the Karkar people live in proximity to the sea, marine resources are of little significance to them, giving rise to the curious paradox of an island population which cannot be regarded as being maritime' (Hornabrook 1977a: 287). Physically they are slight in build and vulnerable to exposure, which matches the ethnographic comments; these are not a maritime people despite their island home.

A hundred miles to the north, the Manus of the Admiralty Islands are a 'sea-dwelling, fishing people who occupy villages built on piles in the shallow lagoons along the south coast off the Great Admiralty Island . . . They practice no agriculture . . . Their entire economic dependence is upon fishing and trade' (Mead 1937: 210). The Manus males display a much better capacity for cold endurance than the Karkar islanders, appropriate to a more maritime existence. This greater endurance is not matched by the females, who are not involved in the sea-fishing/trading existence. 'Their daily economic contribution consists of reef fishing for shellfish and small fish, bringing wood and water . . . making grass skirts and beadwork, helping in the manufacture of lime, and cooking' (*ibid*: 219).

The Halia people from Buka are coastal dwellers with a maritime economy, and a tradition of voyages, out of sight of land, to small offshore islands (Blackwood 1931–32). The males (there are no female data) limit heat deficit under the defined conditions to a little over 1°C/hour.

By contrast, for Bougainville in general, coastal exploitation has not been significant. 'Most of the people of Bougainville itself are concentrated on the plains of the south, which are separated from direct contact with the coast by a broad belt of swamps and poorly drained ground' (Friedlaender 1975: 17). Thus the Nasioi and Nagovisi may be described as inland 'bush' people. So also are the Aita in the north, but they live at a higher altitude, between 500 and 1000 metres. 'It is often cold and very wet in the Aita valley, and warmth, along with protection from the rain . . . is the primary concern of housing' (Mitchell *et al.* 1987: 38). The Nasioi and Nagovisi are slight of frame, with theoretically a poor ability to withstand maritime exposure. The Aita are heavier than the other two Bougainville groups, but the Aita males are also taller and show a resistance to exposure marginally less than that of the Nagovisi. The Aita females show a rather good resistance to exposure. In their relatively windless inland environment this Aita phenotype may have its advantages. Studies in the New Guinea Highlands, at twice the altitude of the Aita, show heat stress, not cold, to be the environmental problem (Budd *et al.* 1974, Fox *et al.* 1974).

The environments of the Baegu and Kwaio of Malaita are similar to those of the Bougainville bush people, and these Malaitans similarly show a rather poor resistance to exposure. At the north-eastern end of Malaita, the Lau people have a different environment, of beach villages and artificial islands built up in shallow parts of the large Lau lagoon. Life in the past has been essentially maritime, with canoes for transport. 'The men often stand on the reef in water chest deep and fish with lines or nets' (Mitchell *et al.* 1987: 42). Physically the Lau are heavier, although no taller, than the bush peoples, and this greater mass reflects in their better cold endurance, with an hourly drop of body temperature under the defined conditions of a little under 1°C for males.

To the north, on the atoll of Ontong Java, life is inevitably maritime. The males have mass, stature and cold resistance similar to the Lau, while the females are taller and heavier than their Lau counterparts, who are primarily tenders of gardens. Finally, the Ulawan people 'divide their time between horticulture and fishing, but there is greater emphasis on gardening here than in either Lau or Ontong Java' (Mitchell *et al.* 1987: 50). Appropriately, with this economy spread between sea and land but emphasizing the latter, the males are placed between the bush and the sea peoples, while the females surpass all bush peoples in cold resistance, except for the relatively high-altitude Aita.

For Remote Oceania at first sight it seems reasonable, on this matter of body temperature homeostasis, to regard the environment within the tropics as similar throughout. We will have more to say on this, but overall the perspective across the Pacific is of an increasing ability to resist cold going hand in hand with an increasing involvement with the sea. In the more comprehensive male sample, the transitional environments and economies of the Ulawa/Manus/Halia groups match their physiological data. In passage from the coast of New Guinea to the isolated atoll of Ontong Java we see the gradual transition to an oceanic physique. Even so, here on the fringe there is quite a way to go to the full-fledged form of Remote Oceania.

An hypothesis for survival

Ultimately, what these figures for heat production and heat loss suggest is that in the oceanic environment, with only neolithic technology, a larger-bodied muscular individual is at a quantifiable and crucial advantage in maintaining body temperature. In addition a thick muscular limb maintains its warmth and function better. In these wet-cold oceanic conditions, only an individual approaching Polynesian proportions and muscularity could maintain body heat and limb function over many hours. For individuals of lesser build the consequences would range from a moderate discomfort to hypothermia and death, depending on the persistence of the windy, wet-cold conditions. For a smaller person with only neolithic technology a few days at sea would often be lethal.

Emerging from these considerations is the selective hypothesis; that the transition to an impressively large and muscular physique with passage from Near to Remote Oceania is a consequence of the selective pressures of the oceanic environment. Phenotypic plasticity can cope to some extent with changing environment, and the point when such plasticity is left behind and genetic selection is occurring may be difficult to define. But the hypothesis is that such true genetic selection has occurred. (Certainly with westernization there has been no return from the large-bodied phenotype of Remote Oceania to something akin to the slighter physiques of Near Oceania.) With passage east across the Pacific the small-island situation also would see the influence of the founder effect, large body size being perpetuated, even if selection pressures eased somewhat as a group became established.

A computer simulation of survival at sea

Thus far in the discussion a particular set of environmental conditions has been used to approximate wet-cold conditions at sea and to assess the ability of different physiques to survive. A crucial question is, just how realistic is this chosen set? Taking 32°C as the body temperature below which people would die, then at the end of a day or night under such conditions only the Polynesians, male and female, are left alive (Tables 3.4 and 3.5). (Recovery from hypothermia below 32°C does occur, and under the highly artificial conditions of a modern operating theatre is routine: but here we have to be realistic about the survival of neolithic people abroad on the Great Ocean.) Is this set of conditions simply too severe, and unrealistic?

Of course the data were used in the first place because they seemed reasonable. However to look more clearly at the problem, Daniel Levy has developed a computer model to simulate a realistic pattern of weather at sea (Levy and Houghton, in press). In this model, two years of data for wind speed, air temperature, cloud cover and (sometimes) rainfall from a number of South Pacific weather stations (Figure 3.2) have been used. Against such a 'natural' environment, the fate of different physiques can be studied over the course of hypothetical voyages. Simulations were run for the November–February period, representing the southern summer, and May–August, representing the winter. In each simulation the voyagers were exposed to a 'natural' pattern of weather derived from the records for a chosen region. In the results presented here they were allowed to remain dry until wind reached a speed of 15 knots (about 28 kph), when spray started to come aboard and the wet wind-chill situation was entered, with all its implications for body cooling. Body heat loss when wet was taken as 1.6 times dry heat loss, rather than nearer twice the rate as is suggested by the wet wind-chill experiment. Shelter from the wind can considerably reduce heat loss. On the other hand it is a basic requirement of seamanship that the craft be always attended to, rather than having everyone in a huddle in the bilges when the wind strengthened. At least one hand would be needed at the helm. Bailing in winds above about 28 kph would be frequently required, and while initially a warming activity it would, paradoxically, lead to increased heat loss from exposure, accentuated by any vasodilation resulting from the exercise. Shelter would never be windless or better than damp. Studies on several animal species show huddling to be an effective method of reducing heat loss by up to one-third for each animal in the huddle up to a maximum of three, compared with that of a solitary animal (Davenport 1992). This doubtless was done, and is allowed for in the simulation. Heat loss from the respiratory system was not allowed for. All in all, to set the onset of wet wind-chill at a wind speed of 28 kph, and then take the increase in heat loss as 1.6 that of the

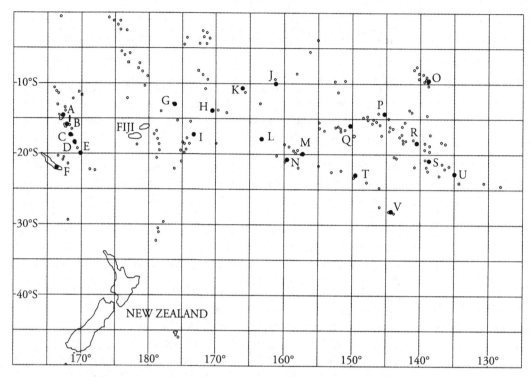

Figure 3.2 Location of 22 named weather stations. A = Sola, Vanuatu; B = Luganville, Vanuatu; C = Lamap, Vanuatu; D = Tanna, Vanuatu; E = Aneityum, Vanuatu; F = Noumea, New Caledonia; G = Hihifo; H = Pagopago, Samoa; I = Vavau, Tonga; J = Rakahanga, Cook Islands; K = Pukapuka, Cook Islands; L = Palmerston, Cook Islands; M = Mauke, Cook Islands; N = Rarotonga, Cook Islands; O = Atuona, Marquesas; P = Takaroa, Tuamotus; Q = Borabora, Society Islands; R = Hao, Society Islands; S = Mururoa, Society Islands; T = Tubuai, Austral Islands; U = Totegegie; V = Rapa. For four stations latitude was available but name and exact location could not be ascertained.

dry state, makes for a conservative or optimistic set of parameters, biased towards survival. Water and food supplies were not considered problems in these survival studies; their shortage could only lessen the survival figures suggested here.

In these 'voyages' the physiques of the 'crew' are the mean weight and surface area for male and female of each of the Pacific groups already considered. The Ulithi male is the only representative from Micronesia, the Maori male the only one from outside the tropics, and the West Nakanai data were obtained too late for inclusion. With 37 people aboard this is quite a big canoe, but that is immaterial to the exercise. The directions of the 'voyages' also are immaterial, but have been taken to continue in the weather pattern of the region in which they started. If required, a voyage may be allowed to progress into the weather parameters of another region, and the consequences are readily assessed

Heat loss from the body was calculated from the wind-chill equation:

$$Kc = 1.16 \ (10v^{0.5} + 10.45 - v) \ (33–Ta)$$
where Kc = wind-chill (Watts/m^2)
v = wind velocity (metres/second)
Ta = air temperature (°C)
with appropriate increase for wet conditions.

Death was taken to occur when body temperature reached 32°C. Temperature decrease of the body (°C) was given by the equation:
$$E/3.5M$$
where E = energy deficit in kJ/hour
M = mass of person in kg
3.5 = specific heat of body tissues

Along with the other biological parameters in the simulation, heat production by the body can be varied. As noted before, a figure of five times basal production seems a reasonable maximum, not sustainable by shivering for long periods, but achievable by moderate deliberate effort such as bailing or paddling.

The equation for heat production is
maximum heat production (kJ/hour) = 4.6.B.M
where B = multiple of basal energy possible to produce
M = mass of person in kg
4.6 = basal heat production in kJ/kg body weight/hour

For most station data at least a hundred simulations were run for any chosen length of time at sea. For each simulation there were only two possible consequences for any member of the crew, survival or death. Survival was scored as 1 and death as zero. All series of simulations were adjusted to 100 for ready comparison. For a given physique, a result of, say, 0.42 indicates that out of 100 simulations (under a given weather pattern, specific number of days, and the set body heat balance parameters for that simulation) survival resulted 42 times and death resulted 58 times.

The results of the simulations suggest that the conditions used initially to assess the survival chances of the different physiques – which left only Polynesians alive – were unlikely often to persist for eight hours at a stretch. However that is not much comfort. After five days at sea in summer in the environs of northern Vanuatu (latitude 14°S) and allowing for heat production by the body of up to five times basal, the slight physique of Karkar Island people sees survival of only 0.65 for males and 0.54 for females. In this locality, or weather pattern, there is not much improvement in survival by physique until the people of the outer limit of Near Oceania are reached – the Malaitan Lau and the Ontong Javanese, who show survival proportions of 0.75–0.80 for males and the Ontong Java females, and 0.66 for the Malaitan Lau females.

A major finding from the simulations was the association of survival with latitude. This was consistent across the range of the weather stations, from the Vanuatu–New Caledonia axis, through to eastern Remote Oceania. Incorporating all groups, the regression of survival on latitude was highly significant. Tables 3.8 and 3.9 give data for survival

Table 3.8 Survival proportions over ten days, by latitude in summer, for Karkar, Lau (Malaita) and Hawaiian groups.

Latitude	Karkar		Lau (Malaita)		Hawaii	
	female	*male*	*female*	*male*	*female*	*male*
9	0.24	0.54	0.54	0.65	0.70	0.74
10	0.48	0.54	0.64	0.73	0.80	0.80
11	0.25	0.41	0.45	0.45	0.56	0.60
13	0.39	0.44	0.46	0.62	0.67	0.79
14	0.27	0.36	0.40	0.52	0.64	0.72
15	0.23	0.23	0.40	0.44	0.59	0.62
16	0.21	0.25	0.32	0.34	0.48	0.54
17	0.24	0.31	0.33	0.43	0.46	0.50
18	0.17	0.23	0.26	0.32	0.37	0.46
20	0.16	0.17	0.18	0.19	0.22	0.22
21	0.00	0.06	0.06	0.10	0.15	0.16
22	0.05	0.06	0.06	0.11	0.16	0.18
23	0.01	0.01	0.05	0.10	0.15	0.15
27	0.00	0.00	0.00	0.00	0.00	0.00

Table 3.9 Survival proportions over ten days, by latitude in winter, for Karkar, Lau (Malaita) and Hawaiian groups.

Latitude	Karkar		Lau (Malaita)		Hawaii	
	female	*male*	*female*	*male*	*female*	*male*
9	0.21	0.28	0.31	0.42	0.50	0.58
10	0.44	0.47	0.53	0.60	0.66	0.71
11	0.33	0.34	0.40	0.45	0.47	0.60
13	0.05	0.07	0.10	0.17	0.25	0.29
14	0.12	0.13	0.14	0.21	0.30	0.33
15	0.02	0.02	0.04	0.08	0.11	0.17
16	0.00	0.00	0.03	0.03	0.05	0.06
17	0.03	0.03	0.04	0.09	0.12	0.14
18	0.00	0.00	0.00	0.02	0.02	0.02
20	0.00	0.00	0.01	0.01	0.02	0.03
21	0.00	0.00	0.00	0.00	0.00	0.00
22	0.00	0.00	0.00	0.00	0.00	0.00
23	0.00	0.00	0.00	0.00	0.00	0.00
27	0.00	0.00	0.00	0.00	0.00	0.00

by latitude over ten days, winter and summer, for Karkar, Lau (Malaita) and Hawaii.

Although in Chapter 1 the noticeable seasons in Noumea and Hawaii at the edge of the tropics were mentioned, and the sea and air temperatures seen to be distinctly lower (Figure 1.5), there is a tendency to regard the tropics as just generally hot. However, being only 10° removed from the equator vastly improved people's survival chances compared with being 20° removed. For example, in summer the five-day survival proportion of Malaitan Lau males and females is 0.75 and 0.66 respectively in the region of northern Vanuatu, but it plummets to 0.2 for both in the region of southern New Caledonia (latitude 22°S). (One continually comes back to Pugh's (1966b: 1281) comment that 'apparently trivial differences in metabolism or total insulation can have disastrous consequences'.) In the east the survival proportions for the Malaitan Lau men across six stations from Atuona in the Marquesas at 9°, south to Rapa at 27°, are 79, 70, 42, 25, 18 and 0. For all groups, whatever their range of survival, there is a rapid decline in survival beyond 15° of latitude. Correlation of survival with latitude was 0.92.

In Figures 3.3 and 3.4, survival by latitude in winter and in summer is plotted for the extremes of physique in the series, Hawaii and Karkar, and for the Malaitan Lau as an intermediate group. We have no data for stations closer to the equator than 9°; while the regression theoretically predicts 100% survival for most groups at the equator there are no data to support this, and it is probable that within 9–10° of the equator a plateau of survival is reached: things do not become much better.

Results of simulations for ten days at sea in summer are presented in Tables 3.10 and 3.11. The hazards of travel are approximately additive, so the consequences of, say, a 20-day voyage under the defined parameters may be roughly determined by multiplying the survival proportion for a ten-day simulation in that latitude, or by the different proportions if the course crosses several degrees of latitude. For example, if a group show survival of 0.70 after ten days, then after 20 days in the same latitude the proportion surviving will be about 0.49.

Two useful rules of thumb emerge from the analysis. For any group there is an approximately 5% decline in survival for each degree movement away from 10° of latitude; and there is an approximately 5% difference in survival between male and female in any group, the male being advantaged.

The weather records for the simulation are limited to two years. It may be argued that against the millennia of prehistory such a sampling is completely inadequate. However as far as the influence on the human frame is concerned this probably is not the case. The crucial influence of wind on body cooling occurs at low velocities (Figure 3.1), far below what could result from any general climatic change. A mean temperature change of 1°C over long periods is regarded as substantial, but such a change to the air and water isotherms of Figure 1.5 makes little difference to the immediate realities of body cooling. In other words, climatic change beyond anything likely to have occurred in recent millennia would be required to significantly influence these results. So interpretations from these simulations should be generally valid for prehistory, and there are interesting matters deriving from them. However in any sequence of years, whatever the mean climatic values, occasional mild and settled periods of weather do occur. (One recalls Herman Buhl's night out at 24,000 feet on Nanga Parbat, an experience no one would usually survive: he happened to encounter a freakishly mild night during which he remained comfortable.) There are good summers and bad

Figure 3.3 Summertime: ten-day mean value survival scatter plot for Karkar, Lau (Malaita) and Hawaii.

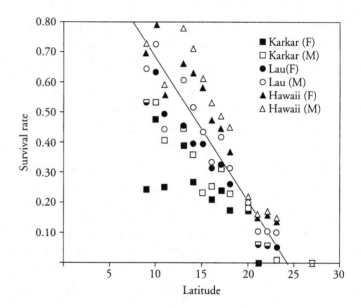

Figure 3.4 Wintertime: ten-day mean value survival scatter plot for Karkar, Lau (Malaita) and Hawaii.

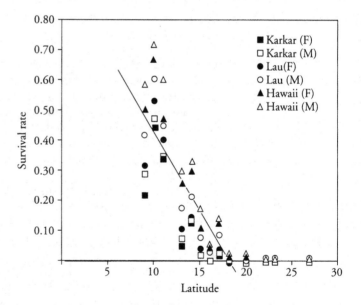

summers, cool winters and mild winters. The simulations suggest that some southern settlement in the Pacific could only have occurred during unusually favourable spells of weather; for example, the model is unkind to people trying to reach a southern outpost such as Rapa, and yet it was reached. Such mild periods with fair winds would not have to last for long – a few days at higher latitudes – if people happened to be at sea at the time. The topic is taken up again in Chapter 5. The immediate purpose has been to pursue the thesis that in the Pacific, environment has been a major determinant of physique.

Table 3.10 Survival proportions for males, over ten days in summer, all weather stations. (For clarity the proportions are given as whole numbers; ie, 51=0.51.)

Station	A	B	D	F	G	J	K	L	N	O	P	Q	R	S	V
Group															
Karkar	51	18	14	2	34	58	39	9	5	46	43	8	2	0	0
Baegu	51	25	14	2	34	58	39	13	5	46	43	10	2	0	0
Nasioi	51	25	14	2	34	58	39	13	5	46	45	10	2	0	0
Aita	51	25	14	2	39	60	39	19	5	46	49	10	2	0	0
Ulawa	60	30	20	2	39	63	44	20	5	51	52	10	2	0	0
Manus	60	30	25	2	39	63	44	20	5	51	52	10	2	0	0
Tolai	60	30	25	2	39	63	44	20	5	51	52	10	2	0	0
Ontong Java	63	33	29	2	46	74	45	20	5	55	58	19	5	0	0
Lau (Malaita)	71	34	29	2	47	76	47	22	5	55	62	19	5	0	0
Ulithi	71	34	29	2	47	76	47	22	5	55	62	22	5	0	0
Fiji coastal	71	34	29	2	47	76	47	22	5	59	62	23	14	2	0
Pukapuka	71	42	29	3	47	78	47	22	5	59	69	23	14	4	0
Tonga (Foa)	87	52	32	3	60	80	57	31	5	65	71	23	20	4	0
Maori	87	52	32	3	60	80	57	31	5	65	71	26	20	4	0
Samoa	88	52	35	3	60	80	63	31	5	65	71	26	20	4	0
Lau (Fiji)	88	52	36	3	60	80	63	31	5	65	79	26	20	4	0
Tokelau	88	52	36	3	60	80	63	31	5	65	79	26	20	4	0
Hawaii	92	52	37	3	64	80	63	33	13	65	79	26	20	4	0

Table 3.11 Survival proportions for females, over ten days in summer, all weather stations. (Proportions given as whole numbers.)

Station	A	B	D	F	G	J	K	L	N	O	P	Q	R	S	V
Group															
Karkar	35	13	9	1	32	52	22	4	0	23	29	4	0	0	0
Baegu	36	13	14	1	33	52	22	6	0	23	43	4	0	0	0
Nasioi	35	13	13	1	33	52	29	4	0	23	29	4	0	0	0
Aita	51	30	20	2	39	62	44	11	5	51	52	10	2	0	0
Ulawa	40	13	14	2	34	52	36	7	0	23	43	7	2	0	0
Manus	35	13	14	1	33	52	22	4	0	23	29	4	0	0	0
Tolai	51	25	14	2	39	58	39	13	5	46	49	10	2	0	0
Ontong Java	63	32	29	2	46	74	45	20	5	55	57	19	5	0	0
Lau (Malaita)	51	26	20	2	39	62	44	20	5	51	52	10	2	0	0
Fiji coastal	63	30	25	2	46	74	45	20	5	55	52	23	5	0	0
Pukapuka	79	44	29	3	47	80	51	22	5	60	71	23	16	4	0
Tonga (Foa)	88	52	35	3	60	80	63	31	5	65	71	26	20	4	0
Samoa	71	34	29	3	47	76	47	22	5	55	62	23	11	2	0
Tokelau	84	62	44	3	64	80	63	37	15	65	79	30	24	4	0
Hawaii	79	49	29	3	47	80	51	22	5	60	71	24	18	4	0

Other explanations

Now it is necessary to question the thesis that the change in physique across the Pacific arises from the need to keep warm, and consider other explanations. Two obvious influences are disease and diet.

Disease

First, the influence of endemic disease, which here really means malaria. I do not think any other parasites – gut worms, filariasis and so on – possess both a defined range, a ubiquitous involvement of those within that range, and such well-catalogued influences on the human frame. In third world countries today diarrhoeas of varying aetiology are signifi- cant in contributing to malnutrition and growth retardation. However these are likely to be problems of acculturation, rather than enduring problems from antiquity.

All four species of the malaria parasite (*Plasmodium*) are found in Melanesia. Until malaria control programmes were set up, *P. vivax* seems to have been the predominant species. The sickle-cell trait that in many African groups provides a degree of resistance to (falciparum) malaria is not found in Melanesia, but there are plenty of other useful polymorphisms, including hereditary ovalocytosis, the thalassaemias, and G6PD deficiency. In Papua New Guinea before the present era of considerable human mobility there were two distinct altitudinal con- centrations of human population: coastal populations, penetrating inland to altitudes of about 600 metres; and the highland populations living above 1300 metres. The upper region was effectively malaria-free while the coastal populations had developed a degree of immunity from constant exposure to the parasite. Any population in the area between was probably subject to more devastating epidemic malaria (Cattani 1992). Further east, it seems that groups such as the Aita and the Kwaio, living in the higher inland parts of Bougainville and Malaita, were also less exposed. Again, with greater mobility of people today this situation has changed.

To assist in examining the influence of malaria on phenotype, nature has provided a reasonably controlled experiment. Through Mela- nesia there are human groups living in malarial zones, and groups living in non-malarial zones. Both are also distributed across oceanic and non- oceanic environments. The slight-built Papua New Guinea Highlanders (Littlewood 1972, Harvey 1974) live above the zone of endemic malaria, so malaria cannot be taken as the basis of their phenotype. Similarly, in Island Melanesia the inland and slight-built Nasioi show a very low incidence of malarial infection both in terms of splenic enlargement and positive blood smears (Friedlaender and Page 1987). By contrast, the Lau of Malaita, living with a fair exposure to the sea and of rather robust build, show a high incidence of splenic enlargement and one of the highest incidences of positive blood smears for malaria. In general the peoples of Island Melanesia with a major maritime aspect to their economy show both infection with the parasite and a rather robust physique. The Ontong Javanese, the most robust of the Island Melanesian groups considered earlier, are an interesting case. Malaria was only introduced to their atoll in historic time, in the 1880s (Hogbin 1957), and while since then there has been substantial depopulation, they remain phenotypically well adapted to their oceanic environment. I know of no suggestion that their actual physique has diminished since malaria arrived. The evidence is that in regions of endemic malaria there is a satisfactory accommodation to the parasite (a matter as much to the advantage of the parasite as to the host) and that prophylaxis does not change growth patterns significantly. For example, reduction of the incidence of malaria in some African societies has not resulted in increase in weight (Tanner 1962) and many groups grow at similar rates despite wide differences in the type and prevalence of the disease. (McGregor *et al.* 1968, Ferro-Luzzi *et al.* 1978). These considerations suggest that the small physique of some groups in Melanesia is not primarily attributable to endemic malaria.

Diet and nutrition

A large body of writing attests to the debate on the relationship in Melanesia between dietary adequacy, body size, health, and efficient biological function. There is no consensus, and for every study or review there seem to be critiques and counter-claims (eg, Rappaport 1968 and McArthur 1974, Dennett and Connell 1988; and, separately, Jenkins, Micozzi, and Nurse, all 1988; and Hornabrook 1977a on the International Biological Programme in Papua New Guinea). Nutritional studies in the field are difficult to carry through, and easy to criticize. Because of the unresolved debate on what constitutes proper size, and thus adequate nutrition, it is not particularly helpful to analyze spoonful by spoonful the various studies that have been carried out and to try to determine their merits and demerits. In discussing the influence of diet on body form in Melanesia, it is probably more useful for the present purpose to stand back and try to gain a perspective on the evidence we have.

The physique and spartan diet of many Melanesian groups when compared with those of the affluent West, and, sometimes, apparent deficiencies when set against such benchmarks as the standards of the Food and Agricultural Organisation/World Health Organisation (FAO/WHO) on nutritional adequacy (1973, 1985), often has steered the debate towards a common arena – an assumption that small size is a result of inadequate diet – and left it as one between those who regard small size as a successful adaptation to sub-optimal nutrition and those who simply believe that small is bad. Other possibilities, such as size smaller than some Western ideal being normal and appropriate to a particular environment, are not always addressed. (This is reminiscent of the old Plunket Society record book for infants in New Zealand. For a child to fall below the red line for weight gained week by week spelled domestic shame and calamity, for here was evidence of neglect. The red line happened to be a running average for infant weight gain.)

There are several points to take up here. Hornabrook (1977b) gives a thoughtful comment on the cultural burden carried by Westerners trying to come to terms with existence in the tropics. The credibility of the FAO/WHO figures on nutritional need and adequacy has been dealt with by Rivers and Payne (1982), whose paper should be read by anyone disposed to take such figures as a benchmark for all *Homo sapiens*. The later revisions of these figures (1985) have not touched their critique. Misunderstandings abound regarding the significance of detailed growth data from Western populations, whose pattern of nutrition carries its own risks in chronic disease mortality and decreased longevity (Eveleth and Tanner 1990). Johnston and Zhen comment on the confusion regarding reference standards of nutritional status, and the misconception that here is some desirable norm for attainment. Rather, the reference data simply provide a position for others relative to the reference population and the interpretation of the nutritional significance of that position is another matter altogether. 'The separation of the anthropometric reference data from the evaluation of abnormality is an important conceptual step in the assessment of nutritional status. First of all it recognises the multiplicity of factors which determines anthropometric dimensions and therefore the complexity of the process of evaluation' (1991: 339).

Amongst this multiplicity, the climatic influence on body build has already been objectively examined. More anecdotal evidence suggests that some physiques may simply be better suited to a particular environment. The hill people of Nepal regard large people as a joke, ill-suited to the terrain, and a study by Strickland and Ulijazek (1990) noted a group of Gurkha soldiers to have more efficient motor function than their British equivalents. Ross (1976) comments on the build of the Baegu of Malaita as being eminently suited to their hill environment. In the Highlands of Papua New Guinea, Littlewood effectively presents the reactions of the rather larger Fore people when asked to carry loads into the rugged terrain of the smaller Awa: ' "the land where the Fore cry": it was an observable fact that Awa carriers moved with dismaying strength and alacrity under heavy loads' (1972: 40).

In a study of strength and motor performance of children of differing body size, a group of relatively small children from Manus Island of the Admiralty Group performed better than did larger American children (Malina *et al.* 1987: 490). The Manus group is described as showing 'chronic protein-energy malnutrition', though the support for this in the paper is limited to the statement 'Conditions in the community suggest mild undernutrition': (this may be unfair to these workers: it is possible that elsewhere more compelling data are set out). Other studies showing 'poorly nourished' or 'low calorie intake' people to have higher mechanical efficiency at various tasks than 'well-nourished' groups are those of Ashworth (1968) and Edmundson (1980).

Mainland Papua New Guinea presents a particularly complex mosaic of dietary variation. Undoubtedly in some areas there have been real deficiencies, but on the other hand some of these are of recent origin and a consequence of acculturation (Malcolm 1970, Lampl *et al.* 1978). In fact the mainland situation is not really relevant here, in that it has not entered into the considerations of physique and the heat balance calculations presented earlier, being removed from the oceanic environment. The relevant pattern of diet and nutrition is that from Island Melanesia through to Remote Oceania.

In an oceanic world the importance of sea foods seems obvious, and certainly they may partly compensate for poor atoll soils. Yet in mature societies throughout Oceania, even with the sea at hand, plant foods are almost always of primary importance. This was brought out in Chapter 1, in Malinowski's comments on the importance of gardening to the Trobriand economy, or in Christiansen's comments on the time spent fishing on Bellona. Further east, through Remote Oceania, 'The most elemental food category, that which minimally constitutes a meal, is the starch staple, product of agricultural labour' (Kirch 1984: 30). The same authority notes in the archaeological record of several islands a decline in the contribution of sea foods to the diet over settlement time, suggesting that a period was required for gardens to be developed and expanded. The only domesticated animals were pig, dog and chicken (unless the rat is considered such) and these were seldom major sources of meat in the sense of being regular components of the diet. Several, workers have stressed the importance of insects in contributing protein to the diet of inland groups.

The significant plant foods in Oceania were mentioned in Chapter 1, along with comment on the sophisticated gardening techniques. These plant foods vary little with movement east into Remote Oceania – many were carried along by the colonizers – yam, sago, breadfruit, bananas, taro and swamp taro, and coconut. Food values for some dietary components are given in Table 3.12. It is the low protein content of many of the traditional Pacific vegetable staples that bothers some observers, particularly when set against such standards as those of the Food and Agricultural Organisation. To repeat, the Rivers and Payne (1982) commentary on these is worth scrutiny. The consistent – if sometimes puzzled – tone of field workers is that throughout Island Melanesia diets, even away from the sea, really do seem adequate for normal and efficient human function.

The traditional diet of the Baegu of Malaita was studied by Ross:

Even with some quibbling about the quality of the diet and whether or not plant proteins can so completely replace animal protein sources, the figures . . . indicate that the Baegu are . . . adequately nourished. Their diet meets or exceeds minimum daily requirements and recommended daily allowances in every nutrient category. Damon's pediatric studies showed no overt evidence of

Table 3.12 Nutritional values of common foods in the Pacific.

Food	water (ml)	calories	protein (g)	fat (g)	CHO (g)	fibre (g)	Ca (mg)	Fe (mg)	Vit.A (I.U.)	thiamine (mg)	ribo-flavin (mg)	nicotin-amide (mg)	Vit.C (mg)
sago flour	12	352	0.5	0	88	0	10	1	0	0.01	0	0.2	0
taro	70	113	2	0	26	0.5	25	1	0	0.1	0.03	1	5
swamp taro	70	111	0.5	0	27	1.2	150	1	0	0.03	0.1	1	0
yam (fresh)	73	104	2	0.2	24	0.5	10	1.2	20	0.1	0.03	0.4	10
yam flour	18	317	3.5	0.3	75	1.5	20	10	0	0.15	0.1	1	0
coconut kernel	45	375	4	35	11	4	10	2	0	0.05	0.02	0.6	0
leaves, dark green	85	48	5	0.7	5	1.5	250	4	3000	0.1	0.3	1.5	100
banana	70	116	1	0.3	27	0.3	7	0.5	100	0.05	0.05	0.7	10
breadfruit	70	113	1.5	0.4	26	1.3	25	1	0	0.1	0.06	1.2	20
fish	70	166	19	10	0	0	30	1.5	100	0.05	0.2	3	0
crustaceans	77	94	18	1.5	2	0	100	5	0	0.05	0.1	2.5	0
molluscs	83	70	10	2	3	0	150	10	200	0.05	0.15	1.5	0
insects	70	134	20	6	0	0	30	1	0	0	5	2.2	0

Values/100g of edible portion

malnutrition of any form . . . Nutritional data presented . . . indicate that even with a starch-rich, meat-poor diet people can get adequate protein, fat, vitamin, and trace mineral intake by eating large quantities of the foods that are available and using a wide variety of species as supplementary or auxiliary foods. Gross caloric intake appears satisfactory. Very few of the Baegu hill people are truly skinny, although body build tends to be thin and wiry. (Obesity is a severe disadvantage in mountainous terrain where all travel is by foot) . . . The Baegu people appear healthy and well fed despite clinical evidence . . . of endemic and chronic malaria and intestinal parasitism at around 25% frequency. (1976: 289–90)

The conclusions on diet and nutrition from the wide-ranging Harvard Solomon Islands study were that 'Diets for all groups are adequate in protein, and for most vitamins. Clinical observations suggest that nutritional deficiencies are uncommon' (Page *et al.* 1987: 86). And 'It is difficult for us to regard the traditional diet of most Solomon islanders as a deficient one. There was no malnutrition, but undernutrition was possible, to judge by growth trajectory changes in recent decades' (Friedlaender 1987: 360).

For Karkar Island, Hornabrook comments: 'In regard to protein intake, the findings suggest that whilst the gross dietary protein was low as was the protein to energy ratio, the amino-acid scores as recommended by FAO/WHO were acceptable, and although there was a low absolute quantity of protein in the diet, both groups probably had adequate intakes' (1977b: 299).

Sexual dimorphism, particularly of stature, is sometimes regarded as an indicator of the quality of nutrition of a group. There are claims that male stature is a more labile thing, and liable to reduce (not in the sense of the life of a single person of course) under poor conditions. By contrast, female physique, exemplified in stature, is constrained at the lower end of its range by the demands of childbearing, and at the upper end by the nutritional drain of lactation (Hamilton 1982). The matter, as always, is complex, and is taken up in Chapter 6. Comment here is limited to the observation that the groups from Island Melanesia show, on this thesis, a healthy dimorphism (see Table 6.6, p. 199) – so perhaps here is another pointer towards nutritional adequacy.

In these matters of diet and nutrition, there also are comparisons possible along the lines taken when considering the influence of malaria – that is, a natural experiment may be viewed. There are groups in Island Melanesia who live on the coast, exist with malaria, yet show contrasting physiques, for through Island Melanesia occupation of a coastal niche does not necessarily mean significant exposure to the oceanic environment. Compared with the Ontong Javanese and the Malaitan Lau who are robust, the Massim of the Trobriands are rather slight in build (Seligmann 1909, Malinowski 1922), for their exposure to the sea has not been of the same order. Many coastal groups have been little more than reef-foragers. For example, on Lesu, off the east coast of New Ireland, Powdermaker (1933) indicates that only reef fishing was carried out. (Haddon (1937) describes New Ireland canoes of varying degrees of seaworthiness but without sails, and cites Stephan as never having seen

New Irelanders make use of the wind.) Reef food provides good protein but does not seem to have directed the New Irelanders (who, from the shape of their island home would have trouble not getting to the sea) towards large physique.

It seems to me that the conclusion to be drawn from these dietary and nutritional assessments is that there really is no compelling case for taking the slight physique of some peoples of Island Melanesia as predominantly reflecting nutritional deficiency. Or, put another way, the change in physique across the Pacific does not seem primarily to rest on a transition from an inadequate to an adequate diet.

Secular change

Secular change may be defined as the change in patterns of growth and development that may occur in a group over generations. Despite the above considerations, there is no reason to think that the people of Island Melanesia will not show, with Western acculturation, the common secular trend towards earlier development and growth and an ultimately greater stature, than their parents and grandparents. Provided a lot of genetic admixture does not occur with the succeeding generations, such secular change reflects the phenotypic plasticity of a particular genotype – that is, the potential for it to be expressed over a phenotypic range. The question now runs along these lines: with optimal (?over-) nutrition, might a Baegu or a Nagovisi turn into a physical Samoan or Tongan? That seems to be the suggestion contained in the earlier citation from Shephard, that 'Most of the small documented differences of body build could easily have arisen through differences in nutrition and physical activity rather than through genetic factors' (1991: 168).

Again, there are problems in teasing out the influences. No factor changes in isolation. That is, no group will simply and suddenly be provided with an optimal Western nutrition (or, one has to keep suggesting, over-nutrition) with no other change in their existence. Outboard motors replace paddles and sails. A simple rainproof and windproof poncho may be worth kilograms of muscle. As Nurse puts it: 'Acculturation is only another form of environmental change' (1988: 286).

Neither are the secular changes always easy to interpret. Most attention is paid to the obvious and most easily-determined change in stature. However this is not inevitably associated with a proportionate increase in mass: 'The positive secular shift in longitudinal growth is not necessarily accompanied by a related weight increase or decrease. Regardless of the age periods, populations of small as well as of large heights were found to be light or heavy in weight for age' (Van Wieringen 1978: 465). Even if increase in mass occurs there is the problem of what makes up this mass. Increase in mass under westernizing conditions, particularly in women, is disproportionately weighted towards

fat, rather than increase in muscle (Finau *et al.* 1983, Friedlaender and
Rhoads 1987, Prior *et al.* 1992).

That said, it is of interest to examine what secular data there are, to
see if such change in physique, interpreted as the phenotypic plasticity of
genotype, is sufficient to throw doubt on the thesis that the peoples of
Oceania show selective adaptation to their particular environments.
There is not much long-term data. From two Bougainville groups are
data for 30–39-year-old males. The first is a series measured by Oliver in
1939 (Oliver 1955), and the second a series measured by Friedlaender in
1965 (Friedlaender 1975). The 1939 group had a mean mass of 56 kg
and stature of 161 cm. The 1965 group had a mean mass of 58 kg and
stature of 164 cm. Using the climatic parameters of Table 3.4, the body
temperature drop/hour only lessens from 1.5°C to 1.44, with body
temperature after 8 hours being 25.5°C compared with 25°C. There is
clearly no worthwhile survival advantage to the later generation.

Other data from which something may be gleaned are those for the
broad age groupings from the Harvard Solomon Islands study (Friedlaender 1987a). When one works through the differences between the
20–34 and 35–49-year groupings, while the younger people usually are
taller and heavier there is no significant difference in tolerance to
exposure. In Table 3.13 is an estimate, using a very crude extrapolation
of some of the Harvard Solomon Islands data, of what secular change
might achieve. The percentage difference between the 50+ and 20–34-
year groupings has been calculated, and the data for the latter group then
increased by the same percentage. It is a crude exercise because the 50+
grouping will be showing the effects of ageing, including decrease in
stature from shrinkage of the intervertebral discs, and decrease in mass
from natural involution of muscle: 'The traditional groups in the Solomons all show a general pattern of weight loss with ageing in the adult
years, most dramatically for females' (Friedlaender and Rhoads 1987:
290). The difference between the two groups must thus be exaggerated
beyond any likely secular change in one generation. (There is also the
matter of how many generations the secular trend might continue for. In
a study of Japanese people in Hawaii, Froehlich (1970) found secular
increase in stature in men to cease with the second generation of
Hawaiian-born, but continue for another generation in women.) However in the Solomon groups the resulting hypothetical generations,
which could be taken as something unlikely to happen three or four
generations later, only manage to struggle up to dimensions approaching
those of the fringe dwellers of Remote Oceania, the Malaitan Lau and
the Ontong Javanese. They do not become Tongans. (The solitary group
that shows substantial secular change is that of the Aita women, for
whom there was a very small sample – nine individuals – of the older age
group. By a quirk of the data, the Nasioi males actually show an
unfavourable trend, with a lessening of resistance to exposure.) The
bodies of these hypothetical generations, probably with a superfluity of
fat, would not simply be bolder versions of the ancestors. They would be

Table 3.13 Heat balance data for Island Melanesian groups using projected secular trend for physique. Column A is the body temperature after 8 hours in shelter at 14.5°C for the extrapolated secular trend physique for each group. Column B is the body temperature for the present physique under the same conditions.

Group	Mass (kg)	Stature (cm)	Rate of change of body temp. °C/hour	A: body temp. after 8 hrs	B: body temp. after 8 hrs
Male					
Aita	62.8	159.6	−0.94	29.5	26.1
Nagovisi	61.0	161.7	−1.13	27.9	25.3
Nasioi	57.7	168.4	−1.62	24.1	25.8
Baegu	63.0	163.0	−1.03	28.7	25.8
Kwaio	63.2	162.8	−1.02	28.9	25.4
Lau (Malaita)	68.3	167.2	−0.83	30.4	29.7
Ulawa	64.3	164.8	−1.01	28.9	27.0
Ontong Java	76.4	170.1	−0.45	33.4	29.0
Female					
Aita	69.2	150.3	−0.22	35.2	27.1
Nagovisi	53.6	154.0	−1.45	25.4	23.0
Nasioi	53.8	155.0	−1.47	25.3	22.0
Baegu	56.4	153.0	−1.17	27.6	23.2
Kwaio	56.0	151.6	−1.16	27.7	23.2
Lau (Malaita)	64.9	158.4	−0.75	31.0	27.2
Ulawa	55.6	154.8	−1.31	26.5	23.8
Ontong Java	62.4	159.0	−0.94	29.5	28.7

something as different as would be the entire new acculturated world from that of the ancestors. The exercise is simply presented to show where theoretical extrapolated trends might end.

The likelihood that secular change might be capable of eliminating differences between groups or converting a slight physique to something as well adapted to exposure as that of the peoples of Remote Oceania may be further examined by looking at the physique of contemporary Western groups who show the full flowering of secular change. Table 3.14 gives data for young adult Dutch, American and Japanese. These are groups for whom it is said that the secular trend has plateaued (Bogin 1988, Gracey 1991, Kimura 1984). The heat balance analysis using the standard conditions of Tables 3.4 and 3.5 finds these contemporary young people to be inferior to the Remote Oceanic peoples in their resistance to exposure. That is, they have put on stature but not the necessary mass. Earlier it was shown that several Oceanic peoples surpass the Westerners in upper arm muscle size.

In the same table are given data for the archetypal cold-climate people, the Inuit, and also the Alacahuf of the Straits of Magellan, both of whom do better than the modern young Westerners. Such data do

Table 3.14 Heat balance data for various groups. Netherlands, US and Japanese are contemporary groups considered to show a plateauing of the secular trend for growth. Alacahuf and Inuit (Eskimo) are high-latitude cold-climate peoples. Spy 1 and Spy 2 are Neandertals.

Group	Mass (kg)	Stature (cm)	Surface area (sq metres)	Heat prod'n kJ/hour	Sheltered heat loss kJ/hour	Total hourly deficit (kJ)	Rate of change of body temp. °C/hour	Body temp. after 8 hrs °C
Male								
Netherlands	70.8	182.0	1.91	1628	1910	−281	−1.13	27.9
US	71.9	177.4	1.89	1654	1887	−233	−0.93	29.6
Japan	64.0	169.7	1.74	1472	1739	−267	−1.19	27.5
Alacahuf	66.0	162.0	1.70	1518	1703	−185	−0.80	30.6
Inuit (Eskimo)	65.3	162.7	1.70	1502	1701	−199	−0.87	30.0
Spy 1	79.0	167.0	1.88	1817	1880	−63	−0.23	35.2
Spy 2	78.8	162.0	1.84	1812	1837	−24	−0.09	36.3
Female								
Netherlands	58.6	168.3	1.66	1348	1665	−317	−1.55	24.6
US	57.2	163.2	1.61	1316	1612	−296	−1.48	25.2
Japan	51.8	157.1	1.50	1191	1504	−313	−1.73	23.2

suggest that the Remote Oceanic groups as well as high-latitude cold-climate groups are specifically endowed with a physique appropriate to their environment. Also in Table 3.14 for light relief are calculations for Neandertal physique, using data from Chapter 2. They do better than anyone alive, including all the peoples of Remote Oceania. What a pity they did not get down to the coast. Perhaps they could not manage to get their tongues around 'Our future lies in maritime technology'.

I do not think the present evidence supports the notion that secular change deriving from a different diet and nutrition, and the elimination of endemic disease, will metamorphose those groups of slight physique in Island Melanesia into something significantly better adapted to the Oceanic environment. Conversely, the lightening of the (here purported) directional selective pressure on modern Polynesians has not seen them drift back to some lesser phenotype. And, in a sense, the matter of secular change is irrelevant to interpretation of changing physique across the Pacific. Pragmatically, what mattered in niche after niche was the appropriateness of physique to environment during the long course of prehistory.

In brief, I think the weight of evidence does lie in favour of a strong directional selection pressure towards large body size in Remote Oceania. The matters of diet and disease and a putative secular trend do not emerge as convincing primary explanations, and while a particular group here or there may bear their mark I do not think these factors have in general made the major contribution to the establishment of the differences in physique across the Pacific. Shephard's view – that we would all look much the same if we all ate much the same – is, I think, unsustainable, at least in the Pacific context. Specifically, to adopt some of his wording, I have tried here to 'assign a numerical value' to the importance of a particular physique in determining human survival.

Voyaging in Near and Remote Oceania

Looking at groups such as the Massim or the peoples of New Ireland does raise the question as to just what constitutes an oceanic environment, sufficient to create selective pressure towards large body size. In the west of Island Melanesia there are some renowned sailors: the Kula traders of the Massim, the Siassi and Tamil islanders of Vitiaz Strait; and on the south coast of New Guinea the Mailu with their double-hulled *orou*, some ten metres in length, with reed-matting crab-claw sails, and an ability to make to windward (albeit at a rather poor angle) at some eight knots (Irwin 1985). At first glance all this activity may seem to resemble much of the sea travel in Remote Oceania. In fact the situation is, or was, very different. The 'voyaging' of these trade networks within Island Melanesia was a much more seasonal and limited thing than in Remote Oceania. In Island Melanesia, barring rain or mist or night, land is seldom out of sight, so navigation is rudimentary. Most legs of a journey are readily covered during daylight hours, and nights are passed ashore. Even in total distance, most of these nightly-interrupted trips are well under 150 kilometres. Malinowski's timetable for a Kula voyage involved nightly pauses ashore over a week to ten days, and he describes with bemusement the way the Kula fleet, after great ceremonial departure, stopped at the first island round the corner, so to speak, in the middle of the day, to have a good feed and a sleep. In this region there is a distinct seasonal cycle of trips to take advantage of fair winds, and in the course of a year few people are likely to have spent more than a few days of daylight hours at sea. Along with the climatic pattern described in Chapter 1, the computer simulations of survival at sea, emphasizing the significance of distance from the equator, suggest that the Coral Sea is a much milder place than the wider ocean. Contrast the Kula sailing conditions with, say, those for the Carolines in Micronesia, where few inter-island distances are less than 130 kilometres, in utterly exposed waters, and where navigation must be equated with survival. When a single isolated small island was vulnerable to loss of crops or resources or people from a typhoon or a tsunami, continued inter-island contact was probably essential for human survival (Hunt and Graves 1990).

None of this is intended in any way as a slight on the sailing abilities of any group in western Island Melanesia, but is simply a statement that the physiological demands on the human frame there were relatively trivial, and the selection pressure towards a larger frame much less significant.

Other matters

The rest of the chapter brings together a few lesser yet relevant matters. I have heard it suggested that selection for large size might have occurred at the start of a voyage, in crew selection. Whether this actually was so can never be more than speculation, and in any case comes to the same thing – selection, and ensuring a founder effect. It is not an argument against the heat balance analysis. If large people were selected for voyages this might have been from the collective wisdom that such people survived. And big people had to be there to be chosen, which implies a reason for being big. People do not grow big just for fun. There is a cost. A big body has to be fed more than a small body. A 75 kg person working quite hard requires about 11,100 kJ energy/day. A 60 kg person requires about 10,000. The difference is only 10%, but in a difficult island environment – and there were plenty of these, as Kirch (1984) has shown – a big person would have been at a disadvantage if there were no other considerations. At sea, if food and water were the constraints rather than the conservation of body heat, then the advantage would have been to the smaller person.

There are sociobiological theses that see large body size in the Pacific as resulting from mate selection. It does seem remarkable that such an inclination to large mates should specifically occur amongst the small (in the sense of numbers of individuals) groups on the fringe of Remote Oceania. And again, apart from the basically speculative nature of these ideas, they are rather beside the point of the cold-adaptation hypothesis. However selection occurred, it enabled people to survive. If only bigger people survived then in turn there were only larger mates to choose. Of course if the heat balance idea is not favoured, then an explanation is needed as to how lesser physiques were eliminated from the normal distribution, or why such remarkable muscularity is evidenced in the region of the world where, in pure locomotor terms, it is least required. The thought, earlier mentioned, that 'salubrious islands' and plenty of food, along with careful mate selection, could see the emergence of the Remote Oceanic phenotype, has its problems. Just eating well puts on fat, not muscle, and for a breeding programme to work for *Homo sapiens* across the vast expanse of Remote Oceania – because that is what the physical record requires – would be a eugenics programme of unusual coordination, success and scope in human affairs. Evolutionary programmes have a better record.

Scepticism has been expressed as to the time available for such proposed selective adaptation to have taken place (Meaney, cited by Irwin 1992). This matter is considered in Chapter 5. However we may note in passing the accumulating evidence that morphological change of a true genetic nature may take place rather rapidly, in a few generations of a species (Lister 1989, Mahotra and Thopre 1991), and that the geography and environment of Oceania so well establish the particular conditions that favour the emergence of distinctions within a species (Mayr 1970). I disagree with Friedlaender's view that in the Pacific 'natural selection has not been a major determinant of summary population relationships in recent times . . . 30,000 years is certainly much too short an interval for major population distinctions to have arisen' (1987b: 357). On the contrary, it is well within that time that the major phenotypic differences have arisen. Indeed, through the Harvard Solomon Islands study runs a pervading contradiction: 30,000 years is seen as being far too short a time for 'major population distinctions' to have occurred, and yet many of the contributing studies are devoted to unsuccessful attempts to discern a fundamental biological dichotomy matching a putative linguistic dichotomy. If the short time-scale of the linguistic model is correct, and if major phenotypic change cannot occur in so short a time, why the difficulty? It can only be – if one accepts the linguistic model – because the original and purportedly distinctive human groups have become phenotypically melded. These are matters discussed in Chapter 5.

In a study of the distribution of the classical ABO blood groups, Beals, Smith and Kelso (1992) have presented a significant positive correlation of Group A, and a corresponding negative correlation of Group B, with latitude (equated with climate), and with mass/surface area ratio. (In their paper they reverse this ratio, presenting it as surface area/mass, so that the correlation is strongly positive.) They suggest some unrecognized interaction here between body and environment, and with climate in particular. I do not know if they used any data from Remote Oceania, which would not have helped in supporting the thesis. However if the peoples of Remote Oceania are viewed as cold-climate people then the idea is well supported: the incidence of Group B declines markedly with passage into the Pacific until in eastern Polynesia it is negligible, and of course the mass/surface area ratio increases.

In a thoughtful paper that represents one of the few anthropological attempts to move craniological data beyond a display of dendrograms, Konigsberg and Blangero (1992) have essayed establishing quantitative genetic differences between two widely separated Pacific groups, the Tolai of New Britain and the Moriori of the Chatham Islands. The selection of groups was dictated more by availability of data (theirs is from the monumental study by Howells 1989) than by any particularly rational model of Pacific settlement. I will mention this paper again, but, as I read their analysis, they conclude that genetic drift cannot even begin to explain the differences between these two widely separated groups. Rather, some major selective force seems to be operating, though as to what this might be they do not comment.

This chapter opened with some discussion of Bergmann's Rule. The analysis suggests that the people of the Pacific illustrate it rather well, and this is largely because neolithic technology could not keep them dry. The theme of strong directional selection for a substantial physique makes sense in terms of matters to be discussed in Chapters 5 and 7.

SKELETAL MORPHOLOGY

4

This chapter describes the skeletal morphology of the people of the wider Pacific, and looks for reasons for any distinctiveness. There is more here about the head than about the infra-cranial skeleton. Partly – and without denying the complexities of locomotor function and adaptation – this is because the head skeleton really is the more complex structure, reflecting its diverse functions. Partly the imbalance simply reflects that not very much research has been done on the morphology of the infra-cranial skeleton of the people of the wider Pacific. Because differences in form between groups are often distinctive, the head certainly has had more anthropological attention.

The head

A succession of anthropological studies has meticulously recorded Oceanic craniofacial dimensions and compared them with data from other regions (eg, Scott 1894; Flower 1896; Duckworth 1900; Thomson 1915–17; Wagner 1937; Shapiro 1943; Marshall and Snow 1956; Shima and Suzuki 1967; Snow 1974; Pietrusewsky 1969, 1983; Howells 1973a, b, 1989; for the extensive earlier European literature see Wagner 1937). Though not always stated, the aim of these studies seems generally to have been classificatory or taxonomic. Earlier studies were limited to simple statistical comparisons, and placed particular emphasis on indices and the form of the cranial vault. Advances in statistical method have led to multivariate analyses of an impressive amount of data, often with equally impressive depictions of cluster analyses.

In general, these craniological studies in Oceania have not sought to establish the reasons for any distinctive features – and some are very distinctive – but have been content to note that they exist. And it must be said that biological reasons for variation in head form between groups anywhere does not seem to have been much thought out. Some limited models have sought to relate variation within particular regions of the head to certain environmental or functional variables. Climate, for example, has been proposed as a determinant of the form of the cranial vault (Guglielmino-Matessi, Gluckman and Cavalli-Sforza 1979; Beals, Smith and Dodd 1983) and of the face (Weiner 1954, Carey and Steegman 1981). Variation in tooth size has been attributed to variation in functional demands on the dentition (Brace and Mahler 1971), and it has been shown that changes in the human masticatory system may alter structures elsewhere in the head (Carlsson and Van Gerven 1977; Hylander 1977). But, as Howells expressed the problem at the outset of a major taxonomic study: 'In cranial comparisons within modern man there are really no general coherent hypotheses to explain the long-recognised differences in head shape (cranial index), prognathism, nasal and facial variation etc' (1973b: 48). The irony in this statement is that there is, therefore, within studies of the species *Homo sapiens*, little theoretical basis for using cranial data as a taxonomic tool.

The first aim of this chapter is to describe the anatomical distinctiveness of the Polynesian head particularly, set out its biological basis, and suggest that it affords insight into the problem of variation in head form between groups of *Homo sapiens* on a wider canvas. Ultimately, the principles that emerge from biological consideration of the Polynesian head allow the formulation of just such a 'general coherent hypothesis' called for by Howells. It is emphasized that this section of the chapter is not an analysis of metric data, but a consideration of the biological principles which ought to precede the gathering of such data for anthropological purposes. In the next chapter the use of cranial data for classification will be examined.

One reason for the focus here on the Polynesian head derives from the general conclusions on physique reached in Chapter 2. In morphological terms there seems little that is unduly distinctive about the peoples of Near Oceania, whereas the peoples of Remote Oceania *are* distinctive amongst *Homo sapiens*. A scrutiny of an extreme variant of the normal range of head form for *Homo sapiens* may be revealing of insights into the bases of cranial variation. A more pragmatic reason is simply that Polynesian morphology has been more closely examined. On limited evidence, Fijian morphology resembles the Polynesian. It would be nice to consider here also the morphology of prehistoric Micronesia, but this I have not had the opportunity to examine in any detail.

Functional components

The complex architecture of the skull results from the diverse functions required of it. Quite a long list may be made of these, but they can be subsumed into three: support and protection of the nervous system, which includes eye and ear; provision of airway; provision of foodway.

Figure 4.1 Division of the head skeleton into neurocranium and splanchnocranium: the latter is indicated by the lined area. The cranial base, the boundary between the two regions, is indicated by the dark shading. N is nasion point, at the bridge of the nose; S is sella, at the centre of the pituitary fossa; B is basion, at the anterior margin of the foramen magnum. The smaller diagram is the standard schematic depiction of the cranial base in profile. S–N is the anterior limb of the base and S–B the posterior limb. CBA is the cranial base angle formed by the two limbs.

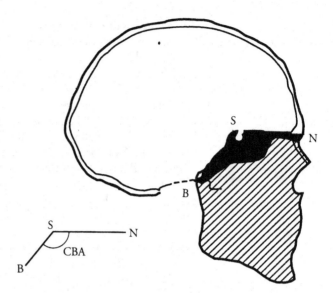

This last term is not in general use, but it should be as acceptable as the term 'airway' and is both less limited and less cumbersome than the usual 'masticatory system'. However the latter term is sometimes more appropriate, being more specific. In the old zoological division of the head skeleton into neurocranium and splanchnocranium (Figure 4.1), airway and foodway come under the latter.

Developmental sequence and skeletal templates

Cranial base

The sequence in which the functions of the head and their supportive structures develop and mature provides insight into the determinants of head form. In this sequence the nervous system takes precedence. As early as the fourth week of fetal life the primitive brain is identifiable as swellings at the rostral or front end of the neural tube (Figure 4.2). In the fifth week are evident flexures of the brain, which relate ultimately to upright posture and the need to keep the senses properly oriented to the environment: in particular, the visual axes align horizontally. At the beginning of the sixth week of fetal life there appears at the undersurface of the developing brain a condensation of tissue that will eventually form the bony platform on which the brain rests – the cranial base. In the same week the cranial nerves are identifiable on the ventral aspect of the developing brain. The cranial base not only mirrors the flexure of the brain, but is traversed by the emerging cranial nerves.

Along with this precocious development of the nervous system goes a cellular conservatism. That is, the cellular complement of the nervous system is determined early, and thereafter it is structurally the least adaptable of body tissues, with a minimal capacity for regeneration and

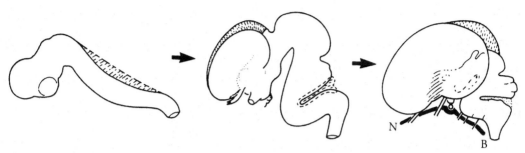

Figure 4.2
Some changes in the form of the developing brain early in fetal life. The original neural tube becomes flexed and the hemispheres enlarge. The cranial base (N–B) develops as support for the brain and is traversed by the cranial nerves.

repair. Essentially we are born with our full supply of neurons, and their further growth is a matter of extension and myelination of processes and the making of new connections. The reason is that for survival of the whole organism neural connections have to be permanent, be they related to suckling reflexes or walking or memory or whatever. This is in contrast to such a tissue as bone, where remodelling may be going on at an almost frantic rate without compromising the function – that is, the strength and stiffness of the bone – in any significant way. This limited adaptability of the nervous system is imparted to the supportive cranial base, whose flexure is effectively set from an early stage of fetal life because of its intimate relationship with the brain and cranial nerves.

This does not mean that the form – and particularly the flexure – of the cranial base is initially set by the nervous system. There is plenty of experimental evidence (Glenister 1976, Bromage 1980) that the cartilaginous precursors of the bony skeleton have their own initial genetic determination of initial form, and the form and flexure of the cranial base are in this category. But soon the intimate relationship with the nervous system ensures that the base changes only *pari passu* with changes in the brain. Clear evidence for initial independence of form of brain and cranial base is seen in cetaceans. These show all the brain flexures of any upright or semi-upright primate, but they rest on a singularly flat cranial base which serves as a fundamental template for the streamlined head.

The flexure of the *Homo sapiens* bony cranial base however is subsequently to be maintained constant by the nervous system. In lateral view (and the limitations of such two-dimensional analysis need to be kept in mind) two limbs to the base may be defined; an anterior limb relating to the cerebral hemispheres, and a posterior limb relating to the brain stem leading down to the spinal cord (Figure 4.3). Though not obvious, nor directly assessable by conventional craniometric techniques, the angle between these two limbs is readily determined from x-ray examination. In *Homo sapiens* in general it averages about 130°, and it is one of the few body dimensions to remain rather unchanged through post-natal life (Riolo *et al.* 1974; Broadbent, Broadbent and Golden 1975); a gradual reduction in angle in the first year of life and some change at adolescence relate more to shift of the end points,

Figure 4.3
Schematic figure showing effect of flattening of the cranial base (ie, increase in cranial base angle) on positioning of the upper facial skeleton. The line D–D indicates the position of the maxillary dentition.

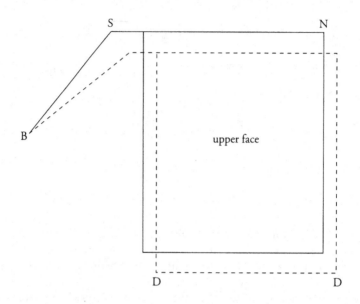

particularly from growth of the facial skeleton, than to any fundamental alteration in the orientation of the two limbs.

In Polynesians, however, the cranial base angle is typically and distinctively large, or open, averaging 140° and often reaching values over 150° in the prehistoric Moriori of the Chatham Islands (Houghton 1978a; Schendel, Walker and Kamisugi 1980). While any group may show individuals possessing such open cranial base angles, the distinctiveness of the Polynesians is that *most* show this considerable divergence from the mean value for *Homo sapiens*. The assumption is that this flat cranial base was a skeletal characteristic of the earliest people in Polynesia. Why this might be is something we will come to shortly.

The structural significance of the cranial base is that, early to develop and stable in form, it forms a major architectural template for other head structures. The posterior part of the base, being the site of spinal and ultimately terrestrial support for the head, is a spatially constant region from which positional variations in head structures may be defined (Figure 4.3). In this region also is the vestibular apparatus for balance, whose orientation in space is constant (DeLattre and Fenart 1960; Moss 1961). The anterior part of the base forms a positionally-variable beam from which the upper face is suspended, and variation in cranial base flexure influences the positioning of the upper face. For example it is positioned more downward and forward when suspended from an open than from a closed base (Enlow 1982; Kean and Houghton 1982; Figure 4.3). Ultimately such variation has a profound influence on individual facial appearance.

Airway

The upper face is dominated by the airway. The increase in airway size that occurs during the growth of any individual is expressed predominantly as a change in its height (Riolo *et al.* 1974) – that is, growth occurs in an inferior or vertical direction. This is because any significant increase in width of the airway would require increase in width also of the supportive cranial base. Because of the neural associations and constraints mentioned above, including the positioning of the eyes, this would be a very complex readjustment. By contrast, vertical growth makes few extra demands on the supportive cranial base, and proceeds into open space, so to speak – there is plenty of open space under the chin. Within *Homo sapiens*, of course, there is variation in nasal breadth between groups, but the geometric reality is that significant change in airway cross-sectional area requires increase in height of the airway.

During growth, increase in airway size reflects the increasing energy demand of the growing body. The graphs in Figure 4.4, derived from various studies, display this. On average, the height of the nasal airway at maturity is greater in males than in females, whereas until about age 12, airway heights are similar, age for age, for each sex. The reason why this should be so lies neither in the nasal cavity nor the face, nor even in the head, but in the greater oxygen needs of mature males, in whom the greater functional demand is imposed on the face by the body. Between six and 16 years body muscle mass increases about threefold in girls and fourfold in boys (Malina 1978), and from about 12 years the curve for male muscle mass rises sharply, whereas that for females is starting to flatten: that is, male and female curves diverge from this age. Similarly divergent growth curves for male and female are seen for measures of lung size (Ferris, Whittenberger and Gallagher 1952; Ferris and Smith 1953). While the graphs depict the results from large studies, they could equally well represent the differences between two individuals, one inherently big and the other small, as they advance to maturity. The larger could well be female and the smaller male. The point is that the

Figure 4.4
Changes in skeletal muscle mass (kg), nasal height (mm) and vital capacity (litres), during growth and maturation. Male data indicated by closed circles and female by open circles.

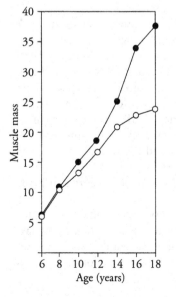
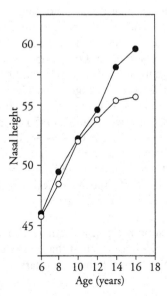
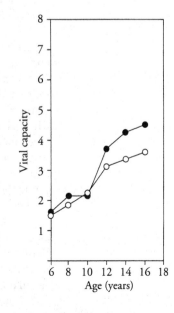

size and shape of the head is not some thing in itself, but is determined to a considerable extent by the demands of the whole body. This, in turn, is influenced by the environment in which the individual grows. And just as the mature head is not of the same proportions as the immature, neither is that of a large adult of the same proportions as that of a small adult.

It is worth observing here that airway height, and vertical measures of the face in general, are not components of stature and inevitably correlated with variation in stature. The splanchnocranium, suspended from the anterior cranial base and with open space below it, is free to make its own adjustments.

Airway size is thus appropriate to oxygen demand and it follows that the extent to which oxygen demand varies between individuals at maturity determines the extent to which the face develops vertically. Amongst adults, differences in anterior nasal height as a simple indicator of airway size are a rough reflection of differing body size, and particularly of differences in muscle mass, for from this tissue comes the great oxygen demand during activity. Such an association, between airway size and body mass, has seldom been sought in growth or anthropometric studies, and in sedentary groups today, where mass may strongly reflect fat, the association may be less discernible. However, evidence of a strong correlation ($p > 0.001$) between body mass and airway height was found in Japanese adults by Miyashita and Takahashi (1971). The individual anthropometric data from Papua New Guinea provided by Dr Robin Harvey allowed a more sophisticated calculation, in that both nasal height and nasal breadth gave a better approximation of the size of the nasal passage. The series showed a correlation coefficient of 0.8 between mass and nasal dimensions, with regression of mass on nasal dimensions giving a highly significant F ratio and a coefficient of determination (r^2) of 0.6. While bearing in mind the tentative nature of the mass estimations earlier described, our skeletal studies gave a correlation of 0.81 between body mass and anterior nasal height.

On the basis of the preceding discussion, nasal height in large muscular Polynesians ought to be in the upper range for *Homo sapiens*, and this proves to be so. For a large global cranial series Howells (1989) gives a range from 42.76 to 56.91 mm for males and 42.86 to 53.33 mm for females. Top of the range for each sex are the Siberian Buriats (a most distinctive, inland, continental, cold-climate people). Chatham Island Moriori are next with values of 55.95 and 52.61. People of Remote Oceania (Hawaii, Guam and Easter Island) take up three of the next five places in the range, along with the Arikara and the Inuit. For the New Zealand Maori, Wagner (1937) gives values of 54.3 for male and 51.4 for female, and for the Marquesas, 57.4 for males – surpassing the Buriats – and 52.7 for females.

The cranial base is an interface between two functional regions, for as well as supporting the brain it also forms the roof of the airway. Berglund (1963) established that flattening of the cranial base led to an increase in volume of the nasopharynx and derived a formula for its calculation:

Capacity of bony nasopharynx = (posterior nasal spine
to basion distance) × choanal width × (perpendicular
from hormion to basion-posterior nasal spine line) / 0.5

Using this formula we established a mean volume of 13.0 ml for the male Polynesian nasopharynx, compared with 10.3 ml for the Suomi, or Lapps (a high-latitude/cold-climate people), 10.1 ml for Norwegians, and 9.5 ml for Australian Aborigines. That is, the Polynesian nasopharynx is on average about 30% larger than that of the northern European, and 35% larger than that of the Australian Aborigine (Kean and Houghton 1982). This is a substantial difference, and the influence on the voice of such a great resonating chamber above the vocal cords (Sundberg 1977; Proctor 1980) may contribute to the well-attested Polynesian singing abilities, and help explain why the minuscule (in global terms) Maori population of New Zealand has produced operatic singers of world renown as well as a host of more popular singers.

The nub of this consideration of the facial and pharyngeal airway is that its development in height and volume is determined extrinsically, by influences beyond the head, and sometimes beyond the body. The body grows the face it needs. The further significance is that variation in airway size influences not only the spatial placement of the upper dentition, but also the overall form and biomechanics of the jaws and teeth – the foodway.

Foodway and mandible

Figure 4.3 shows the influence of cranial base flattening on the spatial position of the upper dentition. The flatter the cranial base, the more inferiorly placed is the upper dentition because the whole supporting upper face is displaced inferiorly. Vertical growth of the airway accompanying the normal body growth of any individual further displaces the upper dentition inferiorly. Maintenance of occlusion, which has been vital in an evolutionary and survival context, requires continuing remodelling of the mandible as the upper dentition shifts its overall spatial position (Figure 4.5). In the head the mandible is the supremely adaptable bone.

The mandibular adaptation involves a progressive lessening of the angle between body and ramus, and an increase in height of the ramus, to accommodate the lengthening upper face. The measure of ramal/body orientation is the mandibular or gonial angle, and its value averages around 140° in young children and 120° in adults. Such maturational changes are universal in the human race, and the mature bone is usually something like the illustration in Figure 4.6. There is a more vertical and higher ramus than in the immature individual, but still evident is an angular process that provides attachment for the medial pterygoid and some of masseter, two major masticatory muscles, and defines a distinct antegonial notch.

In Polynesians in general, but also in any individual with a flat cranial base and large airway because of a large muscular body, the adaptation required of the mandible is extreme. The ramus remodels to

Figure 4.5 Schematic diagram showing variation in mandibular angle with differences in upper face morphology. Lettering as for Figure 4.1. The ramus and body of the mandible are indicated. The continuous lines represent an individual with relatively small cranial base angle and upper facial height. The interrupted lines represent an individual with open cranial base angle and large upper facial height. The mandibular angle is less in the larger individual. In reality the difference in mandibular angle is much greater than suggested in this geometric representation.

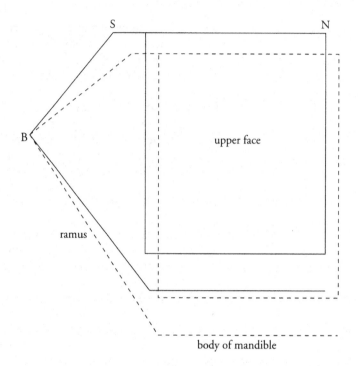

become particularly high and vertical, and this change frequently leads to the disappearance of a distinct antegonial notch as the angular process is absorbed into the general profile of the bone (Figure 4.6). In this form of the bone – the rocker jaw – the mandibular angle is effectively vertical, 90 to 100°, and the relevant masticatory muscles obtain an appropriate spatial position without the need for a distinct angular process. However in x-rays the angular process is still identifiable as a crescent of bone of rather different structure at the angle of the mandible (Houghton 1978a).

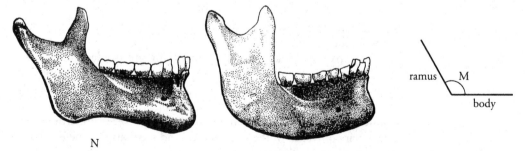

Figure 4.6 Mandibular form. The diagram on the left shows a typical human mandible. The vertical element of the bone is the ramus and the horizontal element bearing the teeth is the body. N indicates the antegonial notch, in front of the angular process.
The line diagram on the right is a schematic representation of the mandible, showing the mandibular or gonial angle, M. The centre diagram shows a typical rocker jaw with no discernible angular process or antegonial notch. The ramus is particularly high and vertical to accommodate the large upper face and large airway.

The immature Polynesian mandible shows antegonial notch and angular process like that of any other youngster, being only rather more robust. The Polynesian child's upper face is still small, but within the craniofacial complex is concealed the flat cranial base. Around eleven or twelve years, as the growth spurt starts and the upper face lengthens, the latent influence of the flat base is brought out, and remodelling of the mandible leads to the development of the rocker form. The mature Polynesian mandible represents the most extreme adaptation of the bone found amongst *Homo sapiens*, and within Polynesia the Moriori with their exceptionally flat cranial bases show the most extreme form of rocker jaw.

The term 'rocker jaw' has no profound meaning, referring only to the instability of this form when placed on a plane surface. It has also become something of a liability. As used here, it is shorthand for the craniofacial complex described, particularly one involving a large airway with flat cranial base and large upper facial height. This complex may be present in individuals of any group of *Homo sapiens*, and the Polynesian distinctiveness is merely that amongst them it is, if not ubiquitous, then in many groups not far off it. Occasional groups of *Homo sapiens* apart from Polynesians may show a high incidence of jaws that rock, but this need not be because of the same craniofacial complex. That is, a jaw that rocks need not be synonymous with Polynesian head form. For example, hypertrophy of the masticatory muscles as a result of considerable masticatory demand may sometimes eliminate the antegonial notch. There may be other craniofacial influences to be identified. However for convenience I shall continue to use the term here as expressive of Polynesian morphology.

The vertical alignment of the ramus of the rocker jaw influences its biomechanical efficiency. It is now generally accepted that during the occlusal stroke of mastication the mandible functions as a third-order lever (Hylander 1975, 1979; Gingerich 1979; Moore 1981; Picq, Plavcan and Hylander 1987). Theoretically, a more vertical ramus actually decreases the efficiency of the mandibular lever by relatively reducing the length of the power arm (Houghton 1978a; Figure 4.7). That is, during

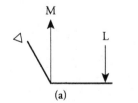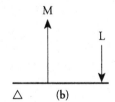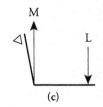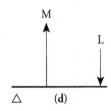

Figure 4.7 Line diagrams of the mandible as a third-order lever. (a) represents an average mandible for *Homo sapiens*, with an angle of about 120°. Δ indicates the fulcrum at the temporomandibular joint, M the mean line of force of the muscles closing the jaw, and L the occlusal pressure or load. In (b) the mandible has been transformed to a standard lever figure: the distance from the fulcrum to the line of action of the muscles (ie, the power arm) is relatively large. (c) and (d) represent the Polynesian mandible. The more vertical ramus leads to a reduction in the length of the power arm, with reduced efficiency of the bony lever. Not to scale.

Table 4.1 Correlations between mandibular variables. *=p<0.01: **=p<0.05.

	r.h.	min. rb.	bic. w.	m. angle	cond.b.	temp.	mass.	load	power	dent.
ramal height	1	0.405*	0.402*	-0.376**	0.262	0.397*	0.402*	0.259	0.192	0.158
min. ramal breadth		1	0.556*	-0.552*	0.764*	0.739*	0.846*	0.499*	0.31**	0.356**
bicondylar width			1	-0.42*	0.705*	0.688**	0.746*	0.707*	0.171	0.738*
mandibular angle				1	-0.425*	-0.386**	-0.612*	-0.161*	0.04	-0.236
condylar breadth					1	0.745*	0.817*	0.564*	0.075	0.643*
temporalis size						1	0.783*	0.622*	0.33**	0.496*
masseter size							1	0.574*	0.171	0.571*
load arm								1	0.609*	0.724*
power arm									1	-0.106
dental component										1

maturation there is actually a set towards increasing inefficiency of function. However this concept of efficiency or inefficiency is one of physics rather than biology, relating to the bony form alone, for the maturation of the masticatory musculature proceeding concomitantly with the bony change ensures biological efficiency.

These principles of mandibular function were examined in a study of mandibular form that included assessment of the size of the masticatory musculature and which established the correlations set out in Table 4.1 (O'Flynn 1992). Of particular interest are the changes correlated with decreasing mandibular angle, this being the natural growth pattern. The robusticity of the mandible as expressed in such dimensions as ramal width and condylar breadth was inversely correlated with mandibular angle – that is, with an increasingly inefficient bony lever. Similarly the size of temporalis and masseter muscles was inversely correlated with mandibular angle, implying muscular hypertrophy with lessening bony lever efficiency. Assessment of changes in the power and lever arms of the mandible with changing mandibular angle is complicated by the fact that in a third-order lever the power arm is a component of the load arm (Figure 4.8). The load arm may thus show an increase in length in the inefficient lever form, but such increase derives specifically from expansion of the power arm contribution (d2 in Figure 4.8) because of the increased size of the attached masticatory muscles. By showing a negative correlation between the tooth-bearing component (d3 in Figure 4.8) of the load arm and the power arm, O'Flynn established that the former becomes relatively smaller as mandibular angle lessens. That is, there is a negative allometric relationship between the two components.

Figure 4.8
The mandible as a third-order lever. The diagram on the left shows the usual action across the jaw during molar chewing: the opposite joint is bearing the load. The diagram on the right indicates the load arm d1, and the power arm d2.

M

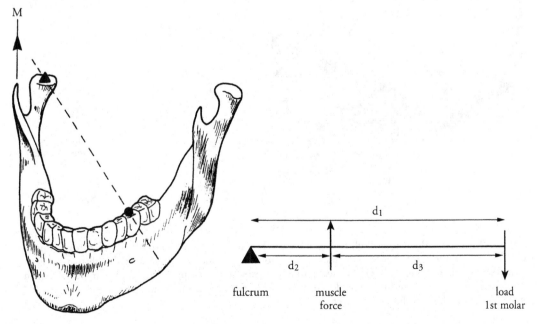

fulcrum   muscle   load
force   1st molar

Thus, in contrast to its broad and high ramus, the body of the Polynesian mandible, bearing the teeth, is relatively short. The apparent massiveness of, for example, a Moriori mandible, is in the muscle-related component, not the tooth-bearing component. In contrast to their physique, the teeth of Polynesians are not in the larger size range for *Homo sapiens*, for the full-blown rocker jaw cannot accommodate a massive dentition. The mandibular body represents a substantial component of the load arm of the lever, and an increase in its size proportionate with the ramus would require an extreme development of the masticatory muscles and related architectural adaptation within the face. In fact, amongst *Homo sapiens* we find rather little variation in size of the tooth-bearing component of the mandible when compared with many craniofacial dimensions (Riolo *et al.* 1974). However on biomechanical grounds, the suggestion is that for a given cranial base morphology, the larger the individual (particularly in muscle mass), then the relatively smaller the dentition. A negative allometric relationship between body size and tooth size has been described in several studies (Henderson and Corruccini 1976, Perzigian 1981, Kieser and Groeneveld 1988). Such a relationship supports the view that the recognized evolutionary trend in *Homo sapiens* to a smaller dentition has an indirect and rather complex biomechanical basis that is likely to be more significant than the cultural influences usually invoked.

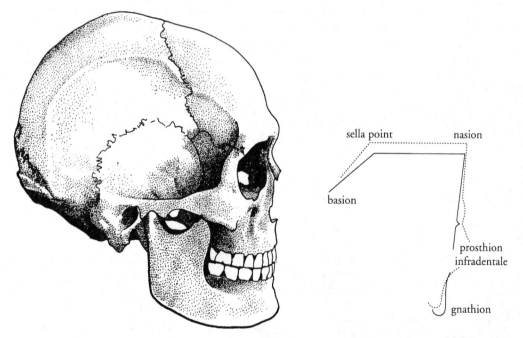

Figure 4.9 Typical Polynesian skull in lateral view. The profile is vertical with no alveolar prognathism. The diagram on the right shows a Polynesian x-ray profile (continuous line) compared with that for an Australian Aborigine (broken line). The less prominent chin and more protrusive dentition are evident in the Aborigine: prosthion and infradentale are the limits of the bony septa between the upper and lower incisors, and gnathion is the tip of the chin.

The necessity to keep the load arm of the mandible relatively short contributes to another distinctive Polynesian feature, the straight or vertical facial profile (Figure 4.9). Polynesians generally lack the alveolar prognathism, or protrusion of the teeth, that is a distinctive feature of some groups. Here they may be contrasted with Australian Aborigines. In general the latter possess cranial bases of rather average flexure for *Homo sapiens*, but they are lightly muscled, and have small upper facial heights reflecting the airway size (Abbie 1966, Brown 1973). With this morphology is associated the pronounced alveolar prognathism associated with large teeth (Brown and Barrett 1964). That is, in the presence of an efficient bony mandibular lever, the dentition can be large and protrusive. By contrast, tall and muscular Polynesians possess very flat cranial bases, and large upper facial heights, summating in an extreme adaptive demand on the mandible, which remodels towards a relatively inefficient bony form, but with a compensatory relatively small mandibular body and dentition.

Mandibular influences on the cranium

Polynesians, with an inefficient mandibular lever, thus show a pronounced development of the masticatory muscles. This of course must be evidenced at both attachments, on the cranium as well as the mandible. Some of the evidence on the cranium is in deep, flattened temporal fossae with high temporal lines, and robust zygomatic arches. Other substantial muscle attachments on the cranium in turn influence the mature width of the cranial vault and face. In width, two components can be recognized in the underlying cranial base (Figure 4.10): a central core penetrated by nerves and blood vessels; and lateral regions related particularly to the masticatory system – the temporomandibular joint and the sites of attachment of the pterygoid muscles. The central core is identified by foramina for neural (and vascular) structures, and its very early development has been mentioned. The lateral regions, though also contributing support to the maturing brain, vary particularly with the degree of development of the masticatory system, whose maturation occurs much later. A flat cranial base, together with the typically large Polynesian body with its consequent large upper facial height, demands extreme adaptation of the mandible and matching development of the mandibular muscles and their attachments. Thus in Polynesians the cranial base is wide and the face overall is wide (Wagner 1937; Howells 1973b, 1989). The form of the face also is influenced by the development of the masticatory muscles behind the cheek-bones (Cachel 1979). Increased muscular development here leads to a vertical expansion of the malar bone and flattening of its facial surface, and a filling-out and flattening of the anterior surface of the maxilla.

The upper face, dominated by the airway, forms the second major template of head form. But unlike the first template, the cranial base, which is set in its flexure from an early stage, the upper face is a template that is changing throughout the years of growth. The mandible, at the

Figure 4.10
Inferior view of the cranium showing two components to the cranial base: a central component (stippled) traversed by many cranial nerves, and particularly related to support for the brain: and lateral areas (lined) varying particularly with extent of development of the masticatory system.

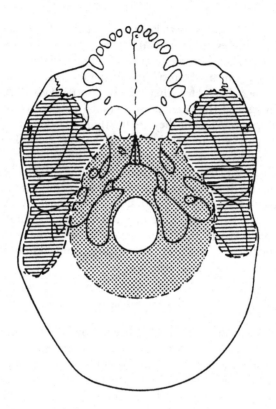

end of the line so to speak, as well as being influenced by the shape of the cranial base, is required continually during growth to adjust to the changing upper facial template. However this is not without exerting its own influence on the upper face, in terms of variation in width of the cranial base, and in the facial profile. By the time of maturity in the Polynesian these adjustments and interactions have become distinctive.

Cranial vault

Only now in this approach to head form can consideration be given to that traditional focus of anthropological study, the cranial vault. Against the complexities of the cranial base the vault is a very simple structure, a mere malleable shell. The view that the cranial base is a fundamental template for craniofacial form supports the idea that long heads should be associated with long cranial bases and round heads with short bases (Lavelle 1979, Enlow 1982). However, such direct expression of cranial base and, probably, brain proportions, may be obscured by the many other influences on the vault. Two factors already discussed particularly affect the Polynesian cranial vault. The very flat cranial base requires development of a high vault to accommodate the brain, for the region below the horizontal plane of the pituitary fossa (particularly the posterior cranial fossa) is rather shallow. This tendency is accentuated by the moulding or flattening of the vault on each side by the substantial temporal muscles, which of course are in part a consequence of the flat base. The result is a very high vault, whether measured by classical craniometric techniques (Wagner 1937, Howells 1989) or by x-rays (Schendel, Walker and Kamasugi 1980, Kean and Houghton 1982), and the effect is compounded by the rather large Polynesian cranial capacity. (In New Guinea in 1871 the Russian anthropologist Miklouho-Maclay (1975) was lamenting, in a rather macabre way, that he could not find a large enough jar in which to preserve the brain of his Samoan assistant who had died of malaria.) On x-ray apparatus designed for examining head form in Europeans, Polynesian heads consistently go beyond the upper margin of the screen. The observation that Moriori crania show a very flat frontal region (sufficient to raise debate as to whether artificial shaping was practised: Thomson 1915–17, Pearson 1921, Howells 1978) is in this group an overt expression of an extreme vault height. The substantial development of the temporal muscles also results in the characteristic pentagonal shape of the Polynesian cranium when viewed from behind and above (Figure 4.11) and explains Larnach's (1976) finding that parietal bosses are most prominent in Polynesians.

Figure 4.11
The Polynesian cranium (right) in posterior and superior view. Compared with most *Homo sapiens* (left), the Polynesian cranium is rather pentagonal when viewed from behind, and shows flattening of the temporal fossae and visible zygomatic arches when viewed from above.

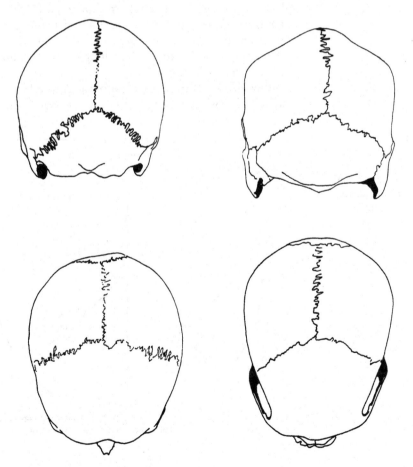

Influences and interactions

Figure 4.12 attempts to chart some of the influences and interactions on and within the architecture of the head. While in the figure the cranial vault is placed centrally, it should already be clear that this does not imply a central place for the vault in determining morphology. Rather, the position reflects the ready plasticity or malleability of the vault. The outer array of structures or influences in the figure are those contained within the body. The inner array are environmental influences of varying importance, from the trivial of ritual headshaping to the substantial influence of nutrition. It may be helpful to follow through some of the interactions, several of which have already been discussed.

On the left side of the figure, nutrition is identified as influencing body mass, which in turn will influence the vertical development of the nasal airway, with flow-on effects on the masticatory system. Vault form may also be directly influenced by nutrition if it is sufficiently inadequate to impair brain growth; however this is an influence rather beyond the normal interplay that is of interest here.

From the top of the figure and moving to the right: a flat cranial base is associated with a rather shallow posterior cranial fossa, which tends to a rather high vault to accommodate the brain. The flat base also positions the upper face more inferiorly, and this positioning leads, as the airway enlarges, to a remodelling of the mandible to a less efficient lever form. In turn this lever requires a more substantial musculature, with widening of the lateral part of the cranial base where these muscles attach. The wider base supports a wider vault, or rather it would if the

Figure 4.12
Flowchart of
influences on
and interactions
between structural
components of the
head. Discussion
in text.

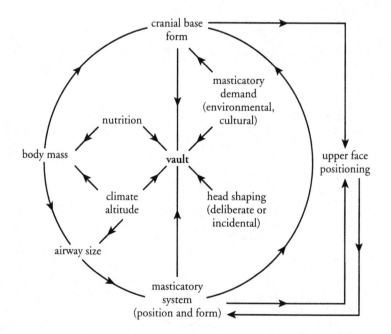

more substantial temporalis muscles did not constrain it laterally, with a further accentuation of vault height. The same muscular enlargement flattens the facial surface of the malar and maxillary bones. The less efficient lever form requires a rather short load arm, so the facial profile is vertical not prognathic, and the dentition tends to be relatively small.

Threaded through all this is the message of the genome. Brain size, from, for example, the difference on average between male and female, clearly has genetic determination. The form and orientation of the cartilaginous cranial base, and thus of its bony successor, appear to have significant genetic determination, and along with the related soft tissues must display as many internal resemblances between relatives as ever appear externally. It is the bony resemblances that taxonomic studies of crania search for. Body mass, and more especially muscularity, has a strong genetic determination (Cotes, Reed and Mortimore 1982: Saltin and Gollnick 1983) with an accompanying demand for a large respiratory tract. It does seem that Polynesians carry with this message for a large body a related message for a flat cranial base to facilitate a large upper airway. The mandibular adaptation then required has been discussed, and there is relatively less room for the dentition than in a smaller individual. Tooth size, or more specifically the antero-posterior length of the dental arcade, must be matched to the potential mandibular support, and because of their differing times of development this can only be achieved by a genetic association. However the genetic influences have to be considered alongside the environmental. In the head these influences are intertwined and perhaps inextricable.

Stress has been laid on the relative immutability of the cranial base because of its intimate relationship with the nervous system, which does not like distortion or disruption. However, unlike the situation with the nerves emerging from the brain stem and passing through bony foramina, a gradual reshaping of the cerebral hemispheres is neurologically as well as socially acceptable, as evidenced by its regular practice in some societies. So while its early expansion is dependent on growth of the underlying brain, the final external shape of the vault is an expression of a spectrum of influences. As with any part of the head, measures or features of it must be interpreted and used with caution in taxonomic studies. We come to this matter in the next chapter.

A factor analysis

In this discussion of the morphology of the Polynesian head, a range of distinctive aspects has been examined from a functional perspective. Another approach to the analysis of biological form is to run collected data through the computer. Analyses of cranial data by multivariate statistics are legion, and there is always the risk of such studies being exercises in what Davis and Hersch (1986) have termed 'rhetorical mathematics', wherein the study and any conclusions may have little to do with any biological reality. Factor analysis purports to reveal significant groupings of shared variability amongst data. We were interested to see if a factor analysis of data from a Polynesian cranial series identified the relationships established on biological considerations.

The data for the study consisted of Cartesian coordinates of cranial landmarks, thus defined in X, Y and Z axes. This approach to data acquisition gives a better chance of defining discrete regions of common variation than does direct use of linear or angular measurements. Figure 4.13 illustrates the major factors that emerged from the analysis (Buranarugsa and Houghton 1981). The figures in parentheses are the

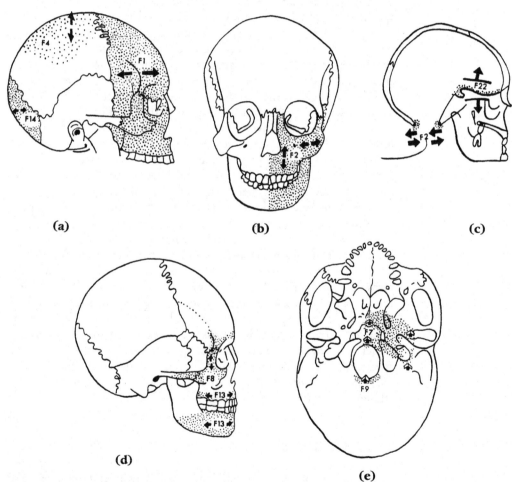

(a) (b) (c)

(d)

(e)

Figure 4.13 Factor analysis of Polynesian skull form.
(a) Skull segments and axes of variation defined by Factors 1, 4 and 14. Factor 14 also displays variation in the X-axis, vertical to the page.
(b) Skull segment and axes of variation defined by Factor 2.
(c) Skull segments and axes of variation defined by Factors 5 and 22.
(d) Skull segments and axes of variation defined by Factors 8 and 13.
(e) Skull segments defined by Factors 7 and 9. The variation is in the Z-axis, vertical to the page.

contribution of each factor to the common variance. Theoretically – that is mathematically – the factors are orthogonal or independent, but however that may be, the biological reality is that this cannot be so. It is encouraging that the analysis has captured many of the distinctive regions of variation in the Polynesian. Not illustrated here are three factors that together contributed 15% to the common variance and which expressed asymmetry of the head. It is generally accepted that a slight amount of asymmetry is inevitable and normal in any ostensibly symmetrical biological structure, at least at the gross morphological level. The vaues in parentheses below indicate the contribution of the particular factor to the common variance.

Factor 1 (16.8%) shows a coordinated antero-posterior variation of the facial skeleton, which relates to the facial profile and extent of prognathism. Factor 4 (5.51%) expresses an association between vault height and lateral prominence of the parietal boss region, features distinctively associated in the Polynesian. Factor 14 (3.54%) expresses variation in flattening of the occipital region; this is a matter brought out in traditional studies of sagittal arcs of the cranial vault, wherein the occipital segment in the Polynesian is found to be particularly short (Wagner 1937, Howells 1973b).

Factor 2 (11.33%) expresses variation in height of the face in association with expansion of the malar region in the lateral axis. This factor expresses the relationship between the vertical size of the airway and the flow-on effect on mandibular form, with consequent variation in the size of temporalis muscle behind the malar.

Factor 5 (3.90%) says that if the anterior margin of the foramen magnum moves in the antero-posterior axis then the posterior margin has similarly to adjust – again a common-sense statement. (Such simple evidence that an analysis is making sensible statements is not to be scorned.) Factor 22 (3.61%) brings out a coordinated supero-inferior positioning of the elements of the anterior cranial base. Here it is encouraging that the analysis has defined a region of biological significance, and the variation probably relates to variation in the cranial base angle.

Factor 8 (4.72%) expresses coordinated variation in height of the malar bone, which is again related to vertical development of the airway and flow-on muscular effect. Factor 13 (3.82%) makes the eminently sensible statement that antero-posterior positioning of the upper and lower dentitions has to be coordinated.

Finally, Factor 9 (3.34%) shows a coordinated supero-inferior positioning of the margins of the foramen magnum, while Factor 7 (3.73%) shows supero-inferior positioning of the central neural component of the cranial base. Again it is encouraging that the analysis has identified a cranial segment defined as significant on biological criteria.

In these studies the biological considerations and the results of the factor analysis come together satisfactorily. It is interesting to consider whether reversing the approach would be as helpful. That is, could a coherent discussion of the biological significance of the resulting factors easily flow from them? The question is worth asking because there are proponents of 'exploratory data analysis' and it is this reverse process that is usual – though in fact biological interpretation of the statistical results of such an exercise usually gets short shrift because the analyst lacks the biological foundations to interpret the results. It seems to me that in the present instance, without a preliminary functional biological model against which to set the results, discussion would be rather incoherent and unable to link up the (mathematically independent) factors. Of the inductive, data exploratory approach, Kowalski, a statistician, comments: 'Has anyone ever found a *useful* morphological factor which was not previously identified on the basis of biological considerations?' (1972: 125).

Some conclusions on Polynesian head form

The particular physique required of the early settlers of Remote Oceania – or at least of Polynesia and Fiji – happens, by its extreme demand on the functions of the head, to illuminate some basic determinants of head form. In determining the morphology of the head, neural influence is significant because of early development and a rather unadaptable nature; but perhaps of even greater import is the demand for an adequate airway, and in these two influences lie much of the basis for the Remote Oceanic distinctiveness. The respiratory demand is expressed in the form – and particularly the flexure – of the cranial base, and in airway size, particularly its height. The templates of cranial base and upper face, being the skeletal support for these functions, in their developmental sequence establish the form to which other head structures must adapt. In particular, the foodway is not some autonomous region. Rather, it is the most adaptable component of the craniofacial complex, last in the developmental hierarchy and to a significant extent shaped by the other components.

By contrast with the discrete hypotheses mentioned earlier regarding form of particular regions of the head, the broader biological view allows establishment of such a 'general coherent hypothesis' as called for by Howells, to account for variation in head form between groups. The model advanced here emphasizes the interplay between the several functional components of this architecturally-integrated structure. There is a problem here of close packaging (Hanken 1983). The form of no part of the head, including the dentition, can be considered in isolation, as though its morphology were in some way autonomous and free of the influence of the whole. This view does not fit easily with some interpretations of fossil evidence which talk of a mosaic of craniofacial features which may be shuffled and reassembled somewhat at random, a bit like an Identikit face (Pope 1992). The model also suggests that is extremely difficult to disentangle genetic and environmental influences on the form of the head. We return to this matter in the next chapter.

Infra-cranial skeleton

The brevity of this section underlines how little has been done. For most individual bones a morphological description in a functional setting is still needed, for the distinctive physique of Remote Oceania will have led to some specific and probably subtle modifications. There are many theoretical approaches to be tried; for example, utilizing beam theory in examining the differences deriving from function or environment between groups (Ruff and Hayes 1983a, b, Ruff 1987). However that is for the future. What discussion there is here is again dominated by the Polynesian morphology, because this has had most attention.

Body proportions and the vertebral column

Body proportions were mentioned in the earlier discussion on estimation of stature. The proportions of body axis (trunk + head) to limbs, quantified in the sitting height ratio, may be a rather old and profound distinction amongst *Homo sapiens*, and it is of some interest that through Melanesia, including the New Guinea Highlands, the ratio seems to be consistently greater than 50: that is, body axis is long relative to the lower limbs. These proportions anticipate that the vertebral column should be rather long, and perhaps show some distinctive segmental proportions or curvatures, or both. Unfortunately the skeletal record is not particularly helpful because much spinal variation is determined by the intervertebral discs, which do not survive. For example the curves of the spine are almost entirely determined by differences between the anterior and posterior heights of the discs, an influence seen at its most extreme at the L5–S1 junction. The only specific study of the spine in Oceanic peoples of which I am aware is one by Christine Barnett on New Zealand material. However this meticulous study was hampered by its being based on a small sample which was, as the author put it, 'riddled with arthrosis'; she concluded that the study had 'posed more questions than it answered' (Barnett 1983: 46). Only slowly, as data accumulate, will reasonable statements emerge on the morphology of the vertebral column in Oceania.

In Chapter 2 overall proportions of limb and trunk were considered. As well, there are segmental proportions within the limbs to examine. These are usually treated as ratios, the standard ones being tibia: femur (crural index), radius: humerus (brachial index) and upper limb: lower limb (intermembral index). Ratios from small series are often suspect; sometimes in such groups discrete individuals, or even male and female, have not been identifiable and the ratios may be much distorted. Gradually data will accumulate, but at present it is wise to accept only ratios derived from larger series – say, more than 20 – and consider smaller series only when the individuals are clearly defined. The group from Mangaia in the Cook Islands (Katayama, Tagaya and Houghton 1988) is marginally suitable on the latter criterion. Here too one runs into one of those little methodological traps. The regression line in Figure 4.14 (after Trinkaus 1981) uses bicondylar (physiological) femoral length and tibial length that takes in the tibial spine as well as the medial malleolus. There are quite a few millimetres tied up in these parts – the femur is reduced in length and the tibia is increased. Before proceeding to ratios it is wise to check the methodology used when the data were collected.

These limb proportions are examined in Figures 4.14 and 4.15, where the Oceanic groups and others are plotted against the regression lines determined for a wide range of *Homo sapiens*. With the exception of the sample from New Caledonia, the Remote Oceanic peoples show a distinctive pattern of a relatively short distal segment to the lower limb and long distal segment to the upper limb. (In New Zealand these distinctive proportions were memorably recorded many years ago by Thomson (1859) in a description of a Maori wearing a borrowed European suit, with the cuffs well rolled up, and forearms protruding from the sleeves.) The large-bodied people from Taumako, the north-eastern outlier of the Solomons, also follow the Remote Oceanic pattern – in fact they show it to a very pronounced degree, as does the single determinable ratio from the Watom series. By contrast, the large prehistoric series from Khok Phanom Di in southern Thailand, dated to about 4000 b.p. (Tayles 1992), shows long distal segments to both limbs; a recent series from Java (Bergman and The 1955) shows relatively short distal segments in both limbs; and Northern Chinese and Ainu plot very close to the regression line for both upper and lower limb for males (Stevenson 1929).

Long bones

Bowing

A distinctive feature of the long bones of Oceanic people is their pronounced curvature or bowing, when compared with groups such as the European or Australian Aborigine. Such bowing is explicable in biomechanical terms. Bone does not like being bent as force comes on it, and counter-bowing of bone in the appropriate plane ensures that the forces of locomotion or whatever, particularly in rather large and muscular people, tend, if anything, to straightening rather than any accentuation of bowing. The appropriate plane for counter-bowing in the lower limb is antero-posterior, with the shaft of the bone being convex anteriorly. Different populations differ in the site of maximum bowing of the femoral shaft. For the Polynesian, Hay (1995) has shown maximum bowing to occur about 45% of the distance along the shaft from the proximal end, which is much the same as in differently proportioned people of African and European ancestry. By contrast, the site of maximum bowing is rather more distal in the shorter-limbed Inuit

Figure 4.14 Plot of tibial maximum length against femoral bicondylar length for several groups. Male means are indicated by the closed squares and female by open squares. The least squares regression line fitting the mean for a larger number of groups of global distribution is shown (after Trinkaus 1981).
AA = Afroamerican, AM = Amerindian, C = Cook Islands, H = Hawaii, In = Inuit, J = Java, K = Khok Phanom Di, NC = New Caledonia, NZ = New Zealand, S = Sigatoka, T = Taumako, Ch = North Chinese, Ai = Ainu. AA, AM, In and NC from Trinkaus 1981; Ch and Ai from Stevenson 1929; H from Snow 1974; J from Bergman and The 1955; K from Tayles 1992; NZ from Houghton n.d.; S from Visser 1995; T from Houghton, n.d.

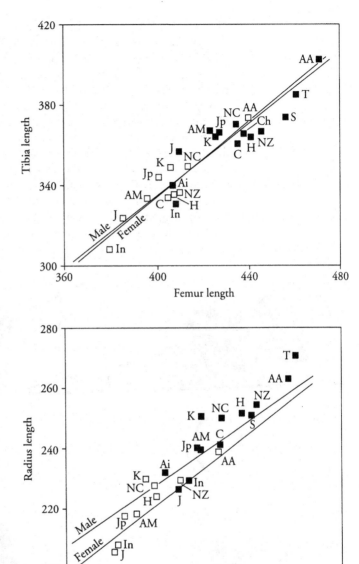

Figure 4.15 Plot of radius maximum length against humerus maximum length for several groups. Symbols and abbreviations as for Figure 4.14. The least squares regression line fitting the mean for a larger series is shown. The separation of the regression lines results from sexual dimorphism in upper limb segment proportions.

and Amerindian groups (Walensky 1965). Biomechanical elucidation of these features does not appear to be simple but the site of maximum bowing presumably is influenced by the length of the limb, the orientation and bulk of the muscles, and associated factors relating to the inertia and pendulum motion of the limb.

Femur and hip

The only Polynesian long bone that has been the subject of detailed study is the femur (Schofield 1959, Hay 1995) and this bone certainly is distinctive to a degree hard to capture in illustrations (Figure 4.16). The shaft of the Polynesian femur is markedly bowed in the antero-posterior plane; shows an extreme flattening or platymeria of the upper shaft, with the long axis of this flattening being oriented in an antero-medial/postero-lateral plane; a large angle of anteversion or torsion, averaging about 25°, which is about twice that of most groups; and some other less blatant features such as an oval fovea – the marking on the femoral head for the ligamentum teres – and a tendency for the gluteal tuberosity to be distinctly enlarged, sufficient sometimes to earn the name 'third trochanter'. In the Polynesian the neck of the femur makes, on average, an angle of 137° with the shaft of the bone, the upper extreme for *Homo sapiens*, in whom it ranges down to 118°.

As with the components of the cranium, these femoral features cannot exist in isolation but must be one with their surrounds, in this

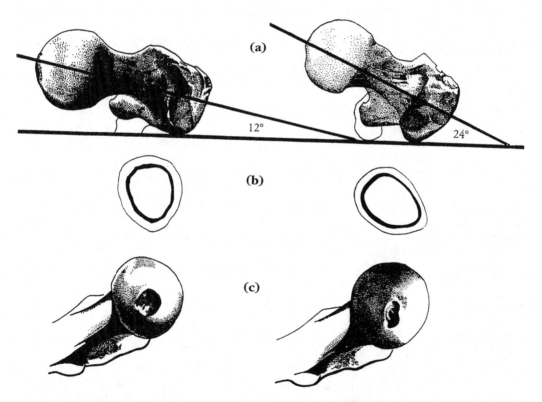

Figure 4.16 Femoral views, contrasting the Polynesian form on right with the more usual form in *Homo sapiens*. In all views the anterior aspect is towards the top of the page.

(a) Superior view, illustrating difference in angle of anteversion or torsion: 12° on the left, 24° on the right.

(b) Schematic cross-section of the upper femoral shaft. Usually roughly circular, in individuals with a large torsional angle there is a flattening or platymeria of the shaft aligned approximately along the angle of torsion. The bone is thicker in the long axis of the platymeric section.

(c) The head of a femur with a large angle of torsion and associated platymeria frequently shows an oval fovea on the femoral head, contrasting with the usual circular depression.

Figure 4.17 Contrasting orientation of the hip associated with femora of small and large torsional angle. In the hip on the left the acetabulum opens at an angle of about 40° to the frontal plane to accommodate the femur with a torsional angle of about 12°. On the right, the acetabulum opens at an angle of about 30° to the frontal plane to accommodate a femur with a torsional angle of about 25°. The sum of the angles in each example is a little over 50°.

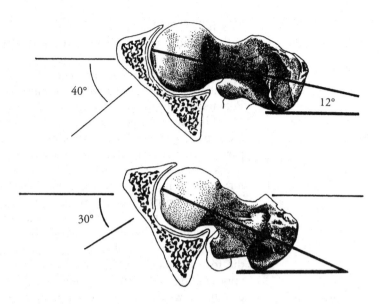

instance with differences in the pelvis and tibia, and ultimately in the whole functional morphology of the lower limb. As an example, a study of the Polynesian pelvis (Baker 1975) showed that by comparison with pelves of Indian and European origin, the Polynesian acetabulum opened or faced more laterally and inferiorly. A study by Domett (1994) showed (the correlation was significant) the acetabulum to open more laterally in harmony with a greater angle of torsion of the associated femur, which makes sense. That is to say, the greater the angle of torsion of the femur, the less the angle made by the axis of the acetabulum to the coronal plane. This is illustrated in Figure 4.17, which contrasts the orientation of the elements of a typical Polynesian hip with that of a more usual form of hip.

   This distinctive femoral morphology may – for these are only suggestions – be related to the generally large body frame and mass, and to the substantial muscularity of the hip and thigh region. Despite the probability that the extensor force capable of being exerted by the gluteus maximus muscle (the largest muscle in the body) is one-third as much again in a Polynesian compared with someone of similar age and sex from the general run of *Homo sapiens*, the pronounced anteversion of the neck tends to alleviate reaction force at the hip joint during extension. (During flexion of the hip in normal movement the weight is largely off the joint and the reaction force there is minimized – that is, alleviation of the flexor force is not necessary.) This force of gluteus maximus and the balancing flexor force of ileopsoas muscle induce the elliptical form or platymeria of the upper femoral shaft. The position of the lesser trochanter (the femoral attachment of ileopsoas) is strongly influenced by anteversion of the femoral neck; the more anteverted the neck, the more the lesser trochanter is carried round the medial margin of the shaft (Lovejoy and Heiple 1972). It has been suggested by

Kapandji (1987) that extreme femoral neck anteversion is associated with a slender femoral shaft and a small pelvis, and favours speed of movement. He contrasts this with the femur with small anteversion which is said to accompany a thicker femoral shaft and broad pelvis; such a configuration is said to favour strength at the expense of speed. The Polynesian morphology contradicts these views.

Usually in *Homo sapiens* the two major forces on the upper part of the femoral shaft tend to be orientated at right angles. The body weight, particularly when standing still, is transmitted medio-laterally from the pelvis through the head and neck of the femur to the shaft, whereas the forces of locomotion are orientated more in the antero-posterior plane. Then the appropriate shape for the upper shaft is approximately circular, indicating that the balance between the laterally-directed force of the body weight and the antero-posterior forces of locomotion is fairly even. By contrast, in the Polynesian with a strongly angled femoral neck, the forces of body weight and locomotion, rather than being evenly opposed, align somewhat together in the long axis of the neck and of the platymeria where the second moment of inertia is much increased.

The platymeria thus tends to follow the anteversion of the femoral neck and is secondary to it. It may be that the features are most pronounced in the Maori because of their different environment, dominated by rather hilly land, terrain which makes greater demands on the extensors of the hip during locomotion.

The large neck-shaft angle of the Polynesian femur favours the direct transfer of body weight to the femoral shaft. A more horizontal femoral neck would accentuate the torque at the neck-shaft junction, with an increased liability to fracture in a heavy muscular person. On the other hand, a disadvantage is that in this axis the more oblique femoral neck leads to an increased reactive force at the joint.

These suggestions regarding the distinctive Polynesian femoral morphology draw again on the theme of environmental adaptation, body mass, and a genetically determined substantial musculature. We do not have adequate data on the juvenile Polynesian femur, partly because the parameters are difficult to quantify before the epiphyses are near-mature; but it seems likely that maturation of an individual must occur for full expression of the femoral morphology to be apparent – a rather similar situation to that for the rocker jaw.

Other limb bones

There has been little scrutiny of Oceanic skeletal morphology distal to the femur. As part of the general bowing of the long bones, the shaft of the tibia is bowed in the sagittal plane, convex anteriorly, presumably for the biomechanical reasons mentioned earlier. The tibial plateau is also markedly retroverted. This has been ascribed to the social/cultural habit of squatting from an early age, but the horizontal plateau of the Australian Aborigine (Quarry-Wood 1920) gives lie to that explanation. (Cultural explanations for morphological distinctions seem rarely to be valid: the explanations are usually more profound and enduring.) It is more probable that such retroversion serves to stabilize the knee as body weight comes on it during locomotion: then the force is more perpendicular to the plane of the plateau and the sheer stresses across the cartilage are minimized. This was proposed by Trinkaus (1975) to explain such morphology in the Neandertal, and it equally applies to the large-bodied Pacific people, who in a very technical sense might be regarded as oceanic Neandertals.

MODELS AND METHODOLOGY

5

This chapter starts by presenting one current view or model of what happened in the Pacific in prehistory. This view and some of the contributing methodology are then scrutinized against the evidence from human biology, including that from gene studies, which over the past decade have greatly expanded our understanding of human relationships in the Pacific. Other perspectives on the Pacific past are then discussed. The chapter closes with a further look at the results of the simulation of human survival at sea, and their implications for Pacific settlement.

Before setting down this view or model of the Pacific past, a few comments are needed on some of the terms, particularly those that derive from physical anthropology and linguistics. One pervading theme in the study of Oceanic prehistory has been to see the wide range of human variation in terms of the movement into and across the Pacific of people of differing physical characteristics. Oliver gives the flavour of these: 'All interpretations of racial differences in Oceania, including the most recent and best informed ones, include at least two or three distinct ancestral strains; and some interpretations include four or five or more' (1988: 49).

The instigator of this approach probably was J. R. Forster, scientist on Cook's second voyage:

We chiefly observed two great varieties of people in the South Seas, the one more fair, well-limbed, athletic, of a fine size, and a kind benevolent temper; the other blacker, the hair just beginning to become woolly and crisp, the body slender and low, and their temper, if possible, more brisk, though somewhat mistrustful. The first race inhabits O-Taheitee, and the Society Islands, the Marquesas, the Friendly Isles, Easter Island and New Zealand. The second race peoples New Caledonia, Tanna and the New Hebrides, especially Mallicollo. (Forster 1778: 228)

Other European observers added to these impressions, and, as mentioned earlier, they became rather fixed by Dumont D'Urville with his partitioning of the Pacific into Melanesia, Micronesia and Polynesia. With increasing awareness of the range of physical appearance across the different groups of mankind, a good deal of nineteenth-century science was devoted to defining these differences. In the vocabulary of racial typology the people of Melanesia were classified, albeit inconsistently, as Australoids, and the people of Polynesia and Micronesia, again rather inconsistently, as Mongoloids. These terms are still used in discussions of Oceanic prehistory, not just in passing but as central elements of the view or model about to be presented (Bellwood 1985a, b, Kirk 1989, Serjeantson 1989). They are the sort of 'distinct ancestral strains' mentioned above by Oliver.

However the migratory concept, of the entrance of different groups of people of differing physical attributes as the basis for the wide range of physical variation in the Pacific, is usually set, not in a biological framework, but in a linguistic framework that stresses the distinctiveness of Papuan and Austronesian languages. Papuan languages, of which there are at least 700, are spoken through much of mainland New Guinea and in Island Melanesia by scattered groups as far east as the Santa Cruz Islands. The embracing term 'Papuan' is deceptive in that it encompasses no coherent grouping of discernibly related languages, and perhaps for this reason is often referred to by the negative term, 'Non-Austronesian'.

The languages of all Remote Oceania and much of Near Oceania belong to the Austronesian family. A current view of the distribution and relationships of this language family is set out in Figures 5.1 and 5.2. It is said to have spread 'from a homeland in the southern China and Taiwan region within the last 6000 years' (Bellwood 1989: 23). From this source it is said to have passed down through the Philippines (Luzon at 5000 b.p.) and into Indonesia, with a subsequent split into western and eastern divisions. The western division, in which here we have no particular interest, has a remarkable outlier in Madagascar, whence it appears to have been carried back along the routes of Indian and Arab traders to Indonesia early in the Christian era – the same route of goods and human contact that brought Islam to South-East Asia. In the east, nearly all the languages of Remote Oceania are placed in the single Oceanic sub-group of Austronesian. The people of the Lapita culture, generally considered to be in their eastward spread ancestral to the Polynesians, are believed to have spoken an Austronesian language: words in the languages of Remote Oceania that relate to pervading elements of maritime existence – canoes and fishing, food plants of the island world and some elements of culture – may be traced to a common source. As the Lapita culture is dated to as early as 3500 b.p. we are looking at a pretty rapid spread of language out of Asia from an origin around 6000 years ago. A most significant point is that in the usual scenario the spread of the Austronesian language family is equated with the spread of people – that is, a

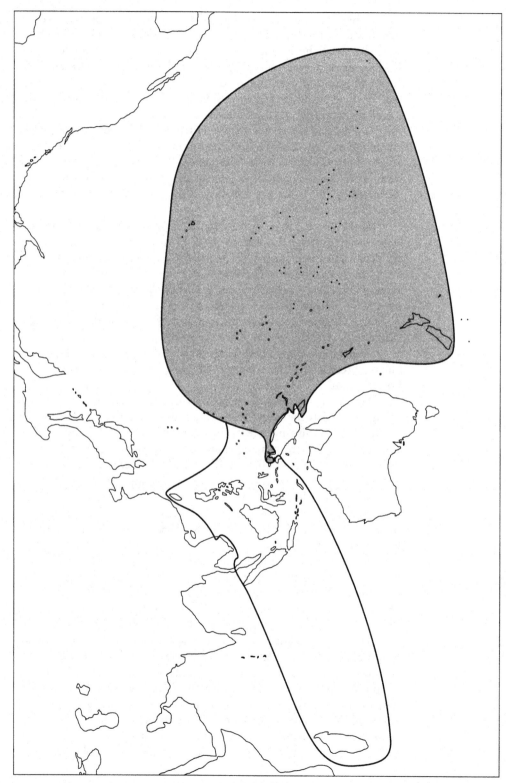

Figure 5.1 Distribution of the Austronesian language group. The region of the proto-Oceanic sub-group is shaded.

Figure 5.2
Sub-groupings of
the Austronesian
languages, with
the Polynesian
relationships shown
in greater detail
(from Irwin 1992).

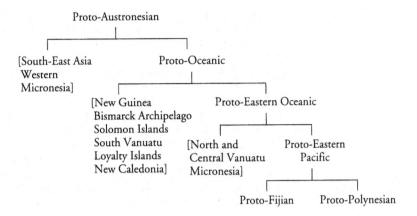

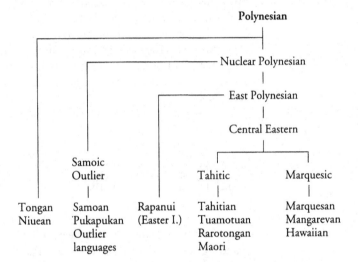

migration, sweeping into Oceania. Thus the linguistic label, Austronesian, is equated with a biological group or continuum, as

the enormous dispersal of the Austronesian languages can only be correlated, over most if not all of its geographical extent, with a dispersal of people ... Around 3500 years ago Austronesian-speaking populations entered western Melanesia from eastern Indonesia. The record of this expansion occurs in the Lapita culture, which involved rapid dispersal eastwards to as far as Samoa, and eventually (by AD 1000) throughout the rest of Polynesia. (Bellwood 1989: 19, 41)

When we stand back and look at this panorama of prehistory there might seem to be a quite satisfying harmony of evidence from physical anthropology, language and the archaeological record. In the words of the same authority:

It is now clear that both New Guinea and Australia were settled possibly more than 40,000 years ago by Australoid founder groups whose descendants are represented by present Australian Aborigines and New Guinea highlanders . . . an Australoid and Papuan-speaking population, probably with horticultural skills and periodic foci of high population density, was established in occupation of New Guinea and some adjacent parts of Island Melanesia at around 2000 B.C. From this time onward small Austronesian-speaking populations entered the region from Indonesia; we see their descendants today as the Polynesians, and of course as the bulk of the Island Melanesian populations whose ancestors intermarried and swapped cultural and linguistic baggage with the earlier inhabitants of western Melanesia. I believe that these Austronesian-speaking populations were initially of a basically Southern Mongoloid phenotype, as befitting their ultimate origins among agricultural populations in Southern China and South-East Asia. Most of them have lost this phenotype today in Melanesia as a result of the genetic success of the previously established Australoid inhabitants of the region. (Bellwood 1985a: 133)

Naturally, as with any attempt at modelling the past, there are a few problems with this proposition, and I would not suggest that every prehistorian concurs with it in full or even in its major part. However, it is widely presented and widely known, and probably for that reason currently forms the background against which the biological evidence on prehistory tends to be set. Along with some contributing methodological approaches that use human biological data, this view or model is now scrutinized.

The linguistic pattern

Linguistic data have been very helpful in the development of an understanding of relationships between Pacific peoples, particularly in Polynesia with its relatively shallow time-depth. In the western Pacific the linguistic picture inevitably becomes hazier and more debatable. When listening to echoes several thousand years old, there may be (but probably is not) at any time a consensus on relationships and origins; but, however thoughtful, that consensus can ultimately only be informed speculation, never disproveable. This demands an extreme caution with dates. When it is stated that Austronesian emerged in Taiwan at 6000 b.p. and was in Luzon at 5000 b.p., a precision is being claimed that is not yet achievable by any physical analysis such as carbon dating or electron-spin resonance, which derive from the rather exact sciences of chemistry and physics. (Of course no one, I am sure, intends ludicrously to adhere to a precise year: but that is how the dates are starkly presented, without any estimate of possible error.) The accuracy and methodology of such physical analyses are subject to intense scrutiny. A linguistic analysis is simply not susceptible to the same order of scrutiny. Such linguistic surety is unanswerable because it is impossible to disprove, and this is the definition of speculation. To base a whole scenario of prehistory on speculation, however informed, seems unwise. These words should not be taken as an aspersion on linguistic models, which would simply be silly, but rather as a statement about levels of testability and levels of complexity.

The linguistic pattern in the Pacific is often presented as a dichotomy, of Austronesian (AN) and non-Austronesian (NAN) languages, with the familiar picture of pages sprinkled with AN and NAN. This is a misleading approach. Rather, under the title of Non-Austronesian are subsumed more than 700 languages that, at best (Wurm 1983) – and still controversially – may be parcelled into five major and six minor groups or phyla, all unrelated. The last couple of words are important. There is no coherence, unless it be patchily geographic, in the Non-Austronesian division of languages. Subsumed in the term is a range of some dozen apparently unrelated language groups, each of which is as taxonomically distinct as Austronesian. The point is that the misleading linguistic presentation too easily leads to an equating with the typological dichotomy of Australoids and Mongoloids.

Nevertheless, this linguistic framework has dominated recent thinking on the early human biology of the Pacific. This is exemplified in two volumes that are probably the most important in Pacific human biology in recent years. Much of the Harvard Solomon Islands study (Friedlaender 1987a, b), particularly the sections on odontometrics and dermatoglyphics, is dedicated to a search for evidence of relationships between biology and language. Interpretation of evidence in a major volume on the molecular genetics of Pacific peoples (Hill and Serjeantson 1989) is similarly influenced. Yet there is simply no inherent reason for apparent linguistic continuity to reflect biological continuity, if we mean by this that, give or take a bit of genetic admixture on the way, a particular human group moved along, in its generations, with a particular language in its generations. Lewontin comments in the foreword to Friedlaender's study of human biology on Bougainville:

Perhaps chief among the defects is the great weight anthropologists must put on a rather too simple measure of linguistic similarity, the percentage of cognates in two languages. It is a peculiarity of the entire method of investigation that one set of characters, the genetic ones, with an extremely sophisticated theory of their evolution, is being compared with another set, linguistic, with an extremely crude theory of evolution. (1975: ix)

I would not suggest that linguistic research has remained static, but since those words were written knowledge of the genome has grown disproportionately. And to anticipate, the genetic evidence from Melanesia shows biological similarities riding roughshod over linguistic divisions.

An immediate example of the fragility of the link between language and people, is given by Oliver, recording in a Bougainville community the almost complete replacement within one generation of one language by another. (Here it happens to record the demise of an Austronesian language.)

When I visited Bougainville Island during 1938–40, there was a large village on the east central coast named Arawa . . . whose residents spoke an Austronesian language closely related to languages spoken in the islands immediately south of Bougainville. They told me confidently and unequivocally that their ancestors had immigrated to Bougainville only a few generations before. On a return visit to the island in 1968, I found the village of Arawa still there, but only the oldest of its residents were fluent in their former language, the rest were speaking the Papuan (i. e., non-Austronesian) language of their hinterland neighbours, the northern Nasioi. (1988: 53)

Because of misgivings over the relationship between language and biology, and the claimed precision of linguistic dates, I have tried through the following discussion to consider the biological data as evidence in itself – which it most surely is, for ultimately we are talking of people not of sounds – rather than against any linguistic model. In some studies however, biological (or, more usually, typological) and linguistic terminology and considerations have been so intertwined that this scarcely is possible.

Australoids, Mongoloids, and racial typology

It has been noted that the people of Melanesia are classified (inconsistently) as Australoids, and the people of Polynesia and Micronesia (inconsistently) as Mongoloids. As these terms are integral to the above model it is important to define them carefully in the words of their proponents. However brevity requires – for often these racial descriptions are wordy – that we focus on the specific physical characteristics that are cited. In passing we may note that some writers, using such features as the shape of cranial sutures, posterior tooth size, and the size of the frontal sinus, have traced this Australoid/Mongoloid dichotomy to the geographic subdivisions of *Homo erectus* (Weidenreich 1936, Coon 1962, Wolpoff 1989). In this view, present day Mongoloids are in the line of descent from *Homo erectus* in China, while Australoids are derived from *Homo erectus* in Java. I will comment further on this idea, but it is not central to the present discussion.

The more substantial volumes devoted to global racial taxonomies of *Homo sapiens*, such as those of Coon, *The Origin of Races* (1962) and *The Living Races of Man* (1965), are now a generation old. Australoids, according to Coon, include Australian Aborigines, Melanesians, Papuans, some of the tribal folk of India, and various Negritos of South Asia and Oceania (1965: 3). They have 'beetling brows, sloping foreheads, concave temples, deep-set eyes, large fleshy noses, projecting jaws and large teeth' (*ibid*: 12). Hair form ranges from tightly curled to straight and skin colour from black to light brown. The specific features of the 'non-Melanesian-speaking Papuans' are tightly curled hair and a high frequency of high-bridged convex noses. Some are of small stature and some have rather light skin. Melanesians are said to differ from the Papuans mainly in having blacker skin, less rugged features and fewer prominent convex noses. Still within Melanesia and Australoid territory, the New Caledonians are thick-set and heavy-boned, with light to medium brown skin and heavy brow ridges and jaws, while some have the Papuan nose. The characteristics of the Fijians, described as the easternmost Melanesians, are tall stature, powerful build, and quite a high incidence of Papuan noses (*ibid.*).

Mongoloids have skin colour said to vary regionally from a 'sallow brunette white to reddish brown'. Hair is black and straight and they have high cheek-bones and 'slant eyes'. Teeth are large and incisors often shovel-shaped, with considerable alveolar prognathism and a 'retreating' chin. Noses are either flat or beaked. The body build tends towards a long trunk and short limbs, with particularly short forearms and lower legs (*ibid*. 11).

Coon then gives a comment on the tall and muscular nature of the Polynesians, and concludes that Polynesians, Micronesian and Melanesians are all of mixed Mongoloid-Australoid origin. Coon did in fact claim an adaptive basis for his racial classifications, but very little support for the claim is given. In terms of distribution and environment it is hard to perceive an adaptive coherence in, for example, his Australoids.

This might seem more than enough on matters of racial typology, but the point has to be made that on these same rather few phenotypic features, other authorities establish very different classifications. For Montague, Australoids comprise the Australian Aborigines, the Veddah of Ceylon, pre-Dravidian people of India, and the Ainu of Japan (1960: 461). Papuans and Melanesians are Oceanic Negroids, and Polynesians are in with the Caucasoids, being a 'far-flung branch of the Mediterranean stock'. Montague was far from being the first with this view. 'That [the Polynesians] are one people is obvious, and that they are an Oceanic branch of the Caucasic division is now admitted by all competent observers' (Keane 1908: 417). Others were less sure. Sullivan (1923) recognized four elements in the population of Polynesia which he called Indonesian, Melanesian, Polynesian, and Polynesian with deformed heads: he was uncertain of the sequence of these types. However Australoids and Mongoloids, more or less as defined by Coon, seem to have won the Pacific day, and with this dichotomy a clear link was seen between the geography of the Pacific and the races inhabiting it.

Some workers have not accepted the old suggestion that a mixing of the primary races has occurred to any significant extent in Polynesia, and much debate about the settlement of the region has been over how the Polynesians managed to get themselves into Polynesia without significant contact with Melanesia. Howells considered that Polynesians were 'simply too different from anything in Melanesia to be derived therefrom by local change in a few thousand years . . . as physical beings the

Polynesians simply could not have emerged from any eastern Melanesian population; they are just too different genetically' (1973: 228, 234).

At this stage one may reasonably protest that much of this was written far away and long ago, which is true. Just as true is the reality that these are terms still being used as central to models of Pacific prehistory. Many still see these concepts as part of the essential vocabulary of pre-history, believing that they serve a useful purpose and that the topic cannot sensibly be discussed without them. For example Bellwood feels that 'for an intelligible narrative of prehistory a concept of race is necess-ary' (1985b: 69). So it really is necessary to try to discern the biological significance, if any, of these terms. The distinguishing features of the specific racial types said to have entered the Pacific do seem rather vague and rather few. Skin colour gets a mention but for Australoids, for example, can vary from light brown to black. Hair characteristics get a mention, but range from tightly curled to straight. There is a comment on body proportions for Mongoloids, and a little on facial features – perhaps a rather slight foundation for a taxonomy of the whole organ-ism. Against the size and complexity of the genome the coding for the phenotypically obvious features of skin colour and hair form is simply trivial. These few characteristics seem a rather shaky base for a classifi-cation of humankind and a model for prehistory. As Jones comments, 'The idea of racial "type" – and, some would argue, of "race" itself – is no longer a very useful one in human biology' (1981: 190).

Certainly in the Pacific arena the variety of presentations of the terms reflects the underlying unease of many who use them. Mongoloid may be presented thus, or as mongoloid, 'Mongoloid', 'mongoloid', *Mongoloid* or *mongoloid*. All these variants are in regular use, and sometimes within the same paper and even the same paragraph. Unfor-tunately the standard subdivisions of the Pacific (Melanesia, Micronesia and Polynesia) have created a sub-order of typologies. For example Howells considers that we may:

use 'Melanesians' in a biological sense, though they can be defined only in the context of comparisons with other Oceanic peoples. The Melanesians are darker-skinned and more uniformly frizzy-haired than populations of Indo-nesia, Micronesia, and Polynesia. It is hard to specify them further, but an experienced observer would recognise them, say, in photographs of samples of populations, and in fact be able to distinguish them from aboriginal Australians, to whom they have a greater resemblance than to other peoples mentioned. (1987: 12)

These criteria are really not sufficient to allow use of the term in a 'biological sense'.

Ironically, if any group of *Homo sapiens* tends to compliance with the outdated concept of a racial type it is the Polynesians, particularly east Polynesians. The Polynesian somatotype is remarkably homo-geneous over one-sixth of the globe, and there is an array of distinctive biochemical and physiological parameters to be considered later. All these are the consequence of a unique geographical and environmental situation in which the full force of selective and drift influences on the genome have been felt. This very distinctiveness ensures that the term Mongoloid (or *Mongoloid* etc.), as a label for the phenotypic characteristics of Polynesians, is not sustainable. In form of head and form of the body overall, Polynesians are distinctive amongst *Homo sapiens*, and they lack significant resemblance to any recent Asian group. This is not the same as saying that their ancestors did not come out of Asia. There is plenty of genetic evidence to show that they did, just as at one time or another did every other Pacific ancestor. The point is that adaptation to the environment of the wider Pacific has substantially changed body and head form, and many biochemical and physiological parameters, so that the successful colonizers of Remote Oceania are likely to little resemble phenotypically their ancestors in Asia, or the present Asian descendants (who themselves will have changed) of those same ancestors.

Until there are more biological data, judgement has to be withheld as to whether Micronesia was sufficiently homogeneous to justify a single epithet along the lines of Polynesia, or even initially, in genetic terms, have been one with Polynesia. The history of western Micronesia is such that prehistoric material will be important in this assessment.

By contrast with the relative sameness of Polynesia, the hallmark of *Homo sapiens* in Melanesia is biological diversity; anthropometric variability goes hand in hand with serological polymorphisms of confounding profusion (e.g., Friedlaender 1975,1987a, b; Lai and Bloom 1982). This is part of the common knowledge of human biology, yet the surprising thing is that the term Melanesian is still used with the implication of some sort of biological unity or homogeneity, and frequently as a foil to the term Polynesian. 'People will inevitably persist in the naming of racial groups based on simple physical and even social attributes, but, at the very least they should be made aware how grossly simplified any such taxonomic system has to be, and the diversity which a name such as "Melanesian" masks' (Friedlaender 1975: 215). The term, Melanesian, is a geographic statement, indicating someone living within the confines of Melanesia. No phenotypic or genotypic uniformity can be inferred from the term.

Because these various terms, Australoid, Mongoloid, Melanesian, are biologically undefinable, they are just cliches that form a real impediment to progress towards an understanding of the early human biology of the Pacific. They make too easy the continuation of thinking in outdated and meaningless racial terms (talking strictly in a taxonomic context). One can have no illusions as to the difficulty in discarding such impedimenta, because they are embedded in the literature and in our own thinking. Even in molecular genetics, the very discipline that offers the chance to discard unbiological racial scenarios, discussion is frequently couched in typological terms.

A couple of examples will do for the moment. A discussion on certain nuclear DNA polymorphisms that opens: 'The three principal races in the South Pacific are Melanesians, Micronesians and Polynesians' (Trent *et al.* 1986: 355). And in the interpretation of HLA antigen data, the Fijiian population is estimated on the basis of the HLA data to be 'ancestrally about 20 per cent Polynesian, 20 per cent Austronesian-speaking Melanesian, and 60 per cent non-Austronesian-speaking Melanesian' (Kirk 1989: 101). This is when 'the parental populations are represented by non-Austronesian-speaking Papua New Guinea Highlanders, Austronesia-speaking Melanesians from New Caledonia, and Polynesians from Western Samoa' (Serjeantson 1989: 164) To complete the confusion over Fiji it is noted that 'Fijiians themselves are not homogeneous ... Lau islanders are closest genetically to Western Samoans ... Koro Islanders are closest to non-Austronesian-speakers in New Caledonia, and ... the Fijians from Nadi are closest to Motu-speakers from Port Moresby' (Kirk 1989:101). The plaint here is not with the accuracy of the genetic statements, which from such authorities will undoubtedly be correct. Rather, it is to ask what the point of the comparisons is. One looks for some direction as to what is being implied – migrations of separate peoples, chance effects of drift, historical or prehistorical mixing of groups – rather that a bald statement.

The linguistic model or framework for Pacific settlement, and studies that adopt it, incorporate typological terms as they move people along with language. A significant problem is that the biology of racial typology is unclear. And to talk of Austronesians in the past is to define morphologies of which we only have glimmerings by a label for sounds that no one living today has ever heard.

Phenotypic evidence

Craniology: metrics and statistics

Into the arena of racial taxonomy the craniologists came early, and, on material more recent than *Homo erectus*, they have had a field century of measuring and calculating and parcelling. This business probably started in any dedicated way with the Swedish scientist Andreas Retzius (1796–1860) who founded a theory of culture and civilization based on head proportions. Retzius perceived round-headed peoples (brachycephalics) as earlier, more primitive types, supplanted in places where it mattered by long-headed peoples (dolicocephalics). Other workers, such as the American S. G. Morton, were more concerned with cranial capacity, for it was clear to some that intellect and thus position on the evolutionary tree was a function of brain size. This sort of thing has been discussed by Gould (1981). Leaving aside the murky ancestry, a taxonomic purpose for measurement of heads was still clearly seen.

The burgeoning data from craniological studies needed ordering, though at the time most anthropologists probably tended to the view of Ernest Rutherford, that anyone who needed statistics should have done a better experiment. In Britain the right man at the right time, around the turn of the twentieth century, was Karl Pearson – a Yorkshireman, despite the Teutonic cast of name, and one of the intellectual heirs of Francis Galton. For Pearson the dictum of Lord Kelvin was truth:

When you can measure what you are speaking about and express it in numbers you know something about it: when you cannot express it in numbers your knowledge is of a meagre and unsatisfactory kind. (Kuhn 1977: 178)

However one must be cautious of spurious objectivity in numerical expression. Kelvin derived, with mathematical assurance, an age for the earth of a few million years, based on the faulty thesis that the earth's heat emanated from a cooling molten core.

Pearson was always alert to defend the objective statement. For example – only one of many crusty responses – in 1915 one of his students, L. M. Thomson, published a study on the crania of the Moriori of the Chatham Islands which was criticized because it failed to discern evidence for artificial shaping. Replying to this presumptive critic, Pearson noted:

It is, perhaps, needless to say that anything produced by the Biometric School is likely to be anathema to an old-fashioned anthropologist of Professor Guiffrida-Ruggeri's type ... The questions of different methods of measurement, the influence of random sampling, the futility of small samples are not difficulties to this type of anthropologist, they are merely shibboleths of the mathematician. ... anatomical anthropologists will agree that the artificial deformation of the Moriori has remained undiscovered up to this day – simply because it does not exist. (1921: 338)

Pearson was fascinated by the statistical problems presented by the steadily-growing work in various fields of biology. The early volumes of *Biometrika*, the journal he co-founded in 1901, are replete with series of measurements from skulls from many parts of the world: India, Thailand, Tibet, Egypt – and the Pacific (Thomson 1915, von Bonin 1931, 1936). Means and variances and estimates of significant differences between groups were presented. Pearson laboured at his statistics, looking towards some method of expression that would enable all these data gathered by his co-workers and students to be sensibly interpreted. In 1926 he produced the Coefficient of Racial Likeness (CRL), the first effective version of multivariate statistics. Soon racial taxonomies from cranial data were everywhere being presented in terms of the CRL, often in misrepresentation of Pearson's own insistence that it was a test of significance: 'it should be a measure, not of how far two races or tribes are alike or divergent, but of how far on the given data we can assert significant resemblance or divergence' (Pearson 1926: 105). Figure 5.3, taken from Wagner's (1937) major study of Oceanic crania, gives CRL determinations for several groups. Problems here include the large distance expressed between the plausibly close Maori and Moriori, and the distance of both these groups from the Cook Islands.

Modifications and enhancements followed. In particular, Mahalanobis, a pupil of Pearson, produced his wellknown D^2 statistic

Figure 5.3
Relationships of
Pacific groups from
skull measurements,
expressed using the
Coefficient of Racial
Likeness (after
Wagner 1937).

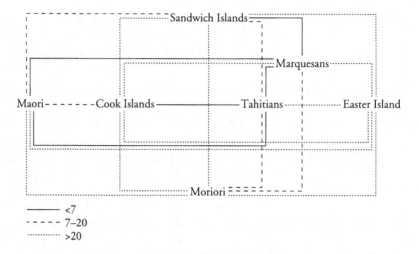

(Mahalanobis, Majundar and Rao 1949). Presented as an advance on the CRL because it eliminates redundant variables, it remains one of the most widely used multivariate approaches in anthropology today. Penrose (1954) showed that the CRL could be divided into two components, size and shape, with much simpler input demands (though I have been instructed that there is no test for significance in a Penrose analysis). Consequently the 'Size and Shape' statistic is quite often used, but the CRL is consistently dismissed in anthropological writings because it does not allow for correlation between variables. On this point I find Gower's views of interest:

> Now it seems to me that the human mind distinguishes between different groups or populations because there are correlated characteristics within the postulated groups. To quantify this idea, some measure of distance such as DD or CRL is necessary; D^2 will not do, because of the way it eliminates correlations. When populations have been established D^2 becomes useful and is essential when building up discriminant functions. These two aspects of distance must be distinguished and carefully considered when deciding why a particular set of data has been collected. (1972: 3)

Cavalli-Sforza and Bodmer comment similarly: 'The search for uncorrelated characters, or the use of analyses that remove such correlations, is valid when the purpose is pure discrimination. In the reconstruction of phylogenies, however, characters may happen to be correlated simply because of evolutionary history. Such correlations should not be eliminated' (1971: 703).

Gower makes another point of interest regarding choice of measures of taxonomic distance (1972: 2–3). He comments that the relations between measures such as the CRL (Pearson), D^2 (Mahalanobis), and the DD of Czeckanowski have been found to be highly correlated, so that one may as well ignore 'erudite arguments' and use DD, the coefficient most simple to calculate. He makes the point that there is no theoretical reason why these measures should always be highly correlated, and suggests that the observed high correlations might imply something about the cranial characters selected for measurement: 'Is the geometry of the skull such that taking standard cranial measurements is equivalent to observing repeatedly a few fundamental, perhaps not directly measurable, variates?'

This thought fits with the discussion in Chapter 4 on determinants of craniofacial form and the establishment of skeletal templates – which may be seen as just such 'few fundamental, perhaps not directly measurable, variates'.

Suggestion has been made that these modes of analysis are outdated (Konigsberg 1992). However it is clear that biological complexity is stretching theoretical statistical analysis when, among the approaches Konigsberg refers to, one assumes that selection (in the evolutionary and adaptive sense) plays no part, and another requires major assumptions to be made in part for mathematical tractability and in part because of empirical justification. In the not-unrelated field of population ecology, Gilbert (1989) – like most who sound warnings, a statistician – simply says that we have no satisfactory mathematical tools for handling the degree of biological complexity. It is clear, though, that some newer methods are making a better job of small samples.

The skull is rich in features on which to place the points of calipers, and large series of landmarks and measurements gradually became defined. Pearson's Biometric School took pride in a meticulous technique of measurement between precisely-defined landmarks. For example, Morant (1936) takes some 300 words to describe the measurement on the mandible between the tips of the coronoid processes. The German School, exemplified in the monumental volumes of Rudolph Martin, did much the same. Today, workers in this field talk of their 'suite' or 'battery' of measurements – their armamentarium, so to speak, derived from traditional sources, usually along with a few of their own. With the advent of computers and the increasing sophistication of multivariate statistics, cluster analyses have been produced in abundance, and the Pacific has seen its share. A typical result is depicted in Figure 5.4. This is work by Pietrusewsky (1983: 73), an exponent of these studies for Pacific groups. Scrutiny of it shows a pretty sensible sorting of people of the Pacific and its rim on the basis of their cranial measurements. Australians are grouped together, Chinese and Japanese are grouped together, and some Polynesian peoples are grouped together. Perhaps the only discordant note is the placing of New Zealand with the southern Moluccas. Pietrusewsky sees these results as contrasting an 'Australo-Melanesian' group with one comprising Indonesia, Micronesia, Asia and Polynesia, which in essence is the Mongoloid/Australoid dichotomy.

Another tree is given in Figure 5.5, this being from Brace and Hunt (1990: 346). These analysts recognize that adaptive influences on the head need to be considered, and the measurements used in their study are stated to include many that are not obviously under the control of specific selective forces. Their view is that in the literature there are

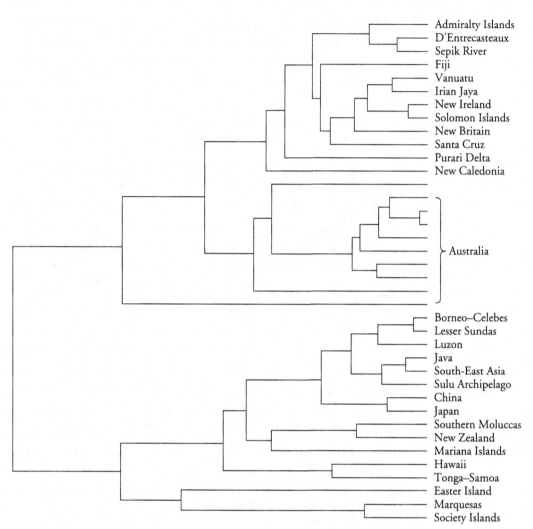

Figure 5.4
Cluster analysis of Pacific, Australian and Asian groups using 28 cranial measurements (from Pietrusewsky 1983).

'countless enumerations of traits that have yet to be shown to have any vestige of adaptive significance'. The specific example they give is the occurrence of a pentagonal cranium. This is an unfortunate example (it was mentioned in Chapter 4), illustrating the dangers of considering any aspect of body morphology in isolation, as though it existed autonomously. Then, when their measurements are scrutinized they are found to include not only such variables as dimensions of the airway, a structure greatly influenced by selective forces, but overlapping dimensions, such as nasal height, nasion-prosthion length, nasal bone height, and piriform fossa height, which have high topographical correlations, so that essentially the same statement is being put again and again into the analysis.

The reality is that for the skull, the theory underlying metric analyses has changed little since the halcyon days of *Biometrika*, and now, as

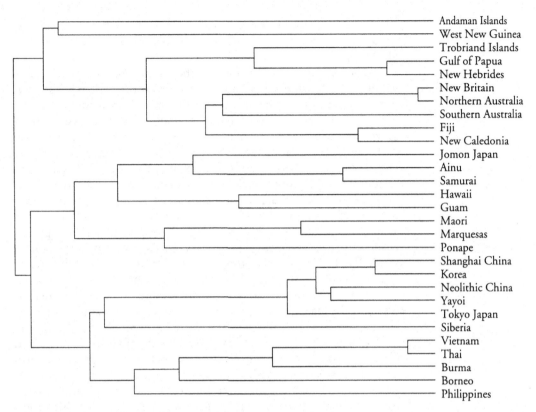

Figure 5.5
Euclidean distance
dendrogram using
skull data from
Pacific, Australian
and Asian groups
(from Brace and
Hunt 1990).

then, there is little evidence that this marvellously complex structure is being considered with any biological perspective, as a once-living thing, responsive and adapted to the needs of the individual and the demands of the environment. Rather, data are collected as though from some inorganic object and scrutinized statistically. The approach is inductive, and it is the advances in statistical technique that are seen as the great step forward.

Yet even at the turn of the century Pearson both promulgated his statistics ('The whole problem of evolution is a problem in vital statistics') and warned of departure from reality ('The danger will no doubt arise in this new branch of science that . . . mathematics may tend to diverge too widely from Nature') (1901b: 320; 1901a: 5). The availability of computers and statistical packages has made this danger real. Most of these skeletal studies undertaken with taxonomic intent 'continue in the old tradition through application of increasingly elaborate quantitative techniques . . . with no major shift in theoretical orientation' (Adams, Van Gerven and Levy 1978: 516). The statistics bring out phenotypic affinities very well, and the results are presented as expressions of 'biological distance' between groups. Because the various influences on morphology are practically never discussed, the undercurrent is that these trees present a sort of phylogeny. Yet, as Howells puts it, 'clustering is not the same as phylogeny, however well the two seemingly or actually correspond. Clustering is simply an arrangement of data' (1984: 6).

The clusters derived from the phenotypic data are presented in various ways. It is usual for slight variation in the basic data, such as dropping of two or three measurements in order to increase sample size, to result in different, even contradictory results. Within any large cluster diagram at least one bizarre juxtaposition can be assured, such as Hawaii with the Gulf region of New Guinea, or the Trobriands with Easter Island (Pietrusewsky 1976). The females of a group may show a different relationship to that of the males (not practical if that source is hundreds or thousands of miles distant) or the interrelationships of groups change with changing statistical methodology (Howells 1989). While a result is assured, the inductive and non-biological nature of such an approach ensures that no judgement is possible as to which presentation, if any, is valid. The hiccups are sometimes said to indicate the need for greater refinement of the statistics. However the problem does not lie there.

The real problem with this inductive approach lies with its dependence on the mathematical process for a result, without any clear idea in any biological context as to why this result has come about, or even why the particular measurements were made in the first place. Davis and Hersh, themselves mathematicians, comment that the scepticism amongst professional mathematicians about mathematical biology is much stronger than it is amongst non-mathematical biologists, a scepticism rarely stated in print (1986: 61–62). They distinguish 'rhetorical mathematics' as distinct from pure and applied mathematics, and define the former as that from which no practical consequences issue: 'either initially or ultimately work in applied maths leads back to the phenomenon being modelled. Rhetorical maths is often incapable of being tested against reality'. That seems a reasonable description of much use of statistics in craniology – which itself, as still often practised, is a field left 'without methodological or theoretical rationale' (Adams *et al.* 1978).

Warnings on the misuse of multivariate statistics have been sounded often enough (Kowalski 1972, Corruccini 1975, Rhoads 1984). The latter comments: 'It is a misuse of the statistical procedures, and a form of neo-Pythagoreanism, to suppose that nature conforms to inductively discoverable arithmetical ideals, and that statistical methods are active agents in an inductive search. We are the agents and must acknowledge the responsibility. The application of a particular method is, or ought to be, a thoughtful act of substantive scholarship' (1984: 248).

This leads to the core of the problem, which is fundamentally not one of statistical method but of biological validity. It is the problem of the fundamental data put into the statistical system (a problem, of course, not remotely confined to craniology). Here, unless there are clear and valid biological reasons for supposing the measurements to carry a reasonably uncluttered and common genetic content between compared groups, or an ability to quantify the other influences, then these studies are untenable as taxonomic exercises. There is nothing original in this view. More than half a century ago R. A. Fisher observed in a discussion entitled ' "The Coefficient of Racial Likeness" and the future of craniometry':

It seems, indeed, undoubtedly true that the theoretical concepts developed in the subject have lagged far behind the mass of observational material which has been accumulated. This may be partly due to the sheer magnitude of the programme which the energy of its founders sketched out, partly to an intuitive confidence, widely held in other fields, though everywhere difficult to justify, that, by amassing sufficient statistical material, all difficulties may ultimately be overcome. Partly, again, to an unconscious minimising of these difficulties. For the establishment of statistically reliable differences between series of skulls from different parts of the world, and from different periods, and the further task of evaluating with precision the magnitude of the measurable differences, does not in itself bring us appreciably nearer the stage of recognising which, if any, among our measurements are of the greatest and which are of the least value as indicators of racial affinity. If, indeed, we knew, of each of these measurements, whether it is much or little affected by purely environmental circumstances; and again, whether it has been often and rapidly, or but seldom and that slowly, modified, without racial admixture, by the selective influences to which human populations are exposed, it might be that some of them, or more probably some particular aspects of the aggregate of measurements, might prove to be of taxonomic value . . . But these necessary and preliminary enquiries seem largely to have been ignored. (1936: 62–63)

It is sobering to note that little has changed in more than half a century. Ironic also is the care taken to ensure precision in measurement and the disregard for influences on these measurements.

The problem is that there is scarcely a part of the body that may not be substantially influenced by the environment. Chapter 4 indicated that this influence extends from the total body to that popular source of data, the skull. The only region where the genetic message may just possibly be relatively unobscured is the central component of the cranial base The rest, vault, upper face and airway, extent of prognathism, dentition and mandible, may all be profoundly shaped by environmental influences. Multivariate techniques reveal resemblances and differences in morphology extremely well, but these associations are not phylogenies. The functional and adaptive demands on the head make a nonsense of attempts to relate peoples across large expanses of prehistoric time and space on the basis of cranial phenotypic data. What is being achieved in these studies is a display of degrees of association of metric data obtained from skulls. It is inaccurate to call such studies biological for biological considerations have not gone into the selection of data. The associations tend to look plausible, for often they are fundamentally geographic, that is, environmental, but they could be obtained more simply just by the use of an atlas.

This is not to deny major genetic influence on head form. However it is a great leap from this statement to establishment from cranial data of relationships across prehistoric time and space.

It might be thought that at least within a particular environment, such as the wide expanse of tropical Remote Oceania, multivariate assessment of similarities and differences in head form should reasonably reflect genetic relationships. However the vagaries of genetic drift and the founder effect may upset this. A quite robust record of archaeological research and dating, and also, in this shallow time span, the evidence of linguistics, supports the sequence of settlement of Polynesia set out in Figure 1.8. The craniological clusterings generally do not follow this pattern. For example, in Figure 5.4, while the Marquesas are linked with Easter Island and the Society Islands, Hawaii is closest to Tonga and Samoa, and New Zealand to the Southern Moluccas. This does not seem very helpful for what should be a fine-grained analysis. The conflicting results from many published cluster analyses defy interpretation.

If these craniometric exercises in taxonomy are ever to contribute usefully to our understanding of Pacific prehistory (and I do not think they can ever achieve a useful precision), then attempts have to be made to untangle and quantify the influences of distance and environment. These are Fisher's 'necessary and preliminary enquiries'. On cranial material there have been very few such enquiries. In a very general study, made difficult of biological interpretation by their working from discriminant functions, Guglielmino-Matessi *et al.* (1977) attempted to 'correct' Howells' (1973b) 'phylogenetic' analyses from cranial data by eliminating climatic influences. However the most significant essay in this field of which I am aware is that of Rothhammer and Silva (1990). These workers looked at the independent contributions of climate, altitude, chronology and geographic location to craniometric variation in a large (1000+) series of South American skulls. The measurements used, gleaned from many studies, were minimal frontal breadth, bizygomatic breadth, nasal height, orbital breadth, palatal length and palatal breadth, a selection providing a fair coverage of upper facial dimensions. (In one of our own unpublished studies we found that, using a maximum of seven head measurements, it did not make much difference to discrimination between groups which seven were selected, as long as the whole structure was reasonably represented; which returns us to Gower's question: 'Is the geometry of the skull such that taking standard cranial measurements is equivalent to observing repeatedly a few fundamental, perhaps not directly measurable, variates?' (1972: 3) – to which I think the answer is 'yes'.)

For the chosen influences Rothhammer and Silva (1990) found an explained variability ranging from 10% for orbital height, to 49% for palatal length, with an average of 30%. (These figures on their own are interesting, showing the conservatism of the nervous system, represented by the orbital dimension, and the adaptability of the masticatory system.) They found geographic isolation explained the highest proportion of craniometric variation (about 23%), followed by climate and then altitude, these each explaining about half this proportion. The contribution of chronology was negligible as their sites were largely synchronic. They concluded that geographic distance partially prevented gene flow from counterbalancing microdifferentiation produced by founder effect.

The study shows a significant environmental influence on craniofacial dimensions, and highlights the problem of disentangling this from the genetic message. Of course the results are particular to the geographic region. A similar exercise for the Pacific, particularly in the transition from Near to Remote Oceania, would probably throw up a different dominance of influences. If the discussion of exposure at sea in Chapter 3 has any validity then simply determining climate on the basis of standard meteorological data would be misleading. Again, because of small numbers and the founder effect, the influence of geographic distance may be difficult to unravel in Remote Oceania.

Craniology: non-metric traits

Measurements are sometimes called continuous variables. The skull is also rich in so-called discontinuous or non-metric variables ('morphological' and 'epigenetic' are, more or less, synonyms). These are the holes and canals, bumps and excrescences which are scattered through it. While some attention had been paid to such features in times past, their intensive study was really launched with the publication in 1967 of a paper by Berry and Berry, 'Epigenetic traits in the human cranium' (which provides an historical resume). They concluded: 'A considerable amount of normal discontinuous variation exists in the human skeleton. This variation is inherited, although it is actually determined by developmental (epigenetic) thresholds rather than by straightforward gene action in the way that gene action is usually understood' (377). Since then, studies of non-metric traits of the human skull and comparisons between populations have possibly surpassed in number even those on metric data. Hauser and De Stefano (1989) document and illustrate just about every such cranial feature that has ever been used.

Analysis of data from discontinuous features is trickier than that from continuous ones, but statisticians have grappled with the problem (Grewal 1962, Green and Suchey 1976, de Souza and Houghton 1977). The result of a cluster analysis for non-metric traits is shown in Figure 5.6. Again, this is from Pietrusewsky (1983) and again the results look pretty sensible – the incongruent note (to some) this time is the placing of New Zealand with Japan. The clusters differ from those in the metric analysis on the same material and it is unclear which is to be preferred.

Unfortunately the use of non-metric traits is not on any sounder theoretical base than when Berry and Berry expressed optimism a generation ago. On the mandible many features, such as those defined on the *Sinjanthropus* specimens by Weidenreich (1936) and subsequently used in Australian studies by Larnach and Macintosh (1971), almost in their entirety owe their variability to functional influences. The genetic

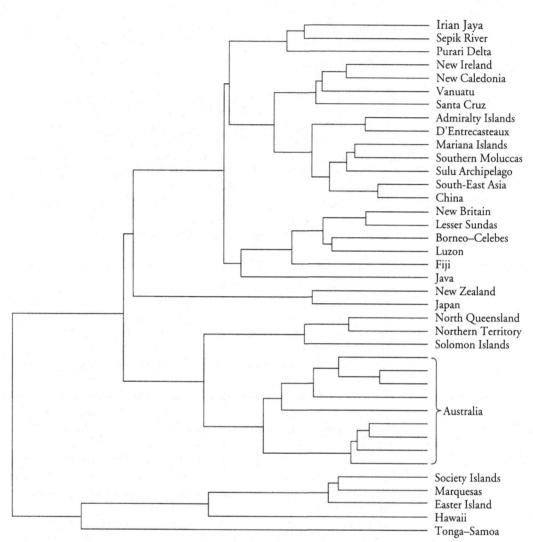

Figure 5.6
Cluster analysis of Pacific, Australian and Asian groups using 29 cranial non-metric traits (from Pietrusewsky 1983).

bases of most of the commonly used battery of discrete cranial traits are unknown and heritability estimates are low and of course are group-specific (Rosing 1984: 319, Sjovold 1984). The former finds the analysis of such traits to be plagued with serious difficulties, particularly a virtually unknown genetic background, and concludes that the commonly used traits appear to be 'unsuited for most populational investigations'. Sjovold analysed the incidence of non-metric traits in crania from an inbred Austrian isolate. In this group, for whom the lineages were known, he considered 7 of 30 traits to show significant heritability. Three of these were related to sutural bones of the vault, one to the maxillary torus, and one each to a foramen for an artery, nerve and vein.

From a study that assessed non-metric traits on a maternally-related population of Rhesus monkeys Cheverud and Buikstra concluded that 'environmental factors are responsible for as much inter-individual variation as are genetic factors' (1981: 47). From a study on random-bred mice, Rightsmeier and McGrath (1986) found only 4 of 35 traits to show statistically significant heritability values.

When such low levels of significance are achieved in controlled studies, the rationale for pan-Pacific comparisons of *Homo sapiens*, stretched across thousands of miles of space and as many years of time, seems tenuous. The increased understanding of phenotypic expression arising from molecular genetics also means that such terms as 'variable penetrance' are suspect and that varying degrees of expression of a morphological trait may have different genetic bases. In arguing a case *for* the use of non-metric traits in palaeoanthropological studies, Wolpoff, Wu and Thorne comment 'Comparisons are likely to be misleading or invalid unless they are between populations that are fairly closely related' (1984: 426). This seems a reasonable statement, and to use such traits to determine this relationship is hardly logical.

In a thoughtful commentary and defence of such studies, Buikstra, Frankenburg and Konigsberg note 'the misconception that because heritability of traits *within* a population is unknown, and in all likelihood not very high, biological distance studies are of no utility when applied to past populations' (1990: 6). (The emphasis on *within* in this quotation is mihe not the authors', and it is an important caveat.) As they see it, if environmental variance is random with regard to the features being studied, then the lower heritabilities simply reflect a lot of random 'noise' in the analyses, and still leave it possible to reach valid conclusions concerning past population structure.

This, of course, is the hope to which we all cling when working through one of these analyses, and empirically there does at times seem to be a fit with other evidence of familial lineages. For example, in a study of groups from a prehistoric Thai cemetery, Tayles (1992), on the basis of features of the vertebral column, presented some support for the archaeological inference that the groups were familial lines. On the other hand, assumptions that environmental variance is random with respect to any trait may be unwise – for example, there are interesting physiological arguments relating climate to the incidence of cranial venous foramina. The present situation may be summed up by saying that within an isolate certain markers, particularly neural foramina, may prove useful, but at present there is no theoretical basis for comparisons between groups separated by large expanses of prehistoric space or time.

Dentition

Analogous to these cranial studies are those made on the dentition. Teeth are a good source of data, being long-lasting – enamel is our hardest biological component – possessing readily-definable dimensions, and quite an intricate crown morphology.

While attempts have been made to discern patterns of size variation across the Pacific, it is really rather difficult to summarize a complex picture. It is said that within Melanesia teeth, especially the molars, are large, although not as large as in Australian Aborigines. With passage east, teeth become smaller and third molar agenesis is more common. However these are only generalizations – for example, the teeth of the New Britain Nakanai, in the heart of Melanesia, are rather small (Turner and Swindler 1978).

One approach to ordering the data has been made by Brace and Hinton (1981), who feel that in Asia and in the western Pacific, long-standing regional differences in selective-force intensity resulted in major differences in tooth size between populations, and that sub-sequent movements brought these disparate populations into contact. (This seems to be the Mongoloid/Australoid dichotomy again.) They set out to show that simple tooth dimensions can provide an index of the extent to which the encountering populations did and did not mix. According to this study, the tooth data show the final peopling of the Pacific to be by groups who had come from a part of the Asian continent where tooth-size reduction had been taking place throughout the late Pleistocene. Because of a sophisticated seagoing technology they were able to move more or less directly into the Island Pacific where they met and mingled in Melanesia with the descendants of earlier, large-toothed settlers. Tooth size is seen as providing a good index of that mingling. Non-Austronesian-speaking people of Bougainville and eastern Highland New Guinea have teeth as large as those found in people of central Australia, suggesting that no appreciable mingling occurred. In eastern Polynesia tooth size is said to be the same as in pre-Chinese Taiwan, suggesting that minimal mingling occurred during the Poly-nesian movement into the Pacific. However tooth size on coastal New Guinea and Island Melanesia is intermediate, reflecting the ancestral contribution of the large and small-toothed groups. I will comment later on this analysis.

Odontometric data from the Harvard Solomon Islands study were analyzed by Harris and Bailit (1987: 250) against a linguistic dichotomy of Austronesian/Non-Austronesian and the implied influx of a new people with the Austronesian language. Discriminant function analysis (though with a rather substantial estimation of missing data to inflate the series) separated, as might be expected, the several groups. In the derived clusters Harris and Bailit see the major split as between the AN-speaking groups of Malaita, Ulawa, and Ontong Java, and the NAN groups of Bougainville. That is, in the Solomons tooth size is seen to parallel the language dichotomy. However it also, and perhaps more realistically, reflects geographic separation and other factors not taken into account.

These authors then subjected the same data to a principal com-ponents analysis in search of more significant morphological differences between groups. They found significant components relating to buc-colingual and mesiodistal dimensions of anterior teeth, to pre-molar size, and to molar size. These are components other authors have extracted, and may reflect the concept of three major morphogenetic fields for the dentition (Butler 1939). After statistical manipulation, any differences established between language groups seem, again, more realistically to derive from geographic distance, and Harris and Bailit accurately sum up their results with the comment: 'Except in extreme cases, size does little to disentangle the web of relationships, either just within the Solomons or in Oceania generally'. In a puzzling and optimistic about-face they conclude, a couple of paragraphs later, that 'The odontometric patterns defined in this study lend strong support to the anthropologic interpretation that NAN-speaking peoples of Bougainville Island have an origin and history in Oceania significantly different from that of the more recent AN speakers' (1987: 259).

On the morphological side of the dentition, a wealth of variation in crown and root features, from standard and accessory cusps to variation in form and number of roots, has been identified: particularly good illustrations of crown variants are given in Littlewood's 1972 study of aspects of physical anthropology in the eastern New Guinea Highlands – which, incidentally, is the fine product of a real anthropological adventure. As with the metric studies it is not easy to make general comments as to the overall patterns. On very limited data Riesenfeld (1956) suggested a west to east (de)cline in the incidence of shovel-shaped incisors across the Pacific. Probably the most concerted attempts at ordering dental morphology in the Pacific region have been those of Hanihara (1969, 1992) and Turner (1989, 1990). For Turner 'The great mosaic of the Pacific peoples actually has only two major pieces: Sundadonty and Sinodonty' (1985: 33). In establishing this he has used a 'battery' of 28 crown and root traits that represents 'at least two dozen separate epigenetic systems and as such is an extremely powerful means for direct estimates of prehistoric population genetic characteristics and phenetic relationships'. The eight traits said to have the most statistical significance in distinguishing Sundadonty and Sinodonty are, in the upper jaw, incisor shovelling and double shovelling, first pre-molar root number, first-molar enamel extension and third molar reduction; and in the lower jaw, first-molar deflecting wrinkle, first-molar root number and second-molar cusp number. Polynesians are placed among Turner's Sundadonts, along with South-East Asians, Micronesians and the Jomon in Japan. Australians and Melanesians may have been derived from early Sundadont groups in South-East Asia. Figure 5.7 is a cluster analysis after Hanihara (1992), who uses morphological criteria similar to those of Turner.

These dental studies, metric and morphological, carry the problems inherent in the skull studies: the genetic control of tooth size and shape is still far from clear, and environmental influences seem often to be significant. Lurking behind most metric studies is the assumption that somehow or other the dentition has a certain autonomy of existence, independent of bone and soft tissue parameters. Harris and Bailit put this explicitly: 'It would appear that essentially separate pattern-generating mechanisms control body size (anthropometrics) on the one hand,

Figure 5.7
Cluster analysis of
Pacific and Asian
groups using tooth
measurements (from
Hanihara 1992).

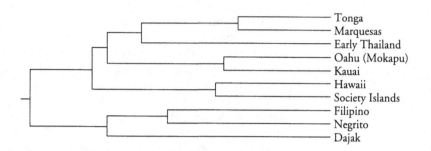

Tonga
Marquesas
Early Thailand
Oahu (Mokapu)
Kauai
Hawaii
Society Islands
Filipino
Negrito
Dajak

and tooth crown size on the other' (1988: 97). However, it would be surprising if in such a tightly organized morphological system as the head, any set of structures was rather free to go its own way. Integration with the bony support of the alveolus of maxillae and mandible are rather obvious requirements, and these in turn must be architecturally compatible with the rest of the facial skeleton and cranial base. Expansion of crown size of the teeth, particularly in a mesiodistal direction, requires a longer, more prognathic alveolus. The trend in larger and more muscular individuals, exemplified in the Polynesian, towards a flatter, less prognathic profile (a consequence of the biomechanical factors discussed in Chapter 4), leaves relatively less room for the dentition. On these grounds one would anticipate that larger people (measured particularly by muscle mass rather than stature) would tend to have relatively smaller teeth. This is borne out by studies showing a negative allometry for tooth size relative to body size (Garn, Lewis and Kerewensky 1968, Henderson and Corruccini 1976, Perzigian 1981, Kieser and Groeneveld 1988). That is, big people have relatively small teeth. Polynesian teeth *are* relatively small on the global range of *Homo sapiens*.

There is good evidence for a strong environmental influence on tooth size. Garn, Osbourne and McCabe concluded that maternal and fetal determinants of tooth crown dimensions might account for as much as half of crown-size variability (1979: 665). For a group of Australian Aborigines, Townsend and Brown (1978) attributed about 64% of total variability of permanent tooth size to genetic factors. For a group of Finns, Alvesalo and Tigerstedt (1974) reported heritability values of about 60%, similar to those established by Goose (1971) for a British group.

This combination of influences – environmental, and a related adjustment in harmony with the facial skeleton – also means that interpretations of tooth size as reflections of cultural differences, such as methods of food preparation, are simplistic. (Through the Pleistocene in the Mediterranean Smith (1977a, b) found wear to increase on the dentition as tooth size decreased. This does not support the 'selective force' reduction concept of Brace and Hinton, 1981.) Even leaving aside the allometric and environmental factors, it seems probable that within *Homo sapiens* the basic dimensions of the dentition lack discrimination for taxonomic purposes. To conclude, as do Brace and Hinton (1981), that because tooth size in eastern Polynesia is exactly the same as it was in pre-Chinese Taiwan, minimal mixing had occurred during ancestral Polynesian movements eastwards, is asking altogether too much of the dentition.

The studies of dental morphology also face the same problems of an uncertain theoretical base. Tooth size itself influences morphology, larger teeth simply having more room in which to exhibit complex morphology (Garn, Lewis and Kerewensky 1963). Environmental factors thus loom immediately as an influence on morphology. For specific morphological features there are surprisingly few rigorous studies. One feature that does come through as having a reasonably high heritability is shovelling of the incisors (Portin and Alvesalo 1974). Nothing else is as encouraging. Some studies maintain a cautious optimism, while affirming how much more is needed to clarify the genetic component (Sofaer, Niswander and MacLean 1972, Palamino, Chakraboty and Rothhammer 1977). Valuable analyses on the morphological aspects of the Bougainville dentition have been done by Harris and Bailit (1980). For the metaconule, an accessory cusp of the upper molars, they established a heritability estimate of 65% for the first molar. However on the second molar, variation in this trait seemed to be (90%) essentially a product of unidentified environmental parameters. This hardly gives confidence in the use of such traits even within a group, let alone across the expanses of prehistoric time and space.

From a complex segregational analysis of dental morphological traits, using casts from 600 individuals of 83 nuclear families in two Amerindian communities, Nichol found 'propositions of major genes' for 13 of 24 traits examined, good evidence for a polygenic origin for two traits, and with a firm conclusion on the mode of inheritance of the remaining nine traits not being possible (1989: 37). Estimates of transmissibility for all characters gave a mean of 0.36. Nichol concluded: 'These findings suggest there is a large amount of environmental influence on the development of dental morphological variants'. Such findings in familial lines emphasize the fragility of theoretical support for the idea of some circum-Pacific dichotomy of tooth form of genetic basis. The Sundadont/Sinodont thesis also has problems of sampling and statistics (Hall 1981, Konigsberg 1992) and of populations that do not fit (Schwarz and Brauer 1990). In general, the genetic bases of the traits used to establish lineages of Sundadonts and Sinodonts, and the heritability of the traits in various groups, are at present unestablished.

Dermatoglyphics

The use of fingerprints as an anthropological tool has a long history (Galton 1892). Finger and palm prints have generally been considered the most useful of phenotypic traits for studying population structure, and some consider them more useful than blood markers, or at least the simple gene frequencies established by the major blood groups. The reasons given for these views are several (Newman 1970, Froehlich 1987).

1. Dermatoglyphic traits are believed to be polygenically controlled, and thus less susceptible to change through genetic drift than are single gene traits.
2. They are unalterable after birth.
3. They are highly heritable and thus relatively independent of environmental influence.
4. They are said to have no adaptive value, and thus to be selectively neutral.
5. Techniques are objective, and consistency of observation is readily attained.

As Jantz comments, 'There is good reason to question some of these assumptions' (1987: 164), and the reasons seem to be growing.

Dermatoglyphic patterns are amenable to statistical ordering and Froehlich (Froehlich 1987, Froehlich and Giles 1981a, b) has been a major contributor to this field in the Pacific. His studies examine the dermatoglyphic patterns against a stated linguistic dichotomy of Austronesian and Non-Austronesian. In a detailed discriminant analysis of finger and palm prints on the eight basic populations of the Harvard Solomon Islands study, Froehlich (1987) obtained a 23.8% correctness of classification to group, which is not a good record for discriminant analysis. Unfortunately, as with the odontometric studies the location of the groups ensures that the claim to separation on linguistic grounds can be as convincingly argued as being simply geographic. The Austronesian-speaking Malaitan groups may be separated, though not particularly distinctly, from the Bougainville Non-Austronesian groups.

Extension of this analysis to 12 populations, incorporating four AN-speaking groups from Bougainville, gave accuracy of classification of 30% for males and 34.2% for females. There were some discrepancies between male and female data, but Froehlich considered that in general the analysis supported a biological separation of AN and NAN speakers. Furthermore, he felt that the consistent grouping together of the AN speakers refuted the suggestion of Dyen (1965) that the Austronesian languages evolved within Melanesia. He concluded: 'Finally, after at least 4000 years of complex prehistory, with extensive gene flow from unrelated groups and genetic drift in small isolated populations, the fact that fingerprint data contain much historical information at all tends to confirm the hypothesis that they are relatively stable genetically. They contrast dramatically with the more labile blood genetic data, which though highly variable have been unable to distinguish AN from NAN groups in the Solomon Islands' (Froehlich 1987: 208).

These are intricate studies, and one is simply in awe of the work and dedication that has gone into them. However, there are problems with some of the underlying assumptions (and the element of circular reasoning in the above citation) and the doubts are growing rather than lessening. Froehlich makes the claim that 'since fingerprint traits are broadly polygenic, any associated selective changes, due to close linkage or pleiotropy, are unlikely to be uniformly directional for the phenotype as a whole' (1987: 177). This relates to an old suggestion, that genetic drift should have less impact on polygenic than monogenic traits, because the effects of drift at multiple loci cancel one another. Hence the between-population divergence of polygenic traits should be slower and less subject to stochastic effects than that of monogenic traits. Thus, polygenic traits, including dermatoglyphics, should be more useful than monogenic blood markers for tracing relationships between populations. However this thought seems not to be correct: there seems no (bio)logical reason why the effects of selection or drift at multiple loci should cancel one another out. (Errors in dead reckoning over long distances at sea do seem to cancel out, but there is no reason why nautical drift in the Pacific should influence genetic drift.) As Jorde comments:

Although this argument has intuitive appeal, it is based on inappropriate reasoning . . . The quantity of interest for this question is the between-groups genetic variance. Wright (1951) demonstrated that the expected value of the between-groups variance is identical in monogenic and polygenic traits. Rogers and Harpending (1983) extended this analysis to show that the variance of the between-groups variance (and thus the precision of the estimate) is also equivalent for monogenic and polygenic traits. Thus, monogenic traits are at least as useful as polygenic traits for studies of population structure. In addition, because most polygenic traits have heritability less than one, and because polygenic traits are more highly intercorrelated with one another, there is probably less unique genetic information in a set of polygenic traits than in a set of monogenic traits. (1986: 210)

Neither does the heritability of fingerprints seem quite as robust as the earlier studies of Holt (1968) suggested. Jantz (1987) estimates an overall heritability of 73.8% from 20 finger ridge counts, and 64.7% for palmar interdigital ridge counts. Pattern elements of the palms and soles may have lower heritabilities, or no evidence of a genetic component (Loesch 1974). The growing body of data on fingerprint association with particular biological problems of fertility has been reviewed by Jantz (1987), who comments on evidence for the importance of intra-uterine developmental factors in variation. Recent studies have shown a significant association of some anthropometric measures with particular dermatoglyphic patterns (Jamison, Jamison and Meier 1990, Loesch and Lafranchi 1990), including palmar features, which were found in the Froehlich study to be useful discriminators. This is an important matter in the Melanesian scene, where body phenotype may differ considerably between groups. Consideration is given by Loesch and Lafranchi (1990) to the possibility of selective advantage for differing dermatoglyphic patterns, but it really need not be necessary to postulate this – it does raise the spectre of the spandrels (Gould and Lewontin 1979). As Mayr comments: 'Some visible components of the phenotype may in fact not be adaptive, but may be the by-product of a genotype selected for its invisible, cryptic contribution to fitness' (1970: 194).

It will be interesting to see how in Melanesia more sampling for

DNA fits with the dermatoglyphic picture, for the gene studies to date are against a biological mirroring of the over-simplified linguistic division. Certainly there is a great deal yet to be learned about the determinants of dermatoglyphic patterns.

Phenotypic evidence in perspective

The tenor of these comments on cranial, dental and dermatoglyphic studies may seem overly, even unrelentingly negative. If so, this is not a problem of the data, which in biological contexts are valuable. The problem lies with the overuse of such data as taxonomic indicators when their underlying determinants are either inadequately known or ignored. The unthinking application of statistics adds to the problem – it is significant that the caveats so often come from statisticians – particularly when matched with an overconcern for fitting biological data to linguistic models for which no *a priori* reason for fit exists. However the allure of data of this nature is such that nothing will halt the inexorable generation of more and more uninterpretable cluster analyses, though Fisher's 'necessary and preliminary enquiries' have hardly been started.

Genetic evidence

Progress in genetics, particularly molecular genetics, has been rapid in the past decade, and the continuing clarification of the Pacific past rests considerably with this discipline. However a problem in the interpretation of some results has been their setting in the rigid linguistic frame, while the persistent use of an antiquated racial typology leads sometimes to a bizarre juxtaposition of the science of the nineteenth and near-twenty-first centuries – analyses of DNA interpreted outside 'perhaps the greatest conceptual revolution that has taken place in biology' (Mayr 1970: 5). In talking of the results, some of these juxtapositions have to be mentioned.

The combination of burgeoning data and a rather indigestible nomenclature has the potential to induce in the non-specialist reader a glazing of the eyes. When this variety is melded with a typological terminology set in a linguistic framework the potential for confusion is great indeed. Thus Kirk asks

whether the peoples of the Pacific originated as a single or multiple migration(s) from a homeland(s) somewhere in Asia. Were the Australoids the original migrants? Were they followed by a wave or waves of Papuan-speakers into New Guinea and some of the nearer islands of the western Pacific? And did the final migrations of Austronesian speakers move down from South-east Asia, or did they evolve independently, both biologically and culturally, in some centre of differentiation in the western Pacific. (1989: 100)

As stated earlier, the approach here is to try to avoid linguistic and typological terms, and to interpret the data in the context of the biological concepts of adaptation and selection, genetic drift and founder effects. Such a context predicts that for Near Oceania genetic relationships will follow geographic and local lines. Human movement through the varied environments of the larger islands of Melanesia saw settlement in various niches. Genetic drift, adaptation, and social influences came to distinguish a group from its neighbours. But even in the face of phenotypic differences (in terms of gross body morphology as determined by anthropometry), geographic neighbours are likely to be genetically closest on the loci commonly assessed. Conversely, phenotypically similar but geographically distant groups need not be particularly close genetically. The reason for this somewhat paradoxical situation is that the polygenetic bases of anthropometric dimensions are still unplaced on the genome, and variants in them are not yet assessable. The determined genes and variants are simpler ones of accessible tissues, such as blood, and geographical proximity means that selective and drift influences on ·these are likely to be minimized.

For Remote Oceania the situation is different. Geography alone suggests that the founder effect is likely to have been influential, and the widespread existence of a distinctive phenotype has been made clear. The genome should reflect limited population origins – that is, a rather close relationship of the people throughout. However the major population changes in Micronesia in historic time, which seldom receive comment, are extremely important in interpretation of genetic data from some parts of the region.

In any genetic distance methodology, the larger the number of loci used the more reliable the estimate (Cavalli-Sforza and Bodmer 1971, Nei and Roychoudhury 1982, Jorde 1985, Stoneking 1993, Szathmary 1993). This is a matter of major significance.

As in any distance methodology, tree estimation is more reliable when a large number of loci is used. Estimated trees are nearly always erroneous . . . if the number of loci is less than 30 . . . If populations are closely related (as in many anthropological studies), even trees based on 100 loci will usually be incorrect. This is because the error variance of genetic distances increases when the distances are small. (Jorde 1985: 349)

The jargon of the trade aside, much of the confusion in interpreting the results of genetic studies in the Pacific arises from taking any one study, or the author's claims for it, too seriously. A few loci do not a phylogeny make. Despite the contributions of some of the most distinguished workers in this field, sampling, both of groups and of loci, is still inadequate for much in the way of firm statements to be made. While of course it is from the aggregate of studies of a few loci here and a few loci there that the substantive patterns will emerge, the present reality is that the occurrence of a few distinctive haplotypes and mutations – particularly in relation to DNA analyses – though falling well short of a phylogeny, do provide the clearest evidence of past associations of present Pacific people.

Gene frequency studies

Assessment of gene frequencies in Pacific peoples goes back to pioneering blood group studies of the 1920s. A succession of such studies established some clear patterns, such as the rarity of blood group B in Polynesia. They were unable to disentangle a larger pattern of overwhelming confusion, leading to the rather wellknown lament of Simmons that 'blood group genetical studies do not tell us the racial components of the Pacific peoples or their paths of migration' (1962: 209). Part of the problem was the relatively small number of loci studied. Later, as a larger range of blood polymorphisms became known and their rather startling variety in Melanesia was revealed, the scrutiny for patterns evidencing relationships revived, and helpful in this has been the continuing development of relevant statistical methods. However for the Pacific, the fine-grained genetic data required to allow reasonable phylogenetic statements to be made are still unavailable.

For the broadest view of affinities of Pacific people in terms of gene frequencies there are studies by Kirk (1989), Cavalli-Sforza *et al.* (1988), and Nei and Roychoudhury (1993). Kirk (1989) analyzed 72 loci for blood group, protein and enzyme systems from six global populations to produce the dendrogram given in Figure 5.8. The grouping of people is very general, and unfortunately no Pacific people east of New Guinea are represented. Amongst Kirk's conclusions are that Highland New Guinea groups are estimated to have separated from Aboriginal Australians 165,000 years ago, and coastal New Guineans from the Australians only slightly later, at 153,000 years. The separation time of Highland and coastal New Guineans is slightly later again, at about 145,000 years. (1989: 108–109). These seem very long times of separation to allow support for the unity of the Australoids, a term Kirk continues to use. Indeed these times are very surprising, for from geography alone one would anticipate a much shorter separation. However they give credence to the mitochondrial suggestion, mentioned below, of separate lineages entering Highland New Guinea and Australia.

The tree of genetic relationships derived in the study by Cavalli-Sforza *et al.* (1988), using 120 allele frequencies from 42 polymorphic systems and the same genetic distance as for Kirk's study, is given in

Figure 5.8
Phylogenetic tree for global groups using data from 72 blood group, protein and enzyme systems (from Kirk 1989).

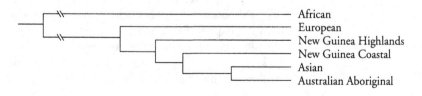

African
European
New Guinea Highlands
New Guinea Coastal
Asian
Australian Aboriginal

Figure 5.9. The tree suggests a major separation of northern and southern Asian populations. The South-East Asian group divides into one branch containing Highland New Guinea and Australian populations; and a second branch that in turn divides into two – mainland and insular South-East Asia, and the Pacific Islands. Thus the regions of Polynesia, Micronesia and Island Melanesia are kept together, with a rather closer relationship between the latter two. The groups used to represent the Pacific peoples are not stated, though of course such information is important for finer-grained interpretation. Insofar as comparison is possible, there is reasonable agreement between these two large-scale studies.

Figure 5.9
Phylogenetic tree for global groups based on 42 polymorphic loci. Adapted from Cavalli-Sforza *et al.* 1988.

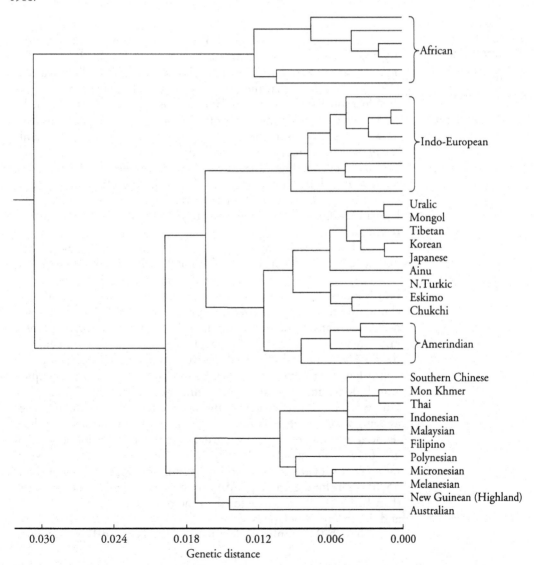

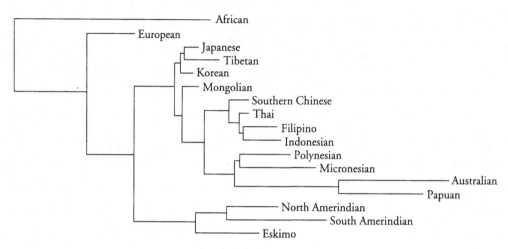

Figure 5.10
Phylogenetic tree for global groups based on 29 polymorphic loci. Adapted from Nei and Roychoudhury 1993.

Nei and Roychoudhury (1993) have similarly analyzed evolutionary relationships of human populations (Figure 5.10). However they argue for the superiority of their analysis over that of Cavalli-Sforza *et al.* (1988) on the basis that their study of 26 global populations not only used loci common to every population, but employed a clustering analysis that allowed for variation in evolutionary rates of divergence. Neither of these conditions were met in the Cavalli-Sforza *et al.* study. Certainly the phylogenetic trees are profoundly different between the two studies. Nei and Roychoudhury's research is the more relevant to the Pacific world as it included data from Samoa, the Caroline Islands and Kiribas, as well as a good representation of Australian Aborigine and Highland Papua New Guinea data. Unfortunately Island Melanesia was not represented.

Nei and Roychoudhury, like Cavalli-Sforza *et al.*, show the first human split as between African and non-African groups. However the second split in the Cavalli-Sforza *et al.* study (Figure 5.9) is between a North Eurasian and a South-East Asian cluster, whereas Nei and Roychoudhury determined the second split to be between Caucasians and non-Caucasians. (One tries to think of these terms as being geographic rather than typological.) Pacific people (Polynesian, Micronesian, Australian Aborigine and Highland Papua New Guinean) separate from a South-East Asian cluster that includes southern Chinese, Thai, Filipino and Indonesian. Australian Aborigine and Highland Papua New Guinean then clearly separate again, which Nei and Roychoudhury see as being a consequence of inbreeding in these populations, which show reduced heterozygosity. And while defining them as a major population group (yes, australoid [sic]) they clearly derive them from a wider mongoloid [sic] group. Nei and Roychoudhury suggest a time of about 55,000 years for their primary African/non-African split, which is far less than Kirk's (1989) estimate for the time of separation of New Guinea and Australian populations.

Though the number of loci used in the Nei and Roychoudhury

(1993) study are still fewer than desirable, it is probably, at this passing moment, the most satisfactory of the wider surveys.

Most studies of course are less ambitious, presenting original data for a few loci. Genetic distances based on blood polymorphisms (ABO, Mn, Hp, Inv and Gm allele frequencies) in the Solomon Islands were determined by Rhoads and Friedlaender, who concluded that the data made 'very nearly a geographic map of the Solomons' (1987:147). Wider Pacific analyses using the ABO, MNSs, Gm and Rh system gene frequencies were presented by Rhoads. Four major clusters were discerned, two of these being primarily Highland New Guinea populations, and with no link between New Guinea and Island Melanesia Non-Austronesian speakers. Rhoads concludes that 'using this set of markers it is futile to seek a distinction between AN and NAN populations which is universal. It is probable that there have been complicated local patterns of admixture, not a simple dichotomy' (1983: 93).

On Karkar Island, off the north coast of New Guinea, the population shows a clear linguistic division into a northern group speaking a Non-Austronesian language, Waskia, and a southern group speaking an Austronesian language, Takia. Gene frequency studies were carried out by Boyce *et al.* (1978) as part of an International Biological Programme study which gave rise to a number of valuable papers, several of which have been cited here in other contexts. Boyce and co-workers concluded that for the populations of the island, selection did not appear as a major determinant of genetic structure – which is to be anticipated in such a limited environment. Despite a rather determined attempt to see any genetic variety in terms of linguistics, they had to conclude that it was very doubtful if the linguistic division would even have been detectable on the genetic evidence alone. Anthropometric data from both language groups were collected by Harvey (1974), who made no comment on any linguistically-related dichotomy in form. As he was happy to use a composite collection as being generally representative of the island, presumably no anthropometric differences were discernible.

Serjeantson, Kirk and Booth (1983) examined linguistic and genetic differentiation in a series of 17 populations in northern New Guinea using an array of blood genetic markers. For the Bogio sub-province a particularly detailed analysis was possible, with the conclusion that gene frequencies in this region of New Guinea were dependent on geographic proximity and not on linguistic similarity. They commented that 'linguistic studies do not provide an infallible guide to ancestral relationships and that postulates of past migrations based on language may be in error unless genetic studies are also taken into account' (Serjeantson *et al.* 1983: 77). This seems an oblique way of saying that genetic data may disprove linguistic models.

In a study of the populations of the smaller islands of the eastern fringes of Island Melanesia – the Banks and Torres groups, Santa Cruz, and Bellona and Rennell – Blake *et al.* (1983) calculated genetic distances using data from 11 polymorphic systems, including the ABO system and several serum protein and red cell markers. Again, geographic distance in this study spelled genetic distance, with the outliers, Anuta, Bellona and Rennell, being relatively removed from the others. The evidence in all these islands was for only a small amount of historic gene admixture from other sources. Bellona and Rennell are classified as Polynesian Outliers, but were shown in earlier studies to have quite different patterns for the major blood groups; instead of high A and little B they show 34% group B and almost no A at all, which is totally different from Polynesia. Perhaps we are seeing here evidence of selective pressure towards larger body size on a rather different genetic pool from much of Polynesia.

Ranford has examined the serum complement proteins, coded by genes in the major histocompatibility complex on chromosome 6. She concludes that the common alleles of the complement genes are the same in all the populations of the Pacific region (a term which here includes the western Pacific rim), suggesting that these now phenotypically dissimilar groups have evolved from the same ancient genetic stock (1989: 188). However she sees sufficient differences in the gene frequencies between mainland Asia and the Pacific to suggest these peoples have been evolving separately for a 'very long time'. Variation in the complement proteins is very restricted in Polynesians, Melanesians and Micronesians. These findings seem compatible with the concept of a considerable period of evolution-in-situ in Near Oceania.

Serjeantson has carried out further detailed studies on the HLA gene system, and some of her findings are summarized in a phylogenetic tree where 'relative branch lengths are indicative of the genetic distance between groups' (Figure 5.11, from Serjeantson 1989: 162–163). Her interpretation is that 'The clear separation of Micronesian and Polynesian confirms that Polynesians did not "island hop" through Micronesia en route to the eastern Pacific, despite anthropomorphic similarities between the two groups . . . The other groups cluster as expected into Australoid, non-Austronesian Melanesian, island Melanesian, and Polynesian groups'. However without reverting to such typological/linguistic terminology, the same phylogenetic tree rather sensibly matches the concept of a gradual spread of *Homo sapiens* through the area. Indeed the figure roughly presents the geography. As is common at the present stage of these studies, sampling is thin: 'Micronesia' actually embraces 22 individuals from Nauru, itself a peripheral and uncharacteristic Micronesian island. The unexpected note, the North American link to New Zealand, presumably is a reflection of the chance influences of drift and the founder effect, and, with this system, the possibility of (undefined) selective forces. There remains the apocryphal thought that here is evidence of the link with the Americas. Others have commented on this possibility, particularly in relation to South America (Salzano 1985).

The persistent message of these gene frequency studies seems to be that in the biological complex of Near Oceania, genetic distance reflects geographic distance, and linguistic and genetic groupings meet somewhat at random.

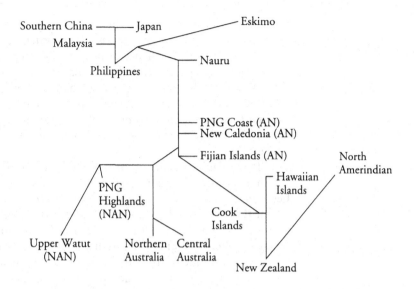

Figure 5.11 Phylogenetic relationships among Pacific and neighbouring populations based on HLA antigen frequencies (from Serjeantson 1989).

DNA studies

We now turn to direct reading of the genome, the fruit of the advances in molecular biology since the elucidation of the genetic code by Watson and Crick in 1952. If these genetic data seem confusing, matters are only going to get worse, for not very much of the nuclear DNA message has been unravelled and there are about three billion base pairs to go. Work in this field proceeds at such a pace that any commentary tends to have an historical air to it by the time it reaches print. The anchor again is that affinities between peoples should be derived from as much DNA as possible and not from a focus on a locus. However the reality is that as sampling from groups and genes in the Pacific is still patchy and inadequate for construction of a phylogeny, the present evidence for past associations of people lies, as already mentioned, in the sharing of distinctive haplotypes and mutations.

Without losing sight of this caveat on overinterpretation of information from a few loci ('in terms of the information contributing to the phylogenetic information, mtDNA is equivalent to a single locus': Nei and Roychoudhury 1993: 932) some of the clearest insights into prehistory have come from mitochondrial DNA, which offers a number of advantages in analysis and interpretation. The mitochondrial molecule is relatively small and limited to about 16 thousand base pairs; it is maternally inherited, and has a mutation rate some ten times that of nuclear DNA. One statistical reality relating to mitochondrial DNA that is consequent on the maternal pattern of inheritance is enunciated by random-walk theory. This indicates that in a stable population of initial size n, then after n generations all individuals in the population will trace their ancestry back to the same single female, other lineages dropping out because of bearing only male, or no children. This situation is rather easily misunderstood to mean the time of origin of the group, which is not the case. What it does mean is that Mother is or was the one of a group who happens by chance to be ultimately represented. The mitochondrial DNA studies received particular attention because of the claim that we are all descended from a woman in Africa who lived about 200,000 years ago (Cann, Stoneking and Wilson 1987), which may be misinterpreted as setting a time for the appearance of the species *Homo sapiens.* Currently there is a lot of jostling in this field that is not of direct concern here.

In the Pacific region, mitochondrial DNA lineages have been helpful in suggesting patterns and numbers of entry, and connections between groups. Stoneking and Wilson (1989) conclude, from studies which they account as preliminary and based on rather small samples, that a minimum of 16–19 mitochondrial DNA types (that is, 16–19 females) initially colonized Australia; and, quite separately from Australia, a minimum of 14 females colonized Highland New Guinea. These mitochondrial DNA studies suggesting different lineages for Australia and for Highland New Guinea, like the Kirk (1989) gene frequency study cited earlier, run counter to the usual Australoid associations. For the Pacific the emphasis shifts to a distinctive 9 base-pair deletion in the mitochondrial sequence. The marker consists of a deletion of one of two copies of a 9 base-pair sequence CCCCCTCTA in the non-coding region V of mtDNA, between the genes for cytochrome oxidase and TRNALys (Cann and Wilson 1983). In studies of DNA variation this has been named haplotype B. It has been found (Stoneking and Wilson 1989) in an incidence of 16% in Japanese, 5% in Okinawans, 18% in East Asians (a mixed group dominated by Chinese and Filipinos), 42% in coastal New Guineans, and not at all in Highland New Guineans or Australian Aborigines. Subsequently Hertzberg *et al.* (1989a) established for this deletion an incidence of 14% in another group of coastal New Guineans, 8% in the Tolai of New Britain, 82% in Fijians and 93% through Polynesia. It has been found at 92% incidence in Philippine Negritos (Harihari *et al.* 1992). It has also been found in both North and South American Indians – for example in the Pima (45%) and Maya (22%) (Schurr *et al.* 1990). An overall incidence through the Americas of 27.4% is given by Balliet *et al.* (1994). Table 5.1 brings together some data on this deletion. For the Pacific the implications are of a distant link with contemporary Asian groups, and evidence of the amplifying nature of the founder effect. The entry of this mutation into the New World has been dated as possibly prior to 21,000 b.p. and certainly much before 10,500 b.p. (Wallace and Torroni 1992). If correct (Szathmary 1993 advocates caution) and representing the same mutational happening, this is a useful estimate, as it provides a minimal antiquity in Asia for the common ancestor of both Pacific and New World groups, and in turn allows a longish time for the mutation to have been in Island Melanesia.

Of course this maternal link with Asia is unsurprising, for the ancestors of every pre-European settler in the Pacific came out of Asia

Table 5.1   Occurrence of the mitochondrial DNA 9 base-pair deletion. Sources: 1 Stoneking and Wilson 1989; 2 Horai and Matsunaga 1986; 3 Harihari *et al.* 1992; 4 Wrischnik *et al.* 1987; 5 Hertzberg *et al.* 1989a; 6 Hagelberg and Clegg 1993; 7 Torroni *et al.* 1992; 8 Schurr *et al.* 1990.

Group		Era	Sample size	% detected	Source
Asia	East Asia	modern	34	18	1
	Japan	modern	116	16	2
	Okinawa	modern	82	5	2
	Philippines Negrito	modern	37	92	3
	Australia Aborigine	modern	21	0	4
	New Guinea Highland	modern	94	0	5
Oceania	New Guinea Coastal	modern	123	24	5
	Watom	2100 b.p.	4	0	6
	Taplins, Efate	2100 b.p.	1	0	6
	Fiji	modern	28	82	5
	Natunuku, Fiji	2100 b.p.	1	0	6
	Sigatoka, Fiji	1900 b.p.	2	0	6
	Tonga	modern	30	77	5
	Tonga	2500 b.p.	1	0	6
	Tonga	300 b.p.	1	100	6
	Samoa	modern	30	100	5
	Samoa	2300 b.p.	1	0	6
	Niue	modern	30	100	5
	Society Islands	prehistoric	1	100	6
	Hawaii	prehistoric	2	100	6
	Cook Islands	modern	30	87	5
	New Zealand	modern	30	100	5
	New Zealand	400 b.p.	2	100	6
	Chatham Islands	400 b.p.	2	100	6
North America	Navajo	modern	48	37.5	7
	Pima	modern	31	45.2	8
	Maya	modern	37	21.6	8
	Athbasca	modern	30	0	7

(accepting the possibility of a few American genes entering from the east, and one way or another these are ultimately out of Asia). The evidence for considerable antiquity for the deletion and its presence through the Pacific is compatible with several scenarios of prehistory, and it did not have to arrive relatively recently or in one migratory rush.

On its own the 9 base-pair deletion is evidently a very broad genetic statement. Identification of other deletions and of base substitutions should allow finer discrimination of the relationships between groups. Using such more widely-founded evidence, Lum *et al.* define at least three distinct mitochondrial DNA groups in Remote Oceania that probably shared a common maternal ancestor more than 85,000 years ago, and their analyses identify 10 or 11 discrete maternal ancestors for Polynesia. They infer 'from the extreme genetic divergence of the Polynesian lineage groups that the original Polynesian colonists must have had significant existing genetic heterozygosity, because the short expansion time of the Lapita peoples would not have allowed sufficient time for new mutations to accumulate' (1994: 584).

Another conclusion drawn from the mitochondrial studies is that geography, not language, is the primary factor influencing the distribution and patterns of variation of mitochondrial DNA types in New Guinea, which is the same conclusion reached from the gene frequency studies.

We turn now to studies of the immensely more complex nuclear DNA. Here, analysis of globin gene variants in particular over the past two decades has helped the understanding of relationships in the Pacific. It has also contributed to an understanding in wider terms of the relationship between genomic coding and the phenotypic picture, including how rather different genetic patterns may give rise to similar phenotypic presentations.

The normal haemoglobin molecule contains 2 pairs of globin types termed alpha (α) and beta (β). The α-globin is coded in four genes, two on the short arm of each chromosome 16. Upstream of the α-globin genes on chromosome 16 lie two zeta (ζ)-globin genes (one non-functional because of an inactivating mutation): this is the embryonic α-like globin chain. One β-globin gene is found on the short arm of each chromosome 11. Adjacent to the β-globin genes are genes for delta (δ), gamma (γ), and epsilon (ϵ) globins, which, like the ζ-globin, are produced in embryonic or early infant life.

Even in normal chromosomes, variation in bases between genes allows restriction enzymes to divide the gene complex into identifiable segments of differing length. These differing segments are termed haplotypes. For the normal α-globin locus, about 30 of a very large number of theoretically possible haplotypes have been described, but only seven are commonly found (Higgs *et al.* 1986). Most of these probably result from single base changes, but others are length polymorphisms. They are designated by Roman numerals I-VII, with subtypes within each haplotype. Generally haplotypes I and II are very common in most populations and III-V are rare. In South-East Asia the haplotypes Ia, IIa and II d/e constitute about 90% of the α-globin haplotypes. In Melanesia these haplotypes account for no more than 7% and haplotypes IIIa, IVa and Vc dominate: to date the latter two are completely absent from South-East Asian samples. The extremely limited sampling that the term 'Melanesian' embraces in the Higgs *et al.* (1986) study is worth noting: 38 individuals from Vanuatu and 24 from Papua New Guinea.

In Polynesia, haplotypes III, IV and V provide 28% of the haplotypes in Tonga, 42% in Samoa, 36% in the Cook Islands, and 37% in Tahiti, with the remainder being haplotypes I and II (O'Shaughnessy *et al.* 1990). In Micronesia overall, haplotypes I and II account for about 64% of the total, the remainder being III, IV and V. The sampling from Micronesia is quite good, taking in Palau, Guam, Truk, Ponape, Majuor in the Marshalls, Tarawa and Nauru. The islands of western Micronesia show the highest incidence of haplotypes I and II, which probably reflects the substantial population change in historic time.

Hertzberg *et al.* (1988) have looked more closely at the α-globin gene haplotypes within Polynesia, finding that Samoans, Niueans and New Zealand Maoris show limited diversity, sharing six haplotypes in all, and they comment that this is consistent with a common ancestral founding population.

Deletion mutations produce globin variants known as thalassaemias, and in the Pacific most of the interest lies with the α-globin variants. While for present purposes severe variants producing distinctive clinical pictures are not of particular interest, the biological or evolutionary background to the origins and perpetuation of some of these blood polymorphisms is relevant. Within Melanesia their incidence – as in most other parts of the world – is related to the presence of malaria. It does seem that these variants offer increased resistance to infection by the parasite, apparently being inimical to the existence of the sporozoite in the red blood cell. The actual molecular or biochemical basis of this effect is still uncertain. For some other red cell polymorphisms which also offer increased resistance to malaria the factors involved are rather clearer. A deficiency of the enzyme glucose-6-phosphate dehydrogenase in red cells leads to a deficit in energy required by the malarial parasite. In sickle-cell anaemia, which is not known from the Pacific but is common in populations long exposed to malaria in Africa, the polymerization of the haemoglobin molecule that occurs is damaging to the malarial parasite.

Much the commonest variant in the Pacific is the $-\alpha^+$ thalassaemia deletion where there is loss of a single α-globin gene resulting in, at worst, a mild anaemia. These single α-gene deletions have been determined to be of two types, $-\alpha^{3.7}$ and $-\alpha^{4.2}$, the superscripts indicating in kilobases the length of the deleted DNA segment. The $-\alpha^{3.7}$ deletions have been further subdivided into three subtypes, $-\alpha^{3.7}$ I, II, and III, according to where the cross-over between misaligned genes occurred when the deletion was being formed. Each subtype is thus the result of a distinct and separate mutation. Both the $-\alpha^{3.7}$III and the $-\alpha^{4.2}$ mutations appear to have arisen in Near Oceania. (The $-\alpha^{4.2}$ deletion in South-East Asia appears to derive from a separate mutation event.)

The distribution of $-\alpha^+$ thalassaemia in the Pacific at present appears thus: Highland New Guinea people have very low frequencies of single α-globin deletions. Coastal New Guinea populations differ considerably from Highland groups in their α^+ thalassaemia patterns, but also differ among themselves. On the north coast of New Guinea the $-\alpha^{4.2}$ deletion is common (>60%). On the south coast the $-\alpha^{3.7}$I deletion is rather commoner than the $-\alpha^{4.2}$, but the overall incidence is only 25%. The $-\alpha^{4.2}$ deletion is also found in the northern Solomons, in Vanuatu and New Caledonia, and in some islands in Micronesia. It is absent from Fiji and Polynesia.

The $-\alpha^{3.7}$III deletion is the commonest type of α-thalassaemia through most of Island Melanesia, and probably the only deletional form in Polynesia. The highest frequencies of the deletion are in northern Vanuatu, parts of the Solomon Islands and in the Bismarck archipelago. In Polynesia $-\alpha^+$ thalassaemia has been shown in recent years to be unexpectedly common: 9% in Tahiti, 12% in the Cook Islands, and 15% in New Zealand.

Of interest is the existence in Island Melanesia of a haemoglobin variant, Hb J[Tongariki], which is known to result from a mutation on the $-\alpha^{3.7}$III deletion: that is, a mutation on a mutation, a chance event on a chance event, both having occurred within Island Melanesia. Tongariki is a small island in central Vanuatu where the variant was first identified. It is now known from throughout Vanuatu (particularly in the north), Ontong Java, New Britain, north coastal New Guinea and Karkar Island. Despite extensive surveys it has not been found in New Guinea Highland populations, Irian Jaya, island South-East Asia or Australia. It has not yet been found in Polynesia; the ancestors of the Polynesians may have moved on from Island Melanesia before its appearance, or its absence may reflect the vagaries of the founder effect.

Zeta (ζ) globin, the embryonic α-like globin chain, shows a distinctive $\zeta\zeta\zeta$ mutation, which occurs at about 9% frequency in South-East Asia, 2% in Melanesia, and is rather common (26% incidence) in Polynesia (Trent *et al.* 1986). The incidence in Tahiti is given as 16%, and in Micronesia as 15% (Hill *et al.* 1987). The shared frequency with Asia appears to be real because of a distinctive haplotype pattern. Of further interest is the appearance in Near Oceania, and through into Remote Oceania, of a superimposed mutation, revealed by the restriction enzyme Bg/II, which has not been recorded elsewhere. That is, there is evidence in the ζ-globin system of an Asian relationship, with a later superimposed mutation unique to Oceania. The Bg/II mutation is known in isolated cases from Vanuatu and New Caledonia, and in the Cook Islands and New Zealand at about 4% incidence (Trent *et al.* 1986, Hill *et al.* 1987). There are apparently no clinical manifestations of these polymorphisms. The single ζ-globin mutation appears to occur only rarely in Remote Oceania in general, but in a relatively high incidence (5%) in Niue – presumably occasioned by the founder effect. (Interestingly, in an unpublished report prepared several years ago, I find I have noted that the morphology of a very limited skeletal series from Niue was not typically Polynesian.)

Variants of the β-globin chain seem to be less significant for the particular purposes of Oceanic prehistory. The β-haplotypes of normal chromosomes in Melanesia appear different from those identified in Asia. Wainscoat *et al.* (1986) have analyzed the haplotypes of the β-globin gene cluster in normal individuals from several regions. The dendrogram resulting from this analysis is shown in Figure 5.12, and the close Melanesia/Polynesia relationship is evident. The only Asian group is moderately removed on the tree. As with α-thalassaemia, the deletion mutations in the β-chain that are of interest here are those compatible with ordinary existence. β-thalassaemia is rare in Highland New Guinea

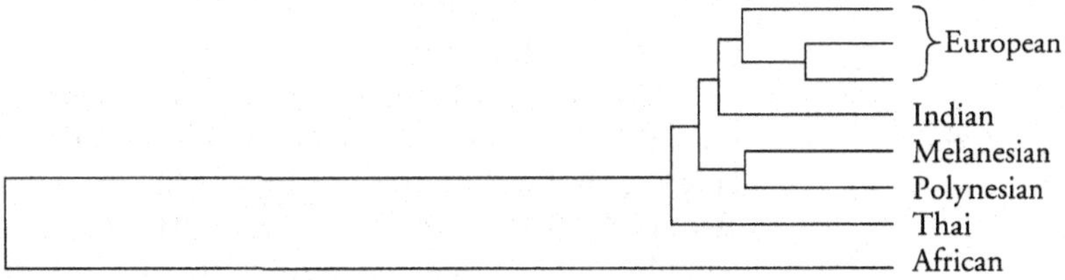

Figure 5.12
Relationships
between global
groups based on
normal β-globin
haplotypes. Adapted
from Wainscoat *et
al.* 1986.

populations, but is found in about 5% of coastal New Guinea populations and in quite high frequency on a few islands of Vanuatu (Hill *et al.*
1989). It has not yet been recorded from the Solomons, New Caledonia
or Fiji. Present evidence is that the common form of β-thalassaemia in
Near Oceania arose in the region and is not related to the form prevalent
further west in Asia. If this is so, it may be significant in understanding
the evolution of malaria in Island Melanesia.

In fetal life haemoglobin contains two α-globin chains and two
γ-globin chains. The genes for γ-globin lie immediately upstream from
the adult equivalent, the β-globin gene, on the short arm of chromosome
11. Cross-over mutations of the γ-gene have occurred, resulting in a γγγ
and a γ arrangement. In Melanesia west of Vanuatu these variants are
uncommon. The γγγ variant is known to occur in relatively high frequency in Vanuatu (3.7%) and New Caledonia (8.3%), while the single
γ variant is also rather common. The incidence of the γγγ is relatively
high in Polynesia (Western Samoa 14%, Tonga 9%, Cook Islands 16%,
Maori 10%) but the single-gene arrangement is rare. The likelihood is
that the mutation originated in Island Melanesia, and that the varying
incidence from island group to island group further east reflects the
founder effect.

We turn to other markers. Phenylketonuria is an inborn error of
metabolism with an incidence in Europeans of about 1:11,000, and in
Asians of 1:16,500. In this condition an absence of the enzyme hepatic
phenylalanine hydroxylase leads to an accumulation of phenylalanine in
the blood with consequent mental retardation or episodes of mental
confusion. (It is not inconsequential in history, as George III of England
is thought to have had the condition and had his thinking been clearer
the American colonies might not have been lost.) The gene for the
enzyme is located on chromosome 12 and a number of haplotypes have
been identified. In Europeans no haplotype achieves a dominant frequency, with the sum of the five most common still being less than 80%.
By contrast, in Asians haplotype 4 dominates with a frequency of almost
80%. Polynesians show a preponderance of three haplotypes, 1, 4 and 7,
with a combined frequency of 95%. This reduction in overall haplotype
number, and relative dominance of haplotype 4 fits with a relationship
with the Asian groups studied, though it has its own distinctiveness
(Daiger *et al.* 1989, Hertzberg *et al.* 1989b).

Another polymorphism of interest relates to the genes affected in
the condition of haemophilia. The British Royal family seems to
accumulate these problems: this is the condition that Queen Victoria's

progeny distributed through the royal houses of Europe. Here factors VIII and IX of the blood coagulation system are deficient. The genes lie on the X-chromosome but the condition only manifests in males who lack a second X-chromosome to override the deficit so to speak. Van de Water, Ridgeway and Ockelford (1991) conclude that the linkage disequilibrium between the restriction sites of the factor VIII gene is consistent with Polynesians originating from a small founder population less than 8500 years ago. The assumptions of the equations used to establish this date are not usually met in studies of human populations, but the time estimate nevertheless fits quite happily with the idea of the founding population(s) of Remote Oceania emerging from Island Melanesia.

Overview of genetic studies on the living

In terms of population sampling it is early days for these invaluable studies of gene frequencies and direct readings of the genome. In setting the present picture from them against differing views on the settlement of the Pacific it is taken that there is no serious argument as to whether or not people came out of Asia. The interesting question is what evidence there is for a longish sojourn for the ancestors of the people of Remote Oceania in Near Oceania. The mitochondrial DNA studies as well as the multiple loci gene frequency studies suggest separate movements of people in the settlement of Australia and of Highland New Guinea. Polynesian groups share maternal lineages with some in coastal New Guinea and Island Melanesia, and in Asia, but the suggestion of an antiquity of some 17,000 years for the 9 base-pair deletion in the Americas (assuming it to be the same event) allows at least such an antiquity for it also in the Pacific. From nuclear DNA there is clear evidence for considerable genetic contact, and presumably – with mutations superimposed on mutations – considerable time-depth of contact between the ancestors of the present peoples of Polynesia and some of Island Melanesia. Amongst the distinctive globin variants shared between Near and Remote Oceania are the $-\alpha^{3.7}$III deletion, originating within Near Oceania; the $\zeta\zeta\zeta$ variant from Asia but with a superimposed 5.5 kb mutation, revealed by Bg/II, originating within Near Oceania; and the $\lambda\lambda\lambda$ variant, originating within Near Oceania. These positive findings, and some of the others mentioned above such as the proportions of α-globin haplotypes, suggest that the Polynesian genome had a considerable ancestral residence in Island Melanesia.

Studies on archaeological bone

Refinements in polymerase chain reaction amplification are enabling extraction of DNA from archaeological bone. One substantial study relating to the Pacific has been published (Hagelberg and Clegg 1993) and the results from this are included in Table 5.1. Basically, the mitochondrial 9 base-pair deletion was detected in all eastern Polynesian samples tested, in a late prehistoric Tongan, and in a late prehistoric individual from Kosrae in the east Caroline Islands. It was not detected in any samples earlier than 700 b.p. (actually more recent, as the date of 700 b.p. was simply given as an extreme range for the New Zealand samples), including those from Samoa and Tonga.

Two individuals separated by the length of the Pacific, from the Chatham Islands and Hawaii, showed both the 9 base-pair deletion and three identical base substitutions in a hypervariable region of the molecule. The two other east Polynesian samples showed the 9 base-pair deletion and two of the base substitutions, and the individual from Kosrae showed the 9 base-pair deletion and one of the substitutions. Neither the 2500–2100-year-old material from the Lapita site of Watom in New Britain, nor the 1900-year-old material from Sigatoka in Fiji showed the 9 base-pair deletion. The base substitutions of the eastern Polynesian material were not present in Watom, and not ascertainable from Sigatoka.

Hagelberg and Clegg suggest that their results 'fail to support current views that the central Pacific was settled directly by voyagers from island Southeast Asia, the putative ancestors of modern Polynesians. An earlier occupation by peoples from the neighbouring Melanesian archipelagos seems more likely' (1993: 163). However their results do pose fresh problems in suggesting a very recent movement of people, or at least women, into the far reaches of Oceania – matters such as phenotypic change and even logistics have to be fitted into this scenario. Of concern in the results is the temporal cut-off for detection of the 9 base-pair deletion, none being found in bone older than (at the very most, and probably 200 years less) 700 years. Contamination by recent DNA does seem a real possibility for the older, much handled, and often treated material. On the other hand the evidence from studies on the living (Lum *et al.* 1994) of multiple female lineages entering Polynesia may link with these archaeological results. Again, there are apocryphal thoughts of American mitochondria. All in all, while taking this study as a landmark in its detection of significant molecular variation in archaeological material from the Pacific, it is probably wise in the meantime to suspend judgement on its results.

Other possibilities

At present there is something of a polarization between those who accept in its generality the model presented at the beginning of the chapter, and those whose views might be closer to Swindler's words, written a generation ago:

Melanesia's populations as seen today are the products of generations of racial churning, in which the evolutionary processes of mutation, migration, natural selection, genetic drift and selective mating have effectively contributed to the racial diversity so demonstrable there today . . . The original ingredients which went to make up this mixture could have been represented by many polytypic populations slowly wandering out of Asia into Melanesia, populating the islands as they came, undoubtedly mixing with neighbouring bands and at the same time establishing rules of endogamy with respect to others. Whether this exodus was represented by three or four separate migrations or simply was a slow gradual dribbling of small groups is not known. However the latter assumption seems more plausible. (1962: 48–49)

This is something of the approach that I have been trying to take here, when avoiding interpretation of results in typological terms and linguistic models. A model that embodies scepticism over linguistic models for prehistory has been presented by Terrell (1986). In arguing for the immediate origin of many of the peoples of Remote Oceania within the Pacific itself, he suggests:

(1) people began to colonize the islands thousands of years ago (a popular estimate today is that at least New Guinea Island was settled by 30,000–50,000 years ago);
(2) patterns of human communication vary and have varied from place to place and from time to time;
(3) when people are out of touch with each other they may go their separate ways in different directions of change and development; and
(4) there has been time enough since human colonization of the islands began for similarities and differences to arise within the Pacific that may owe little if anything to customs, linguistic conventions, and old biological distinctions carried out to the islands from other parts of the world. (244)

Terrell observes that 'many of the assumptions behind this alternative model are simple and far from controversial in themselves' (*ibid*.). He does question the very fundamentals of the linguistic dichotomy of Austronesian and non-Austronesian languages, which is something I am unable to comment on, but it is worth noting that not all linguists have originated Austronesian in Taiwan: Dyen (1965) placed the origin of the Austronesian language family within Island Melanesia on the basis of its greatest diversity in that region. However on one point Terrell seems to have been misunderstood. Because he mentions a figure of 30–50,000 years for settlement of parts of Melanesia, his model seems to have been taken by some to presume an effective human isolation in the region over this time, which I do not see to have been the intention.

Of an earlier statement of Terrell's views, Bellwood commented, somewhat acerbically: 'If those . . . who claim a totally Melanesian origin for the Polynesians wish to argue that Polynesian origins lie entirely within these long-established Australoid populations of the western Pacific, via the founder effect and genetic drift . . . they will perhaps need more evidence than their present blind faith' (1985a: 133).

In this statement, aside from the recurring typological problem, there is, in the polarization of view expressed in the words 'totally Melanesian', an adherence to the linguistic model. Yet this model, claiming an origin of a language group at 6000 years b.p. in the region of Taiwan, with speakers of derived languages sweeping down through the Philippines and Indonesia, into Island Melanesia and out to Remote Oceania, all within 3000 years, does seem to have some problematical elements. (The figure of 3000 derives from a date in Fiji of 3000 b.p., which in all likelihood has not caught the earliest settlement.) Questions have to be asked as to what numbers of people we should suppose to have been involved in these movements over such time and distance, and even why the relentless onward pace? I have never seen an estimate of the numbers involved or an explanation of how they moved apace while purportedly retaining their distinctiveness. I do suspect we are talking of logistical as much as biological unrealities in this supposed movement of people and language. (Of course such a rapidity of evolution of a technology of sea-craft and navigation makes nonsense of any talk of a voyaging nursery in Island Melanesia.) The numerically rather small Austronesian-speaking populations that entered Island Melanesia are said to have become, within 2500 years, the bulk of the present population there, though having lost their 'Southern Mongoloid' phenotype because of the genetic success of the previously-established 'Australoids' (Bellwood 1985a) – which is perhaps surprising if they, the 'Southern Mongoloids', form the bulk of the present population, and at present there is no evidence for 'previously established Australoids' east of north Bougainville – yet their Austronesian language prevailed. And these are the same people of whom Howells could conclude that 'as physical beings the Polynesians simply could not have emerged from any eastern Melanesian population; they are just too different genetically' (1973a: 234). Then there is their physical evolution. If the adaptive concept is valid, time has to be allowed for phenotypic change to take place. While I take the selective pressure to have been intense, it was not so until the fringes of Remote Oceania were reached. I actually do think that such change could have emerged within a few hundred years, but I suspect most would want a few thousand. If the concept of selective adaptation suggested here is wrong, then where are these big pre-Polynesians back in Asia, and particularly in the region of Taiwan (Taiwan aborigines are today slight: Chai 1967) and why were they there in defiance of biogeography?

This postulated rapid progress out of Asia has been termed the 'Express train to Polynesia' theory by Diamond (1988), who presents a remarkable map of the stations along the way – of which the first five, from Taiwan to Island Melanesia, are speculative. It may be a little early to get on that train, the 6000 b.p. from Taiwan. We see also the speed with which the linguistic framework is discarded when convenient. That Polynesians speak Austronesian languages is evidence for their Mongoloid origin, but the use of the same languages is not evidence for a Mongoloid origin of some Australoids (whatever they are). On the other hand, the lack of affinity between languages of Australia and Highland New Guinea does not prevent the people being lumped together as Australoids, though the biological evidence for this is unclear.

If Swindler's (1962) view on the movement of *Homo sapiens* into

Melanesia is valid, and when in turn we consider the few and small founding groups likely to have entered and spread through Remote Oceania, then the pattern of the genome of these latter people is likely to owe a good deal to chance – the chance of just which of many Near Oceanic groups and Near Oceanic gene complexes happened to move on. From that perspective, it is likely to be an impossibility to link, in genetic terms, the descendants in Remote Oceania with any particular extant group in Asia. Perhaps the only thing the Remote Oceanic genome was required to carry was the message for a large body.

Something has been said about misleading and meaningless typological terms. The alternative to their use is simply to specify groups by geographic origin. This does have to be reasonably precise. To state, as so often in titles, that Melanesians and Micronesians differ in some polymorphism, when 'Micronesians' refers to a group from Nauru, and 'Melanesians' to a group from Vanuatu and another from New Guinea, neither does justice to the complexities of Pacific human biology nor helps in clarifying it. Similarly, a statement such as that some Polynesian beta-globin gene haplotypes are predominantly southern Chinese in type carries with it, wittingly or unwittingly, the idea that Polynesians derive from southern Chinese, perhaps as evidenced in the predominant Han population of today. This tends to reinforce the idea of the great Austronesian migration, though it is unclear what the biological link is between the aborigines of Taiwan of 6000 years ago and the dominant people of the region today. I think the more helpful wording here is along the lines 'Polynesians and southern Chinese possess similar beta-globin gene haplotypes', which leaves open the possibility and time of a common ancestor.

In summarizing the results of some of the gene studies, Serjeantson and Hill comment:

The extreme view taken by Terrell (1986) and White *et al.* (1988), that Polynesians evolved within Melanesia from a population resident there for at least 30,000 years, is untenable in the light of the genetic evidence. It seems quite implausible that a group evolving within Melanesia could have acquired, by chance, so many non-Melanesian genes! Rather, it seems likely that the Austronesian speakers were the source of those genes found commonly in Polynesia and sporadically in coastal New Guinea. (1989: 287)

There is the old mix of problems here, with people being defined by language, and the typological Melanesian/Polynesian partitioning wherein Melanesia is seen as some discrete and distinctive biological entity. To say it all again, Melanesia is a geographic statement, and at one time or another the ancestors of every person in Melanesia came out of Asia. The presence today in Asia and Polynesia of certain shared and distinctive genetic features does not preclude the subsequent continued evolution within Island Melanesia of human groups bearing these markers for many thousands of years before their spread into Remote Oceania. The genetic markers and mutations shared by both regions, and probably also the phenotypic change between, seem to require this pause.

Hill *et al.* similarly address Terrell's model of Pacific settlement, in considering that the 'high' frequencies of the α-globin haplotypes Ia and IIa in Polynesians today, haplotypes which are 'very rare' in most of Melanesia, makes a view of a 'wholly Melanesian origin for the Polynesians – untenable. The data argue rather for a basically Mongoloid population which has interbred to some considerable extent with Melanesians' (1989: 276).

Again, the typological terms create their own confusion, the term 'wholly Melanesian' with its misleading idea of total isolation makes for polarization of views, and 'untenable' seems too strong a word. Considering the small and few founding groups likely to have entered Remote Oceania several thousand years ago, it is hardly untenable that a pattern at a few loci that is 'very rare' in Island Melanesia today could not by the chance of the founder effect have attained dominance in the further reaches of the Pacific. This is just one argument proposed by molecular geneticists O'Shaughnessy *et al.* (1990) to explain the conflict between HLA and globin gene data regarding Melanesia/Polynesia relationships. I have commented earlier on the often scanty sampling – the data are thoroughly inadequate to allow assertions about 'most of Melanesia' – and the need to consider many loci has been stressed. Rather, the overall evidence from genotype and phenotype does point towards people evolving for quite a long time within Island Melanesia before moving on to Remote Oceania. It is fair also to add that the approach in many studies in Polynesia to eliminate sampling of foreign genes is often rather innocent.

The genetic evidence overall – gene frequencies, and mitochondrial and nuclear DNA – shows that in general linguistic boundaries are not mirrored by the biology, and from this vantage the model claiming a rapid spread of Austronesian languages, with such well-defined dates, seems hard to equate with a synchronous movement of a biologically distinctive group of people out of Asia. The evolutionary time required for the change in phenotype may be too great to match the claimed linguistic dates. Several shared molecular features suggest a longish sojourn in Island Melanesia by the ancestors of the settlers of Remote Oceania. The evidence is against a grand and recent migration of people with their language. It is for biological and geographic realism. (The fact is that the seemingly opposed views are not all that far apart. If the typological concepts could be abandoned and a modest and realistic flexibility admitted to the linguistic chronology then there may not be too much in contention.)

In the end the reality is that the human biology within the distinctive Pacific environments conforms well to evolutionary principles and expectations, and a pattern of progress of settlement such as envisaged by Swindler (1962). There happens to be a separate, intriguing, but much simpler problem, as Lewontin (Friedlaender 1975) indicated, of the origins and distribution of the Austronesian languages. It should not be necessary to try so strenuously to fit the biology with brittle tools (a term not applicable to the gene studies) to some non-biological model for which there is no *a priori* reason for fit – language and culture are but changeable veneers on the biological substance. Nor to resort to explanations involving migrations from elsewhere to account for biological differences, when, in general, thinking has so turned against such concepts. Bringing in different people of different form merely shifts the biological and evolutionary problem elsewhere. The differences still have to be explained. The study of Pacific prehistory must be just about the last area where racial typologies are taken seriously. My own feeling – speculation I suppose, but there is much evidence in support – is that the Austronesian linguistic model of a grand and relatively recent movement of people out of Asia and inexorably on to the outer reaches of Oceania, is in several ways too simplistic and naive a concept to sensibly frame the human happenings through Oceania in prehistory. And I suspect that the present jumblings of linguistic, typological and genetic terminology of many writings will look as quaint in a few years as some of the craniological musings of a generation or so ago (albeit lingering on) now do.

Simulation of survival at sea

In the last part of this chapter on models and methodology in the study of Pacific prehistory we return to the computer simulation described in Chapter 3. There the highly significant association of survival with latitude was mentioned, and here we look in more detail at the implications for the settlement of Remote Oceania. This is a change of perspective from the genetic considerations, but the simulation indeed serves as a model and another methodology, and more needs to be said than was appropriate earlier.

The first point to be made is really an extrapolation from the findings. The study did not incorporate any data closer to the equator than 9°, nor any North Pacific data. Nevertheless the significant regression obtained strongly suggests that the South Pacific findings should be relevant across the equator. The water isotherms and mean wind speeds indicated in Figures 1.2 and 1.5 show that a wider band of the western Pacific is as mild as the near-equatorial latitudes of the eastern Pacific, which in turn suggests that Near Oceania was a voyaging nursery not only in terms of development of Oceanic seagoing skills and technology, but in the adaptation of the human frame.

It is likely that the simulations do present a rather (but not drastically) gloomier picture of survival because human judgement is not allowed for. Craft push off and take whatever weather the computer throws out. In reality, experience and judgement on favourable and persisting weather patterns would be significant in determining when a voyage would start. This is important when, by and large, we are not considering long periods – many weeks – at sea. On Irwin's (1992) 'search and return' thesis, embracing the most reasonable idea that people actually were interested in getting home safely again, then seven days out and five days back might be taken as a long voyage.

Recent studies of voyaging and Pacific weather patterns (Finney *et al.* 1989, Irwin 1992) have examined the possibility that winter would have been a favourable time for travelling east. However the simulations suggest that beyond about 12° of latitude voyaging in winter could not have occurred. Summer was the time of exploration and colonization.

Much of Micronesia – the Carolines and some of the Marshall Islands – lies within the relatively mild equatorial zone. The rather late dates presently known for settlement of these places may reflect the problems of finding and then surviving on very small islands. The earlier dates for the Marianas are interesting, and DNA analyses on archaeological bone from there may be informative.

Beyond Near Oceania, it is not only distances, wind direction and target angles of islands that need to be considered when looking at the pattern of settlement (Irwin 1992), but also latitude as an expression of human survival. Through Remote Oceania the different phenotypes now established on the islands – or at least those visible at time of European contact – probably are indicative of the form of the first colonizers. For example in Vanuatu it seems probable that a rather slight physique could have reached the north through the Santa Cruz Islands and then filtered south through the archipelago. However despite the relative proximity of New Caledonia, the higher latitudes are sufficiently harsh as to postpone expansion of settlement there until development of a more robust physique.

Much of Fiji descends into the harsher latitudes, but it does present a rather large target, and presumably a more favourable environment for settlement than the small atolls of Micronesia. It is likely that earliest settlement occurred in the north, from where movement south would have been simple, at least by the standards of the explorers of the Pacific. Some of the Tongan archipelago lies even further south, but the proximity to Fiji, with small intervening islands, enabled relatively speedy settlement. Once the relative positions of these islands had been ascertained, taking advantage of fair winds in summer, a trip between the Lau outliers of Fiji and the Tongan archipelago of only a couple of days could be anticipated. There is not too much risk to the Tongan physique over this period, at this season and latitude. Samoa lies in kinder latitudes, accessible from either Fiji or Tonga.

Beyond the Fiji–Samoa–Tonga axis, distances to the major island groups of the South Pacific become greater, and the latitudes to be passed through, higher. It is 1100 km from Tonga to the southern Cook Islands, with only tiny Niue intervening along latitudes 18–21°, and these latitudes lie beyond the survival precipice, even for Polynesians in summer. Even in the best conditions ten days is probably the least time in which that distance could reasonably be covered. Palmerston atoll, in the Cooks, on latitude 18°, gives 10-day survival proportions of 0.31 for both males and females, tumbling at Rarotonga, not too much further south, to 0.05 for both sexes. It is unlikely that settlement could easily occur along this axis. This offers an explanation for the relatively late dates now available for settlement of the southern Cooks. Of course this is not to say that these islands were not reached earlier, perhaps much earlier, but demographic realities may have prevented successful colonization.

On the other hand the simulations show much better survival along the low latitudes between Samoa and the Marquesas, taking in the northern Cook Islands along the way. That is, the simulations support the suggestions of early settlement for Pukapuka (A.D. 0) and the Marquesas (A.D. 300). The language of Pukapuka is in the Samoic lineage. From the Marquesas, exploration south would be a much safer exercise, with successive discovery of the relatively close Tuamotus, Society Islands, Australs, Gambiers and southern Cooks, which matches available carbon dates on sites of human occupation. These studies on the resilience of the human frame support such a pattern of settlement over the systematic, radiating expansion postulated by Irwin (1992), but which in turn is supported by palaeoenvironmental suggestions (in the form of charcoal layers suggestive of burning) of much earlier human occupation of the southern Cooks (Kirch and Ellison 1994). However Irwin's (1992) study and the present simulation results are complementary, not contradictory. Systematic, radiating exploration with an eye to survival seems rational. The biological considerations merely suggest that the more prolonged and successful probes were initially in low latitudes, a view supported by the available archaeological dates.

Settlement of a southern outpost such as Rapa is likely to have been a marginal happening, with little communication with the islands to the north thereafter. Such a statement of course applies much more strongly to Easter Island, which probably was a chance discovery by a group *in extremis*, and from which return voyaging can be discounted. That leaves New Zealand as the ultimate southern settlement. One can only postulate that this was reached during a particularly favourable period of summer weather by a motivated group directed by generations of observation, such as the route of migrating birds. New Zealand does have the advantage of being a particularly large target as the land is closed, and having undreamed-of resources for survival. Return voyages to tropical Polynesia seem improbable – perhaps once or twice in earlier generations. Though Tonga is close, the genetic and linguistic evidence favours settlement from the southern Cook Islands. It does not make survival sense to settle New Zealand directly from groups further north, such as the Society Islands or the Marquesas.

Hawaii at first glance presents the same formidable distances from the Marquesas as does New Zealand from the Cook Islands, but the situation is very different. Most of the Hawaiian course is in low latitudes, across the equator, with only a final five or six days in higher latitudes, and the likelihood of a beam wind. However one must be sceptical of ideas that return voyaging was a regular occurrence.

If the American coast was reached by Polynesians it was north of

15° south latitude, and not from Easter Island, which on distance alone might seem the obvious departure point. A northern encounter with South America makes sense from the evidence of the kumara. An encounter with the Galapagos Islands is a possibility.

On this matter of human survival at sea, of interest are the impressive achievements of the Vikings in reaching North America across the high latitudes of the Atlantic. These voyages were undertaken during a rather mild climatic period, over shorter distances than those traversed in the Pacific, later in time, and, in the sense that there are sometimes advantages in metal and what it can fashion, with a more advanced technology than had the peoples of Remote Oceania. The Icelandic sagas, with their accounts of storms and icebergs and sea-monsters, are good reading, but I take them to be the equivalent of the Hollywood epics of the day, and not representative of everyday life. It is significant that the North American colonies did not flourish. Buck (1938) pictured his ancestors as 'Vikings of the Sunrise', but the Polynesian maritime achievement preceded and surpassed that of the Vikings, who might, more aptly, be termed 'Polynesians of the Sunset'.

At present two significant pauses in Pacific prehistory are apparent. One is between the Solomon Islands, with dates of 28,000 years b.p., and the earliest in Remote Oceania of about 3500 b.p. (Though sea-level changes may have covered much, there must be a high likelihood of earlier dates emerging further east in Near Oceania.) The second is the 'long pause' of perhaps 1000 years in the Fiji–Samoa–Tonga axis of central Polynesia, before settlement extended east. A further cause of perplexity has been the apparent earliest settlement, when this pause ended, of the northern islands such as Pukapuka and the remote Marquesas. The simulations suggest that both these gaps, as well as the pattern of settlement, may be explicable in terms of human survival. An evolutionary adaptation of the human frame was required before people could cross the distances, first between the Solomon Islands and Vanuatu (eased by the scatter of the Santa Cruz Islands) and then the more formidable gap between Vanuatu and Fiji. The greater distances beyond central Polynesia and the higher latitudes of most eastern islands meant that even with maximal human physical adaptation, discovery and colonization of new land took many years, and then first succeeded along the line of best survival, closest to the equator. Of course all these statements are hostage to the next crop of carbon dates, and settlement but not survival of generations must have sometimes occurred. And through all the settlement of the Pacific must run the thread of chance, the chance of a particularly kind spell of weather when people were afloat, and the greatest vagary of all, human motivation and resilience. The apparent early settlement in Micronesia of the distant Marianas may evidence just such chance. However the simulations point up the limits to resilience.

HEALTH

6

An approach to matters of health in prehistory that I have found useful is to consider first the general health of a group, and second the specific evidence of disease. In Table 6.1 are listed several general indicators of health. Some, such as age at death and stature, may often be assessed with minimal equipment, which does not necessarily mean they are always easy to determine. Others, such as radiological assessment of bone structure, require more complex technology. Whatever the technology, these indicators often may be determined on most members of a group from the past and therefore contribute to a general statement about the health of the group. Usually they all point in the same direction; that is, they are consistent in suggesting that a group was relatively healthy, unhealthy, or whatever, and thus work well in comparisons between different groups. By contrast, the occasional pathological lesion, if sometimes spectacular, simply indicates the expected – cells can run amuck in anyone. An example would be the isolated finding of an osteosarcoma; its existence tells us nothing about the particular conditions under which the group as a whole lived, or even the general state of health and nutrition of the

Table 6.1 Parameters providing general evidence of health.

1. Age	at death	
	of menarche	
2. Physique	*a.* outer parameters	stature
		mass
		robusticity
	b. inner parameters	general bone structure
		cortical thickness
		lines of arrested growth
3. Dental parameters	enamel hypoplasia	

individual with the tumour. Of course the finding of bony indicators of a particular disease widespread in a group *is* informative (and that could include osteosarcoma), but the point is that while that euphonious term 'palaeopathology' often claims the fascination, more basic matters relating to growth and age and diet usually tell a lot more about the existence of people in the past.

Another problem encountered in palaeopathological studies is that dry bone can only show a limited range of changes: a loss of substance, an accretion of substance, or a general or local alteration of architecture. This means that the diagnosis of a bone lesion in an individual from the past often means reduction to a rather unsatisfactory list of possibilities, and it may be speculative to plump for any one of them.

The comprehensive classifications of disease and pathology of the living do not provide a very useful basis for a readable account of disease from the past. (Compendia of palaeopathology such as the splendid volume by Ortner and Putschar, 1981, may be interesting reading because all conditions can be discussed and illustrated.) For example, primary bone tumours occur rarely in any group, ancient or modern, and to wade through a lengthy classification of these with a reiteration of negative findings would be tedious. Secondary bone tumours (that is, from the spread of soft tissue carcinomas) are common in the modern world, but for demographic reasons they also were uncommon in the past. Yet simply to present the lesions that do occur without any ordering is unhelpful. So some adaptable ordering is desirable. The classification of pathologies used in the second part of this chapter derives from the old 'surgical sieve'. When confronted with an obscure clinical problem (particularly in the abdomen) one mentally works through a series of possibilities (see Table 6.10). With a few additional headings for the particular problems of prehistory, the classification is capable of catching everything, yet allows concentration on the essentials. It makes discussion of lesions from the past manageable, perhaps even readable. However we begin with health.

General evidence of health

Evidence from assessment of age

Age at death, and demography

Just as the mean age at death of a modern population provides one index of its state of health, so should that of a group from the past. However the circumstances of prehistory pose particular problems. Sampling is one. In the Pacific an increasing amount of data is coming from controlled excavations of large prehistoric sites, but much still derives from small and chance discoveries. Young bones, being less durable, are less likely to survive unfavourable soil conditions or even to be noticed, and most skeletal collections are dominated by the more robust adult material. Burial practices may contribute to such disparity, but by and large it is likely that environmental processes are more often responsible. Therefore when comparing ages at death across groups it is often necessary to consider adults only – those above, say, 15 years, from which age reliable sexing of the remains also is possible.

After about 30 years the placing of individuals even within five-year age brackets is sometimes optimistic. There is also the problem of different ageing criteria adopted by different workers. However I have the impression that those doing such studies in the Pacific arena have tended to follow fairly similar standards, and the results across the various studies are, I think, comparable. This is not the place for any prolonged discussion on ageing criteria. The sequence of development and maturation of tooth and bone is the basis for the first two decades or so of life, with accuracy sometimes of the order of plus or minus one year. Ageing of mature individuals is much trickier, and there is a range of methods, some (or most, one could argue) of which are group-specific – that is, criteria have to be defined for each group. Pubic symphysis change, tooth wear, and changes in bone structure assessed on x-ray are some of the armamentarium. Bone histology is probably best of all (Kerley 1970, Ortner 1975), but the bone has to be in quite good condition. Always, beyond about 30 years an adult age must be regarded just as an observer's estimate as to which five-year span the person may have belonged.

There are problems piled on problems here. It is one thing to make an estimate of the age at death of an individual or to establish the average for a group. It is another thing to interpret its meaning. The age distribution of the series may be (it usually is) distorted by a clear under-representation of juveniles. The significance of the mean age at death is complicated by the insistence of demographers that, rather than simply equating with life expectancy, it contains an expression of the rate and direction of growth in a population, with the mean age at death decreasing when a group is expanding, and increasing if it is declining (Johansson and Horowitz 1986). The basis of this is the concept is that if there are more young people in a population they will be duly represented in the skeletal record. This complicates interpretation of rather slight differences in mean age at death. A group with a lesser age may have been thriving in its environment. Here, consideration of other factors indicative of general health will be useful.

For all the difficulties in its assessment, age at death remains a – perhaps *the* – crucial parameter in much interpretation of the human past. Table 6.2 gives the mean age at death of the adults of various Pacific groups. The results range from 29 to 37 years for males and a tighter cluster of 29 to 30 years for females. Micronesia gives the highest figures, and I have heard anecdotes of a series of 400 prehistoric individuals from Guam said to show a mean age at death in the late thirties. On the basis of what was said above, this greater span in Micronesia may be a reflec-

Table 6.2   Mean age at death of adults (ae > 15 years) for several prehistoric Pacific groups. Figures in parentheses are the number of individuals in the group.

Group	Region	Date (b.p.)	Male age (years)	Female age (years)	Source
Watom	New Britain	2100	35.0(6)		Houghton 1989
Sigatoka	Fiji	1900	31.3(18)	30.3(28)	Visser 1993
Taumako	Solomon Islands	700	32.0(80)	28.6(57)	Houghton (unpub. data)
Afetna	Marianas	500	37.3(9)	30.5(10)	Roy & Tayles 1989
Mokapu	Hawaii	500	31.9(295)	29.0(336)	Snow 1974
New Zealand		2–400	29.2(159)	29.0(182)	Houghton (unpub. data)
Tepoto	Tuamotus	150	29.9(7)	30.3(6)	Dennison 1992

tion of a stable or declining population in a world of very small islands, rather than demographic success. For nearly all groups females show a lower average age at death than males. This is a consistent finding in studies of prehistoric material, and is usually interpreted as expressing the risks of childbirth. However it may sometimes reflect relative female success.

Population pyramids for two Pacific groups, portraying age at death by 5-year intervals, are given in Figures 6.1 and 6.2. The final age groupings are open-ended: I am unable even sensibly to guess at discrete ages beyond about 50 years. The first pyramid is for 176 individuals from a single site in Taumako in the Solomon Islands, dated to about A.D. 1500. The infant representation in this group is quite good. Figure 6.2 is a composite histogram derived from all available New Zealand prehistoric data and the unrealistic representation of juveniles is clear. Leaving this aspect aside, the contrast of both with the biblical three score years and ten or the average life span in modern westernized countries is clear. So although these Pacific data do follow the pattern of

Figure 6.1   Population pyramid for Taumako (Solomon Islands), about A.D. 1500. The five-year age clusters are indicated on the right-hand scale. Adult males are distributed to the left of the central axis and adult females to the right. Individuals below age 15 years have been distributed equally to either side of the central axis. Discussion in text.

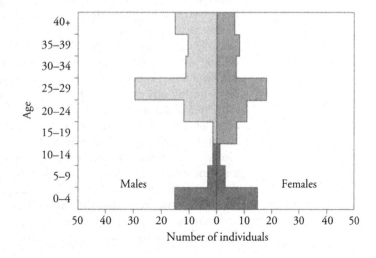

Figure 6.2 Population pyramid for New Zealand, about A.D. 1600–1800. Arrangement as for Figure 6.1

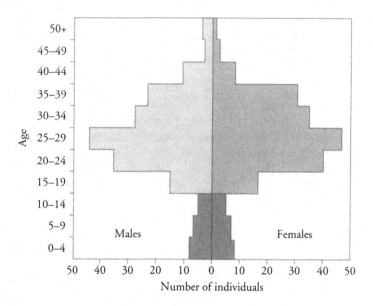

prehistoric groups in other parts of the world, they are sometimes greeted with scepticism. For example, Pool feels that the New Zealand samples, with no one dying over the age of 55 years, portray a situation which is entirely unrealistic. He suggests that levels of life expectancy at birth of 25–30 years, and at age 15 of 13–17 years, 'do not fit with what is known about the epidemiological transition. Moreover, these figures imply a modal age at death in the late 20s–early 30s, which would have played havoc with the reproductive capacity of the populations . . . Perhaps here the analyst is a victim of the difficulties inherent in the age estimation of skeletal remains' (1991: 34, 36). Elsewhere, Howell talks of the 'unusual and implausible features of the population structure' (1982: 263) of a prehistoric Amerindian group.

The epidemiological transition mentioned by Pool I take to be more or less equivalent to a central paradigm of demographic studies that relates patterns of demographic change and related patterns of disease in the transition from traditional societies (characterized by high fecundity and mortality) to modern technological societies (characterized by low fecundity and mortality). It is fair to say that such theory has not been much concerned with the conditions of prehistory, and while I do not think one can always be completely at ease with some of the low estimates of age at death, there are, for all the problems in estimating adult age from the skeleton, good signs that these profiles from the past are on the right lines. The matters being assessed are common to the biology of all *Homo sapiens*. The 20–29-year age bracket dominates most assessments, and much of this bracket simply is not tradeable. That is, the biological criteria for ageing the last stages of maturation, through to about 25 years, are unequivocal and not for debate, beyond a bit of give and take of 2 to 3 years. Examples of such ageing criteria include the fusion of the medial clavicular epiphysis and the last of the ring epiphyses on the vertebral bodies, completion of the root of the third molar, and elimination of the circumferential bone lamellae in histological section. In this group also, criteria such as change at the pubic symphysis or age estimation from tooth wear are likely to be at their most accurate. Further, many of these criteria have initially been established on recent material, particularly of European or American origin, against whom Polynesians (and presumably their ancestors) display an accelerated development and maturation. This means that the tendency is to age these prehistoric people in the Pacific as rather *older* than they really were; the error is in the direction of a greater than real age. After 30 years, the vigorous life of the past accelerated ageing features, such as joint degeneration and tooth wear (which are always considered on a group-specific basis anyway), and the important histological changes within the bone (Currey 1984, Martin and Burr 1989). Again, the error is in the direction of a greater than true age. (In passing, it is of interest to note that the allometric prediction of life span for animals of our average mass is 27 years (Calder 1984: 153).)

Even if we accept that there may be problems in justifying the upper limit of this first group (that is, confining it to 29 years) and arbitrarily decide to extend it to a 20–34-year span – to then spread the remaining individuals out for 30 or 40 years, to age 65 or 75, results in an unusual mortality curve. With the New Zealand data I experimented with age distributions that assumed 33% of the individuals in the 25–29 group to belong to the next span (30–34 years), and so on for each group. The result from this empirical exercise was only to establish a mean age at death for adult males of 30.6 years rather than 29.2, and 30.9 years for adult females rather than 29.0 years. Unfortunately the population pyramid then developed a suspicious double peak.

Scepticism has been expressed as to how populations could expand when the mean age at death of the adult females was around 30 years. However, while there are significant differences in survivorship curves between recent groups and people from prehistory, Weiss (1973) suggests life expectancies at birth between 19 and 24 years for recent groups leading a traditional non-westernized existence. For the Dobe !Kung, Howell (1979) determined mean age of final birth as a little over 34 years. While accepting the value of two further years of reproductive life, even without further analysis the figures are not too far apart. More importantly though, these criticisms of the age profiles may be misplaced if it is assumed that mean age at death of the skeletal group, or any sub-set of it, equates with life expectancy of that group. We will return to this matter. In the end, whatever demographic theory may suggest, the skeletal data are primary and cannot be adjusted to fit preconceptions as to past demographic realities. After allowance for sampling error and with a considered estimate of the rate of growth of the population, if the data do not fit the theory then the theory requires adjustment, not the data.

For a few groups from the past the numbers have been sufficient to provoke attempts at reconstruction of life tables. I say 'attempts' because the data are rarely likely to meet the criteria of classical population theory, which postulates a stable, self-contained population. 'Stable population theory is about hypothetical populations closed to migration that have experienced constant growth rates (zero, positive, or negative) based on unchanging age-specific fertility and mortality rates' (Johansson and Horowitz 1986: 235). Also, sampling in space and time for skeletal series is generally crude and gross, and there is almost always the problem of infant under-representation. There is the risk of the oft-times fragile data on age and sex being converted into expansive statements and comparisons between groups in the past, and of the presentation of seductive arrays of numbers carrying a spurious objectivity. Most professional demographers seem to regard palaeodemography as a speculative exercise. 'On the whole, anthropologists in general find more to admire and trust in attempted prehistoric demographic reconstructions . . . than formally trained demographers' (Johansson 1982: 135). Peterson (1975) and Bocquet-Appel and Masset ('Farewell to paleodemography' 1982) have written in similar vein. Others have defended some of the reconstructions, and palaeodemography in general (Van Gerven and Armelagos 1983). It is fair to say that the sceptics are too sceptical as to the worth of skeletal ageing and sexing methods. As palaeodemography deals with nearly all our existence as a species, persistence seems worthwhile.

Kirch (1984) considered four Pacific skeletal populations to be suitable for palaeodemographic analysis, and provides life tables for them. Two of these groups are Hawaiian. From Mokapu on Oahu a census of 1163 individuals was established by Snow (1974); and from the Pu'u Ali'i Sand Dune site on Hawaii derives a census of 92 individuals (Underwood 1969). From the Marquesas a population of 41 individuals excavated at the Hane Dune site was analysed by Pietrusewsky (1976), and from 'Atele on Tongatapu a sample of 61 individuals was also studied by Pietrusewsky (1969). The numbers for the Marquesan and Tongan series must render them marginally appropriate for demographic analysis. Two others worth examining are those of the population pyramids of Figures 6.1 and 6.2. The New Zealand sample might seem a metaphorical millennium away from the coherent, stable population of actuarial theory, but justification has earlier been offered for putting New Zealand data together for many purposes. The 176 ageable individuals of the Taumako series constitute one of the more satisfactory samples in the Pacific for palaeodemographic purposes, being derived from a single site with a time-depth of about 150 years, and showing what may be family sequences of burial.

For palaeodemographic reconstruction, Johansson and Horowitz (1986) suggest that the (rather mechanical) approach via the formulation of standard life tables is inappropriate. They emphasize the effect of population change rate, positive or negative, on the interpretation of the determined mean age at death for a skeletal series. They also stress the importance of striving for some estimate of the degree of isolation of the group. When these points are considered it can be seen that the particular circumstances of an island world may be rather favourable for establishing these parameters, and for the application of demographic theory. Island populations sometimes were closed or largely closed: and sometimes we have a reasonable estimate of population at time of contact and an archaeological assessment of time of first settlement. Sensible statements about size of the initial colonizing population can be made – they can be nothing other than small multiples of canoe loads.

For New Zealand, a rigorous scrutiny of carbon dates has not provided clear support for settlement earlier than about 800 years ago – that is, around A.D. 1200 (Anderson 1991). This is despite various murkier dates back to A.D. 750, and suggestions from the fringe of much earlier settlement still. An antiquity of no more than 800 years is embarassingly close to that of popular tradition for the time of the Great Fleet, from which most of Maoridom strive to trace their genealogy. One says embarassing because the usual date for the fleet, A.D. 1350, derives from a degree of mangling and amalgamating of local traditions by nineteenth-century anthropologists (Simmons 1975). For the last couple of scholarly generations such versions have been on the discredited list. Tradition may be more robust stuff than often credited.

So, there is scientific support for first settlement of New Zealand being about 800 years ago. In terms of European contact, and taking that in round figures as A.D. 1750 (ignoring Tasman of 1642, who had very little to do with the country), the settlement time before European arrival may be 550 years. From now on many permutations of the data are possible; a couple are set out in Table 6.3. If this was a single colonization by a single canoe then we may postulate the arrival at that time of 14 people, with the sexes balanced (though a ridiculous precision is not intended by fixing on this number). Whether there were other arrivals at other times is something debated. In Chapter 3 it was suggested that multiple settlement by large numbers or any significant amount of return voyaging to tropical Polynesia was unlikely to have occurred. What we are left with are figures for a settlement time and founding population that have some substantive basis.

The next estimate needed is that of New Zealand's population at time of European contact. This has been debated at length (Davidson 1984), but in the end most seem to come back to that of Cook, of about

Table 6.3 Some population growth estimates.

Island	Possible size of founding group	Estimated population at European contact	Years of settlement	Mean annual population growth (%)
New Zealand	14	100,000	550	1.61
New Zealand	50	100,000	500	1.38
Chatham Islands	6	2000	400	1.45
Hawaii	14	250,000	950	1.03
Easter Island	14	7000	800	0.78

100,000 people. Some would place it a bit higher – around 125,000. On the other hand Anderson (personal communication), on the basis of a range of archaeological evidence, suggests that it was no more than 80,000. Fox (1983), on the basis of estimated numbers occupying pa sites, has also suggested deflating the estimates. It seems best to settle for Cook as observer and arbiter.

We now have reasoned estimates of founding population and achieved population after a defined time interval. The rate of population increase needed to achieve one from the other is set out in Table 6.3. Starting from a population of 14 individuals, an average annual growth rate of about 1.6% is needed to reach about 100,000. Such a rate of increase is very modest. In recent times there are figures exceeding 3% in some developing countries. An average annual growth rate of 3% has been determined for Pitcairn Island between 1790 and 1856: for the period 1808–1830, after the initial ructions of settlement had subsided, the rate was 4.3% (Refshauge and Walshe 1981). A suggested growth rate of 1.6 for New Zealand, a relatively vast new environment abundant with resources, does not seem excessive. The point of the exercise is not to assert that this was just the situation, but to demonstrate that the short settlement time of New Zealand provided by present dating evidence is compatible with a very small founding group.

The Maori tradition of the Great Fleet, with the date of A.D. 1350 ascribed to it by Percy Smith, has been reinterpreted as the diaspora to other parts of the country from some early focus of settlement in northern New Zealand (Simmons 1975). After 150 years of settlement, from A.D. 1200, our settling group of 14, given a growth rate of about 2%, increases to 300, quite enough to provide for a general diaspora. Again, the story is compatible with modest growth rates.

Other possibilities are given in Table 6.3. A founding group of 50 requires a mean annual growth rate of about 1.4% over 550 years to reach the estimated contact population, while 14 people require a growth rate of 1.2% to reach it in 1000 years.

Another calculation is possible in the New Zealand region. The Chatham Islands at time of European contact (A.D. 1792 – we can round it to 1800) have been estimated to have had a population of about 2000 people. There are no firm carbon dates from the Chathams before about A.D. 1500, but we may reasonably push settlement back a little, to A.D. 1400. The evidence seems now to be that the Chathams were settled from New Zealand. This was probably after the time of great voyaging, and it may be that the arrivals had neither the abilities, the means, nor the incentive to return – the Chathams in prehistory would have been a pleasant place. (The previous sentence was written before the simulations were run; it is highly unlikely that the lucky survivors had any of the attributes listed.) The most likely possibility is that the Chatham Islands were accidentally reached by a canoe blown offshore from the South Island. If this founding group is taken to be a mere 4, with the sexes balanced, then the required annual average growth rate is a modest 1.55%.

What these various estimates support is a population growth rate of 1.5–2% for New Zealand in general through prehistoric time. They support the idea that in northern New Zealand in the later prehistoric era population pressure was quite quickly becoming a cause of strife, with an inordinate amount of human energy being poured into warfare and its associated fortifications and earthworks (Davidson 1984). The estimates also suggest that we should not anguish too much over the precise population at contact; a couple of decades could readily take it from a lower to a higher estimate, from say, 80,000 people to 100,000.

With a statement on rate of population change that has as firm support as any in prehistory is likely to have, we can return to the basic age-at-death data. The focus is on the adults, here defined as fifteen years and above, as the juvenile data are inadequate. Mean age at death in New Zealand for both male and female is 29 years. Figure 6.3, from Johansson and Horowitz 1986, and in turn derived from the life tables of Coale and Demeny 1983, sets out the relationship between mean age at death and life expectancy for varying rates of population growth. With a growth rate of 1.5–2%, a skeletally-derived age at death of 30 years translates into a life expectancy of 45–50 years for adult New Zealanders in pre-history. That is, if an individual reached the age of 15 then that person could expect to live into her or his late forties. Previously (Houghton 1980) I have taken the mean age at death for adults as assessed on the remains – about 30 years – as equating the mean life span, and this has always seemed a puzzling and uncomfortable figure. The figure of 45–50 years seems a much more realistic figure for the New Zealand past, given what must have been a rather favourable environment. It is a life expectancy still to be found in some countries today, though of course patterns of disease and mortality are different.

Figure 6.3
Adjustment of life expectancy from mean skeletal age at death for different rates of population change (after Johansson and Horowitz 1986).

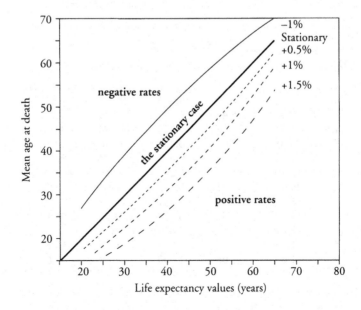

The importance of an informed estimate of population growth can be seen when standard life tables derived for the Taumako and New Zealand populations are examined (Tables 6.4 and 6.5). The juvenile representations are inadequate for comparison, but for the adults (over 15 years) the indication is that Taumako has the edge in longevity. Taken at face value this would suggest a marginally better environment on that small tropical island than in the southern latitudes. Yet a knowledge of the human material makes this an uncomfortable conclusion.

Table 6.4 Life table for Taumako. x = age or age interval in years, Dx = absolute number of dead of age x, dx = number of dead of age x as percentage of total, lx = number of survivors to age x out of radix of 100, qx = probability of dying in succeeding age class for those reaching age x, Lx = years lived by survivors in the age group, Tx = person-years lived in this and all subsequent age intervals, ex = average number of years of life remaining at age x.

x	Dx	dx	lx	qx	Lx	Tx	ex	Age interval (years)
0	19	10.80	100.00	0.108	94.60	·2458.81	24.58	1
−4	11	6.25	89.20	0.070	344.32	2364.20	26.50	4
−9	6	3.41	82.95	0.041	406.25	2019.89	24.35	5
10–14	3	1.70	79.55	0.021	393.47	1613.64	20.29	5
15–19	9	5.11	77.84	0.066	376.42	1220.17	15.68	5
20–24	23	13.07	72.73	0.180	330.97	843.75	11.60	5
25–29	48	27.27	59.66	0.457	230.11	512.78	8.60	5
30–34	18	10.23	32.39	0.316	136.36	282.67	8.73	5
35–39	18	10.23	22.16	0.462	85.23	146.31	6.60	5
40–44	13	7.39	11.93	0.619	41.19	61.08	5.12	5
45–49	5	2.84	4.55	0.625	15.63	19.89	4.38	5
50+	3	1.70	1.70	1.000	4.26	4.26	2.50	5
Total	*176*							

Table 6.5 Life table for New Zealand: legend as for Table 6.4.

x	Dx	dx	lx	qx	Lx	Tx	ex	Age interval (years)
0	6	1.58	100.00	0.016	99.21	2711.21	27.11	1
−4	10	2.64	98.42	0.027	388.39	2612.01	26.54	4
−9	12	3.17	95.78	0.033	470.98	2223.61	23.22	5
10–14	10	2.64	92.61	0.028	456.46	1752.64	18.92	5
15–19	31	8.18	89.97	0.091	429.42	1296.17	14.41	5
20–24	75	19.79	81.79	0.242	359.50	866.75	10.60	5
25–29	91	24.01	62.01	0.387	250.00	507.26	8.18	5
30–34	62	16.36	37.99	0.431	149.08	257.26	6.77	5
35–39	54	14.25	21.64	0.659	72.56	108.18	5.00	5
40–44	19	5.01	7.39	0.679	24.41	35.62	4.82	5
45–49	5	1.32	2.37	0.556	8.58	11.21	4.72	5
50+	4	1.06	1.06	1.000	2.64	2.64	2.50	5
Total	*379*							

The problems of the people of Taumako are laid out later in the chapter, but it can be said that this was an island population devastated by yaws to an extent not recorded, to my knowledge, in any other skeletal series. That, in the confines of such a small and isolated place, makes it unlikely that this was a burgeoning population. The chances are that at best the population was stable, but just as probably it was in decline. If population growth was nil, then the life expectancy of adults was just that set out in Table 6.2 – under 29 years for women and 32 years for men. If, very reasonably from the gross skeletal pathology, a 0.5% annual decline is postulated, then, from Figure 6.3, the life expectancy of women was about 27 years and that of men about 29 years. It may not even have been as good as that – the continuing drain of pregnancy and lactation on a woman burdened with ulcerating and secondarily infected yaws lesions suggests an image that is not pleasant to contemplate.

A similar exercise may be attempted for the Hawaiian islands. A considered estimate of population at contact is 250,000, and settlement time according to the chronometric cleansing exercise of Spriggs and Anderson (1993) is about 950 years before 1770. The annual rate of population growth required to achieve that figure from a minimal 14 settlers is only 1.03%. This low rate is probably indicative of a high density of population having been reached at least a century earlier, with natural and social constraints on growth then having effect. The situation on Easter Island was probably similar. Spriggs and Anderson (1993) suggest settlement in the latter part of the first millennium A.D. If the founding population is taken as 14,800 years of settlement before European contact are presumed, and a population at contact of some 7000 (Kirch 1984), then a growth rate of only 0.78% is needed to achieve one from the other. This does support Kirch's suggestion that the population may actually have been considerably higher, and helps explain the degree of environmental degradation that did occur. The likelihood is of higher growth rates in the early years and much reduced rates as crowding occurred. In fact this exercise of estimating growth rates and then reinterpreting the skeletal mean age at death, such as attempted here for the Chathams and New Zealand, cannot be so clearly applied to most of the islands of the tropical Pacific. Smaller land areas and a longer period of settlement must mean that population dynamics have been more complex and are now less amenable to analysis. I would hazard the guess that on smaller islands periodic natural calamities and food shortages meant that populations were rarely static and that the mean age at death of any small skeletal series is unlikely often to be an accurate reflection of the actual life expectancy of the group. This may also often have been the situation in continental environments.

Age at menarche

Another perspective on the matter of age in relation to health and demography relates to reproduction. Pregnancy is said to leave its mark on the bony pelvis in the form of distinctive pits on the dorsum of the pubic bone adjacent to the symphysis and on the pre-auricular groove of the ilium (Houghton 1974, 1975b). (In passing, the term 'scar' seems to be creeping in to describe them. This is not correct. A scar results from the replacement of specialized soft tissues, particularly epithelium, by fibrous tissue, which is not the situation with pits in bone. Palaeo-pathology has enough interpretive problems without adding to them the misuse of standard terms.) The basis for the changes is the stimulation of osteoclastic erosion of the bone at the sites of attachment of the pelvic ligaments during pregnancy. The hormone relaxin produced by the corpus luteum of the ovary underlies this response and the associated softening of the ligaments.

There have been several studies of these features in skeletal and recent autopsy series, offering varying degrees of support for the thesis that they are indicative of past pregnancy. Along the road to interpret-ation there are hazards. Medical records may be far from accurate, something we found in our own small (n = 20) autopsy series. For example one person recorded as having had no children was confirmed by her family doctor as having had four. This was not an isolated example in case notes from a teaching hospital. A significant factor until very recent time has probably been the social stigma attached to preg-nancy outside marriage. There are other, more biological factors. Nowadays activity and therapy during pregnancy are very different, and more varied, than in times past. These factors may influence cellular activity at the ligamentous attachments.

As I have said, different studies have given differing results. Suchey *et al.* (1979) did find a statistically significant association between pubic pitting and past pregnancy in autopsy specimens from a modern popu-lation. From radiographic films, Spring *et al.* (1989) did not find an association for the pre-auricular groove, but a two-dimensional glimpse of a deeply-situated feature of variable orientation cannot be very sat-isfactory. Cox and Scott (1992) found no significant correlation between records of pregnancy and the bony imprint in a series from a cemetery at Spitalfields in London with associated burial records. The population was what I think the English would term upper middle-class – it con-tained, for example, an erstwhile Lord Mayor of London, and the ancestors of some still-prominent names. The environment was very different from that of prehistory. The people were well-nourished with the women probably letting their children out to the wet-nurse soon after birth. An appendix to the paper outlines how the crucial matter of reproductive history was established. Certainly it was not a simple matter of reading the details on a tombstone, and it is not detracting from the quality of the paper to suggest that the archival trail might have its false scents. By contrast, Igarashi (1992) has shown a significant association between these pelvic markings and recorded pregnancy in a recent Japanese series.

I think in all this the jury is still out. I remain biased to the view that in the active existence of prehistoric time a reproductive history is being displayed in these markings, and the discussion continues on this assumption. Anyway, not too much is being hung on it here. We have found in prehistoric material that about one in forty to one in fifty mature females fail to show pitting at the pre-auricular sulcus, which equates with a 'natural' rate of sterility (Short 1976). Pitting becomes deeper over the third and fourth decades of life, suggestive of the cumulative effect of pregnancies. In the few surviving women in their fifties the pits tend to shallow again, filling out after the cessation of childbearing.

If these marks on the female pelvis are an indicator of past pregnancy then they provide a further clue to the health of a group. Age of menarche and subsequent fertility are influenced by nutrition (some references are given below). The better the nutrition then the younger the age of menarche and fertility. If first evidence of pregnancy in a large group is at age 19, we may consider that nutrition and general health were probably not particularly good. Evidence of pregnancy at 16 suggests menarche at age 13 or 14 years and a rather good nutrition. Both for Taumako and for New Zealand, the youngest on whom I have seen evidence of pregnancy markings are aged 16 years: a pointer for these groups towards a good diet. In prehistoric Thai groups Wiriyaromp (1984) found evidence for menarche at 14–15 years in one individual from Ba Na Di, while Tayles (1992) placed the age at menarche for two individuals from Khok Phanom Di at 15–16 years.

The possibility arises that these purported imprints of pregnancy might also indicate the number of children borne. This is something that Ullrich (1975) has attempted in detail to quantify. Here I am sceptical. I think that the link with the likely number of children borne – which is of course a very important parameter – must derive from different, physiological sources. The natural reproductive pattern – natural in the sense that it has been the pattern for most of womankind for most of our time as a species – has been splendidly discussed by Short (1976), who has shown that in the light of biology neither the modern planned family of one or two children nor the Victorian achievements of 12 to 16 are normal.

The possible reproductive span of modern woman covers some thirty-five years, from about fifteen years through to about fifty. This last figure is optimistic, for it is clear that even in a contemporary well-nourished group, fertility starts to drop away quite quickly after about 30 years of age. From the earlier discussion on life span in prehistory it is evident that with truncation of life (mean life expectancy of adults) to, say, 40 years, ten years may immediately be lopped off the reproductive span of the past: however these are probably not significant years in terms of population success.

At the other end of the reproductive years there is good evidence from Europe that in the past the age of menarche, now 12–13 years, was later (Tanner 1973, Laslett 1985, Helm and Helm 1987), about 17 years for much of the population. This would still be followed by an average (and physiologically normal) two years of anovulatory cycles, suggesting that the first child would be borne when the woman was about 19 years old. That such a pattern may have been general is suggested by the finding of Howell (1979) that this also was the average age of first birth for the !Kung Bushwomen.

The reproductive life of modern women may thus be reduced 'in nature' to about 20 years. This might still allow a Victorian with a good wet-nurse to produce as many children. But probably not most prehistoric women or even women in recent traditional societies. Several factors intervene. First, there may be the continuing influence of marginal nutrition, which as well as delaying menarche has a rather depressing effect on fertility. Second is the post-partum amenorrhoea and anovulation induced by an unrelenting lactation in an environment when the foods for easy weaning were not always to be had. There is little doubt that suckling-induced anovulation has been the great contraceptive for the species for most of our time (Short 1976, Dobbing 1982, Galdikas and Wood 1990, Gray *et al.* 1990).

This combination of a sometimes marginal nutrition, intensive lactation and, in places, societal proscription on too-ready reproduction, suggests a spacing of 3 to 3.5 years between children – the sort of spacing observed in traditional groups both in Africa and in New Guinea (Howell 1979, Galdikas and Wood 1990). The mean value of about 34 years for age at birth of the last child established by Howell seems also reasonable for prehistory. About three-and-a-half years into a reproductive life of 20 years suggests that in many groups a woman would bear at most five or six children. Again, this is where seemingly trivial differences in life span may be crucial in determining the reproductive success and, thus, the survival of a group.

The situation in at least some parts of Remote Oceania may have been better. In Tahiti, Ducros and Ducros (1987) determined age at menarche nowadays to be at just under 13 years, and they cite evidence to suggest that it may have been rather earlier in the past. William Ellis, a missionary in Tahiti, commented that 'The age of puberty amongst Tahitians is about ten or eleven', which was greeted sceptically by his medical enquirer (Roberton 1832). However 'Data from different sources on food consumption in Tahiti show that Tahitians have had both a high caloric intake and a diet rich in proteins for a long time . . . and could . . . explain the relatively early age of menarche in Tahiti' (Ducros and Ducros 1987: 561). For Fiji, Thompson (1940) suggests that in traditional society puberty occurred as early as 10 or 11 years. It seems likely that reproductive potential in the more favoured parts of Remote Oceania in prehistory was that of the modern world. The Pitcairn study (Refshauge and Walshe 1981) supports that view. A selective push towards early fertility seems probable in a world of small groups settling small isolated islands.

These matters of life span and fertility lead on to the matter of small-group survival so well discussed by McArthur and her co-workers (McArthur *et al.* 1976, McArthur 1982). In a series of computer analyses, using a range of mortality and fertility schedules that from consideration of existing populations seemed realistic in the oceanic scene, they concluded that a group of 14 young adults, evenly balanced between the sexes, had a 78% chance of being ancestral to a viable population following settlement of an island. The criterion for survival was attainment of a population of 50 individuals, a figure mirrored in studies of survival of other animal groups and which might take many generations to achieve. Below 50 a population is vulnerable to natural disasters such as loss of a canoe at sea or the impact of a tsunami or a cyclone, or simply to chance imbalance in male and female births. Even after much analysis McArthur was not able fully to consider all variables that might have had influence, but another estimate was that a group of six young adults stood an almost 50% chance of survival. For two females and a male the chances dropped significantly; even with an (inadequate) allowance of only one post-partum year of infertility, after 250 years few such founding groups had achieved 50 descendants. In survival following small-group settlement of islands clearly there were many precarious early generations. If islands such as Hawaii or New Zealand were settled by small groups, then when the vagaries and fluctuations of small populations are considered a considerably longer settlement time than yet detected might be the reality. And for any island it is possible that earlier arrivals than any yet discerned might not have survived the first few precarious generations.

Evidence of health from assessment of physique

External parameters

Stature, mass and robusticity have been sufficiently discussed in Chapter 2. The dimensions of the Remote Oceanic people suggest a good environment and diet – the proof of the pudding is in the eater. The additional topic to be examined here is that of sexual dimorphism. The discussion of this complex matter is limited to dimorphism of stature.

In human populations, sexual dimorphism in stature is usually of the order of 6–8%: that is, within a group males are on average that much taller than females (Eveleth 1975, Gray and Wolfe 1980). Values for Europe and Africa have been presented by Stini (1982); for Europe the range is 109.0 to 105.7, and for Africa 108.8 to 104.6. As Stini notes, these differences, even at the top of the range, are really quite modest.

The basis of dimorphism is one of those unending debates in biology, which reflects its far-from-simple background, embracing both genetic and environmental factors, including possible social influences. There seem to be obvious selective reasons why males should be bigger, such as protection for the group and possibly greater success in some aspects of food-gathering. A sociobiological thesis (Alexander *et al.* 1979) suggests that male/male competition in polygynous mating systems leads to greater dimorphism than in monogamous systems, because the former exclude some male lines whereas the latter do not. However this view does not receive robust support (Gray and Wolfe 1980). The constraints on female size have been cogently discussed by Hamilton (1982). The lower limit to female size is set by the requirements for childbirth, where there is good evidence that smaller women are significantly more at risk than others (Niswander and Jackson 1974). Childbearing also places constraints on the upper limit. The extra nutritional demands of pregnancy are up to an extra 2500 kJ/day, and 5–50 extra grams of protein (FAO/WHO 1973, 1985). Lactation demands even more energy, at least a further 2000 kJ/day (Gopalan and Belavady 1961). As the state of lactation accounts for most of a woman's reproductive life span in non-technological societies, it represents a selective force that has operated during human evolution to constrain female size. If females started off with the body size, muscle mass and energy needs of males, it would in the past have been difficult to obtain that extra energy. 'As with lactation, the physical requirements of birth create selective forces that limit the extremes of phenotypic response to stress and favour larger women in the absence of dietary limitations. For each population the actual size that marks the balance between selective forces favouring large and small females will differ' (Hamilton 1982: 138–9).

On the environmental side, a copious literature suggests that chronic malnutrition or infection may cause individuals to fall short of their genetic potential for stature, and it is males who seem particularly sensitive to such pressures (Tanner 1962, Stini 1969, Garn *et al.* 1973, Gaulin and Boster 1992). These observations might seem to provide a basis for using dimorphism as a guide to nutrition in the past. As so often, things are lamentably less simple. Gray and Wolfe conclude that 'the degree of sexual dimorphism of stature cannot be used to gauge nutritional status of a society [as] the greatest and least degree of sexual dimorphism in height is found in those societies with high protein availability' (1980: 455).

It is necessary, then, to proceed with caution, looking at the extent of dimorphism but balancing it against the other indicators of health. Dimorphism in stature for some living Pacific groups is set out in Table 6.6. The peoples of Remote Oceania show a lesser dimorphism than the peoples of Near Oceania, even when the low figure for Pukapuka is disregarded. This is most unlikely to be on the basis of malnutrition. Rather, there may be here a profounder, older statement about survival, with the female being closer to the large male size because of the stringency of the Remote Oceanic climate – and able to be that size because of ample food. Conversely, the data on dimorphism do not support notions of growth impairment in Island Melanesia resulting either from malnutrition or endemic disease such as malaria.

This anthropometric evidence of reduced dimorphism is often in harmony with the historical record. In the Marquesas and in Hawaii the sexes are put together in the descriptions of physique. In Tonga 'the women are comparatively quite as sturdy of body and limbs as are the men' (Tasman 1968: 164), a comment echoed by Wales: 'The women

Table 6.6 Sexual dimorphism in stature in living Pacific groups. Marquesas data from Sullivan 1923, other sources as for Table 2.1.

Near Oceania	M/F ratio	Remote Oceania	M/F ratio
Baining	106.8	Ontong Java	104.8
Karkar	106.6	Fiji coastal	106.3
Nasioi	105.4	Tonga (Foa)	104.3
Manus	106.8	Samoa	107.3
Nagovisi	107.2	Marquesas	106.8
Baegu	109.5	Tokelau	106.6
Kwaio	106.6	Pukapuka	102.1
Ulawa	107.9	Hawaii	105.8
Aita	108.7		
Lau (Malaita)	107.1		
average	107.3	*average*	105.5
S.D.	1.19	*S.D.*	1.72

also are tall, well form'd' (1967: 808). A conflicting note is sounded by Cook in Tahiti: 'The superior women are in every respect as large as Europeans but the inferior sort are in general small' (1955: 123–124). However it is from New Zealand that there comes the recurring suggestion of women being of relatively lesser physique. 'The women are rather short and not nearly as well built as the men' (Montesson 1985: 243). 'The women of the country are generally small and fairly badly built' (Crozet 1985: 14). Unfortunately we have no anthropometric data for Maori women, and it also has to be remembered that these observations represent a very slight sampling of the country. There is the skeletal record for help. Calculated stature cannot be used, for in the Polynesian female stature equations (Houghton *et al.* 1975) a dimorphism of 7% was assumed. One approach is to look at male/female ratio for lengths of femur or tibia in various Pacific groups. Data are sparse as a series even of 40 individuals may provide only six bones from each sex for comparison. Dimorphism in tibial length in three large series, as a percentage, is: New Zealand 9.1, Mokapu 8.5, Taumako 7.7. So the suggestion remains that for one reason or another New Zealanders did display a rather large dimorphism in stature. Such a difference could be environmental or genetic or, of course and most likely, a combination of influences. On the genetic side, it is possible that by chance there happened to be a greater than usual dimorphism in the numerically small founding population(s). If so, then one might anticipate that a few centuries would see a more usual value assert itself. However on this Rogers and Mukherjee comment:

The additive genetic covariances between male and female length measurements are extremely high, suggesting that genes for such characters tend to affect males and females in the same way. The result is that the mean of the two sexes responds to selection many times faster than does sexual dimorphism. (1992: 243)

That is, a particular setting of dimorphism might be very tardy to change.

As any environmental influence for greater dimorphism in New Zealand is likely to be nutritional, there may be a hint here of a different and poorer diet for women in New Zealand's prehistoric past. If there is ethnographic evidence for this, I am not aware of it.

Internal skeletal parameters

Anthropometric or osteometric dimensions and standard indices provide an external statement about the body and its interface with the environment. While such data may be rather simply acquired, major conclusions may be drawn from them, size being one of the most significant characteristics of any animal (Calder 1984). It can be argued that basic anthropometry still provides the best assessment of nutrition (Himes 1991). To examine the internal fabric, more complex technology is required. The focus is structural, looking at the quality of the bone as framework and support. (The important isotopic and trace element studies on bone, directed at dietary interpretation, are not under discussion here, though of course diet may well ultimately be reflected in bone structure.) The discussion centres round the findings from x-ray studies because these have been most used. Techniques such as photon absorption or CT scanning have yet to be used to any extent on Pacific material, being newer, more costly, and less available, whereas basic radiological equipment may be available in very isolated places and almost as a field resource. Even so, radiological studies of Pacific archaeological material are still sparse.

From x-rays, general statements on bone structure may be made, particularly relating to the cancellous pattern at the ends of long bones. Within a group such changes are a useful ancillary method for relative ageing, though this is rather subjective. For quantification of bone structure the most ready measure is cortical thickness of the second metacarpal, for which a wide range of data is available, both because of its use as a clinical index of osteoporosis and in surveys of nutrition and health in human biology. The measures on the x-ray are whole shaft width, medullary cavity width, and the derived cortical thickness, all measured at the mid-point of the length of the bone (Figure 6.4). The use of a single (second) metacarpal has been shown to be practically as good as a whole hand array (Garn *et al.* 1991) and in large series there may be significant correlations of these measures with body mass and stature. However these are not particularly useful for small groups or individuals. Of more practical use is the standardization of the measures against the length of the bone, expressed as Nordin's Score (Barnett and Nordin 1961). One of the most helpful aspects of this approach is that results are comparable between present and past populations; on such a small bone any change in dimensions as a result of drying is much smaller than interobservational differences in measurement. Tubular (long) bones do show an expansion of the medullary cavity with ageing, but

Figure 6.4
Derivation of
Nordin's Score from
measurements on
the second
metacarpal.

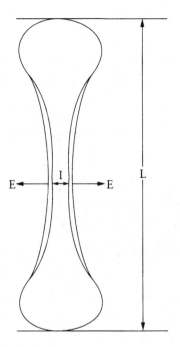

usually no significant change in the external dimensions of the bone. Nordin's Score thus decreases with age. To avoid this bias, and on the assumption that most of the prehistoric people are younger adults, the modern groups used for comparison are drawn wherever possible from the under-40s bracket. In measurement and calculation the metacarpal is regarded as being tubular or round in cross-section. In many prehistoric peoples this is not really so, the interosseii shaping the shafts; however the approximation is probably reasonable.

Nordin's Scores for a range of living and prehistoric groups, both within the Pacific and, for comparative purposes, beyond, are given in Table 6.7. The prehistoric data are limited, but using the recent Western data for well-nourished people as benchmark and thereby taking a score of 57 as the lower end of the optimum range, they suggest that bone structure/nutrition was particularly satisfactory in Afetna (Marianas) and Taumako, both small islands in the west. For New Zealand in prehistory the suggestion is that bone structure was a little deficient, particularly in females, which is an interesting finding in light of the historical observations on Maori female physique. Similarly the small protohistoric series from Tepoto in the northern Tuamotos (Dennison 1992) also show rather thin cortices though they were of tall stature. The suggestion is that size, the body's interface with the environment, is being preserved at the expense of internal structure. Of the outsiders introduced for comparison, the Khok Phanom Di population from Thailand show thick bones, but this finding is complicated by a probable high incidence of thalassaemia in this group, with expansion of bone because of increased activity in the marrow (Tayles 1992). The Ban Na

Table 6.7 Nordin's Score on the second metacarpal for prehistoric and living groups.

Group	Female				Male				Sources
	n	*age*	*mean*	*S.D.*	*n*	*age*	*mean*	*S.D.*	
N.Z. Maori	36	20–40	49.2	6.8	27	20–40	52.9	8.9	Houghton (unpub. data)
Taumako	19	20–40	61.6	–	31	20–40	59.7	–	Houghton (unpub. data)
Afetna	5	20–40	59.5	7.17	4	20–40	63.4	3.9	Roy and Tayles 1989
Tepoto	7	20–40	49.2	7.69	4	20–40	50.1	7.08	Dennison 1992
Khok Phanom Di	26	20–50	59.2	10.3	21	20–50	63.0	7.6	Tayles 1992
Ban Na Di	4	20–40	64.0		6	20–40	62.6		Wiriyaromp 1984
Indian Knoll	–	17–29	51.0	–	–	17–29	51.0	–	Cassidy 1984
		30–49	36.0	–		30–49	47.0	–	
British	–	20–40	63.0	–					Nordin n.d. (cited Garn 1970)
U.S. White	–	20–40	64.0	–	–	20–40	62.5	–	female, Saville *et al* 1976 male, Skrobak-Kaczynski and Anderson 1974
Norwegian					51	30–39	57.9	–	Skrobak-Kaczynski and Anderson 1974

Di people from northern Thailand show good bone structure without any stigmata of thalassaemia. The bone structure appears particularly poor in older females of the Indian Knoll population.

Another internal skeletal parameter of health is the occurrence of lines of arrested growth, radiologically-observed transverse lines across the shafts of long bones. Such a line in reality is a flimsy plate of cancellous bone stretching across the medullary cavity. (As with the purported pregnancy markings on the pelvis, the term 'scar' sometimes used for these features is inappropriate; a scar arises when a soft tissue is replaced by fibrous tissue.) The sheet is readily destroyed if the medullary cavity has been opened by post-mortem breakage or deterioration and the interior reamed out by earth or roots.

The theory is that as a result of an episode of malnutrition or illness, cellular activity in the epiphyseal cartilage (growth plate) is reduced for a while. The rather thickened, quiescent cartilage shows changes of ossification on its proximal surface. When the cartilage begins to grow freely again, the ossifying component is left behind, ultimately as fine cancellous bone. (The final sites of the growth plates near the extremities of the bones are often marked by an enduring, obvious, but non-pathological transverse line.) It is possible to estimate the age at which a line was formed from a knowledge of the mean length of the bone at different ages and the differing rates of growth of the two ends (Simpson 1979, Hummert and Van Gerven 1985). The 'growing end' of a long bone is the end at which the growth plate persists longest, and thus the end at which most growth of the bone occurs. In the upper limb the growing ends of the long bones enter into the shoulder and the wrist joints. By contrast, in the lower limb they are at the knee: the distal growth plate of the tibia contributes 40–45% of the total diaphyseal length of the bone, while the proximal growth plate contributes 55–60%. From these proportions, the age at which a line was formed may be calculated by measurement on the diaphysis.

During growth and maturation there is a lateral drift of the long bones as a consequence of the general widening of the pelvis and the whole body. That is, deposition of bone is occurring on the lateral aspect of the shaft with removal of bone along the medial aspect. The residual transverse lines often clearly show this, being absent in their lateral part where bone has been added since the time of their formation. Probably for the same reason the lines tend to show best – and are certainly most interpretable – in the rather straighter, simpler bones such as tibia and radius.

A considerable literature on transverse lines supports, on balance, the view that they are a result of an interruption in the continuous normal growth of the bone because of a moderate or severe illness, or episode of malnutrition (Park 1964, McHenry 1968, Gindhart 1969, Blanco *et al.* 1974). A one-to-one relationship cannot be determined – that would be too much to hope for. But, as Hummert and Van Gerven put it: 'transverse lines may be analytically useful as subsidiary criteria for more fully understanding the biological well-being of prehistoric populations' (1985: 297) – and that is their place in the present discussion.

Data on the incidence of lines of arrested growth in Pacific populations and some comparative data from other regions are given in Table 6.8. It is clear that the data are too scanty to allow much to be said. There is some support for the suggestion from the cortical thickness (Nordin's Score) data that the New Zealanders inclined towards nutritional problems sufficient to affect internal bone structure. On the other hand, the New Zealand series is much the largest and consisted of bone in particularly good condition; in archaeological material in poor condition the plates of bone across the medullary cavity are readily destroyed.

Dental evidence of general health

Enamel hypoplasia

The dental equivalent of the skeletal line of arrested growth is a developmental defect of the enamel. It differs in being externally visible and assessable. Illness or dietary deficiency during amelogenesis may lead to a recognizable enamel defect. The association of enamel hypoplasia with difficult periods during the biological development of an individual seems firmer than for lines of arrested growth (Rose, Condon and Goodman 1984). The canines often show hypoplasia most clearly, probably because of their length and thus extended development period. Scrutiny of both deciduous and permanent teeth assesses the developmental period from late intrauterine to adolescence. At its simplest and commonest, hypoplasia manifests as a transverse groove in the enamel. More grossly, small and large-scale pitting may occur with full-thickness loss of enamel. Gross lesions are sometimes hard to distinguish from caries. Some minimal data on enamel hypoplastic defects in the teeth of various Pacific groups are given in Table 6.9. The incidence ranges from near-ubiquitous in the people from Nebira in Papua New Guinea, who appear to have been afflicted with malaria and who show other evidence of a general failure to thrive, through to a low incidence in early New Zealand. The larger series show incidences of between 50 and 80%.

Review

In the end, what can be said in general terms of the health of prehistoric people in the Pacific? The short comment is, nothing definitive, because sampling is inadequate, methodology is inadequate and not uniform, and reports, like the remains, are often buried again in obscure places. At present New Zealand probably provides the clearest picture, with some regional statements being possible.

The age-at-death data (Table 6.2 above) provide a remarkably – or

Table 6.8. Incidence of lines of arrested growth in several prehistoric groups.

Group	Female				Male				Sources
	n	% with lines	mean	S.D.	n	% with lines	mean	S.D.	
Taumako	7	86	5.9	4.00	28	86	3.2	3.00	Houghton (unpub. data)
Khok Phanom Di	20	100	2.9	2.70	24	92	2.0	1.10	Tayles 1992
Ban Na Di	2	100	9.0	–	5	80	1.0	0.00	Houghton (unpub. data)
Dickson Mounds	43	63	1.2	1.20	65	74	1.5	1.30	Goodman and Clark 1981
Nubia	84	60	–	–	67	66	–	–	Hummert and Van Gerven 1985
NZ Maori	101	45	5.9	5.10	81	32	5.5	5.20	Simpson 1979
Regional NZ									Simpson 1979
northern North Island	29		6.1	5.51	15		5.9	4.89	
southern North Island	22		8.0	4.38	16		8.3	6.28	
northern South Island	8		5.9	5.64	13		5.4	5.95	
southern South Island	26		4.3	4.63	24		4.3	4.32	
Chatham Islands	5		5.0	4.53	5		4.8	4.32	

Table 6.9 Enamel hypoplasia in Pacific groups. The New Zealand group is late prehistoric, while Castlepoint and Wairau Bar are early New Zealand prehistoric (pre-A.D.1500).

Group	n	% affected	% male	% female
Tonga	64	53		
Taumako	140	67		
Sigatoka	18	52	67	30
Watom	2	50		
Nebira	34	97		
Taplins	6	33		
New Zealand	81	79	74	84
Chatham Islands	61	50		
Castlepoint	5	0	0	0
Wairau Bar	37	26		

suspiciously – uniform picture, and reference has already been made to scepticism over these data and to problems of interpretation. I suspect that seldom in studies of these prehistoric groups are we seeing the stable, stationary populations that analyses tend to assume, wherein the skeletal age at death equates with the life span. For New Zealand and the Chathams an adult (> ae 15 years) life span of 45–50 years is suggested – good for prehistory, and appropriate to a good environment. I suspect the Sigatoka series also follows this pattern. By contrast, for Taumako the skeletal age at death is likely at best to be a realistic statement of life expectancy, but may rather inflate it. Perhaps each group on each island (or any prehistoric group anywhere) needs to be considered against its whole environment and settlement history before judgement can be made on the likely direction and rate of population growth. Such considerations require rather a lot of quite sophisticated information, much of it from the fields of archaeology and prehistory, and I would not attempt them here.

The available skeletal series are largely from Remote Oceania or outliers in Near Oceania. The gross morphology of these series, with the estimates of stature and mass generally indicating tall and rather large people, implies an ample nutrition. However the studies of internal bone structure suggest that the important parameter of size, being the interface with the environment, has been achieved and maintained somewhat at the expense of internal quality: the bone inclines to the thin side. The rather high incidence of enamel hypoplasia supports this view. With some parameters it is likely that different workers are using different methods of assessment, and a move towards consistency and uniformity is necessary. Within New Zealand it is suggested that environmental pressures were rather less in the early prehistoric, that women were less well nourished than men, and that life just possibly might have been more favourable in the south.

Specific evidence of disease

The classification of disease and pathology used in this discussion is given in Table 6.10. Justification of this classification was given at the beginning of the chapter. It is of course not remotely comprehensive as set out, but it does have the potential to be so. One or two of the parameters assessed might reasonably be placed in the previous 'general indicators of health' category (Table 6.1).

Congenital diseases

These are discussed more fully in Chapter 7. Congenital diseases or congenital lesions are defined as those present at birth as a result of an inherited genetic defect or of something going awry during the developmental process, perhaps as a result of an environmental agent. From the modern world thalidomide is an example of such an agent. Congenital defects range from the innocuous, such as an extra finger or toe, through to those barely compatible with life, such as a major heart defect.

Congenital lesions from the past are to be expected simply because they are in the range of possibilities along the developmental way. The evidence is obviously limited to the skeletal record and further limited by the relatively poor preservation of young bones. The nature of Pacific settlement, with small founder populations, favoured distinctive incidences of congenital lesions in Pacific populations, and some contemporary figures are given in the next chapter. From the past, examples of club foot have been recorded from prehistoric burials in Hawaii (Snow 1974), but not to my knowledge elsewhere. This may simply be from a lack of attention, for foot bones tend to get rather short shrift in skeletal assessment. Cleft lip is a soft tissue problem so the record is lost. While there are scattered reports of cleft palate in adults from prehistory, a severe form of the condition was generally lethal because an affected infant has problems suckling. Spina bifida is a lesion arising from failure of the spinal cord to completely separate and close off from the skin surface – the two tissues sharing a common developmental origin. There are all grades of spina bifida, from the gross, where the tissue of the spinal cord opens widely on to the skin of the lower back, to innocuous deep dimples in the same region. The skeletal manifestation of a minor degree of spina bifida is a defect in the posterior arch of (usually) lower vertebrae. This may mean that the neural arches of the vertebrae are open; or there may be variable completion of the posterior sacral arch, generally called sacral hiatus. Minor degrees of this are common, to the extent that it is hard to say what is normal morphology, but a complete sacral hiatus may be regarded as a defect and there may be a familial incidence. Though the genetic basis is still unknown it is one of

Table 6.10 A working classification for skeletal pathology and disease. The conditions listed are given as examples of the various categories, and mention of a condition does not necessarily mean that it was present in Oceania in the past.

1. Congenital	club foot cleft palate spina bifida cleft lip, with or without cleft palate torticollis congenital dislocation of the hip	
2. Traumatic	fractures dislocations	
3. Inflammatory	acute infectious disease	zoonoses
	chronic infectious diseases	yaws leprosy tuberculosis
	local infections	osteomyelitis periostitis septic arthritis urinary infection
4. Parasitic	malaria filariasis intestinal worms	
5. Vascular	atheroma cribra orbitalia and spongy hyperostosis	
6. Neoplastic	benign	auditory exostoses
	malignant	primary secondary
7. Degenerative	osteoarthritis ankylosing spondylitis spinal degeneration spondylolisthesis	
8. Iatrogenic	trephination headshaping	
	dietary	kava amyotrophic lateral sclerosis
9. Dental	attrition evidence of dental abscess alveolar erosion caries mandibular joint degeneration	
	tooth modification	industrial social

the features that might be of some use in suggesting familial lines in burial sites (Bennett 1972).

From prehistoric Hawaii come craniofacial asymmetries that have been ascribed to congenital torticollis or wryneck (Douglas 1991). This obscure condition appears to arise from a gradual contracture of one of the sternomastoid muscles in the neck. The head is gradually tilted obliquely, with compensatory remodelling of facial structure in an attempt to preserve particularly the visual axes. The sternomastoid lesion is believed to arise in utero during the development of the muscle – perhaps because of interference with its blood supply – with the condition only manifesting as fibrosis progresses in the muscle during the first few years of life. I do not know of reports from elsewhere in the Pacific; again, it may be missed through lack of awareness.

Trauma

This means injury, and under this heading come fractures and dislocations. A vigorous physical life might seem likely to lead to such injuries, but they were relatively far less common in prehistory than in mechanized modern times, to the extent that in the general pattern of health of a group they were insignificant – evolutionary selection has ensured that the skeletal framework of the body is pretty robust.

Archaeological remains are of course often much broken, and there is a need to try to distinguish fractures occurring during life from breaks occurring after death. Generally, multiple transverse breaks in bone can be regarded as having occurred after death. With loss of organic content the bone becomes brittle and may simply snap across from the weight of the earth or from surface pressure. By contrast, spiral fractures are suggestive of the torsional forces inherent in locomotion and of bone with a high organic content, and are much more likely to have occurred in life. The classical 'greenstick' fractures of childhood are unlikely to be recognized, unless death occurred within a few weeks of the injury, as remodelling is usually near-perfect. This means that the epidemiological pattern of peaks of fracture incidence in childhood and in old age are not likely to show up in the record of the past as there were not too many old people around.

Some classical patterns of fracture have their prehistoric representation. The fragments of an untreated fracture of the shaft of the femur angulate and overlap because of the pull of the powerful thigh muscles. Healing in this position may be sound but inevitably is accompanied by shortening of the leg by several centimetres so that the person would thereafter walk with a limp. While some traditional splinting of forearm fractures is quite successful, the thigh is too massively muscular for splinting to work. By contrast, lighter bones, such as the ulna with a classical 'parry' fracture, might heal in more satisfactory alignment. All these fractures seem to be uncommon in Pacific groups from the past, certainly too few to allow any pattern of occurrence to be discerned within a group. Notions of mass carnage from warfare are not supported.

Until recently the consequences of a fracture hinged on whether it was simple or compound. A simple fracture is one in which the overlying skin surface is unbreached. In a compound fracture the skin surface is broken, exposing the fracture to infection. Until the antibiotic era the difference in prognosis between these two types was absolute. Simple fractures had a good chance of firm union, albeit with overlap and shortening of a bone in many situations. Lighter bones surrounded by lighter musculature, such as the ulna or fibula, might heal in alignment. Compound fractures, by contrast, have been in the past either a death sentence or a condemnation to living with a suppurating wound with all its influences on health. A compound fracture might be identifiable if death was delayed more than a few weeks. This would allow time for resorption of the sharp bone edges (a finding of much significance in assessing whether a lesion occurred before or after death) and also, with the open wound, changes indicative of bone infection.

Similar statements apply to dislocations. I do not know what ancient expertise might have been applied to reducing dislocations of shallower major joints, such as the shoulder, but it seems unlikely that a dislocated hip – which would anyway have been a rare occurrence – could have been reduced. I have seen one unreduced hip dislocation, where the distorted head of the femur has shaped for itself a new and crude socket on the hip bone (Houghton 1980a: 139). This was a lesion of many years' standing, and could possibly be an example of a dislocation present at birth – congenital dislocation of the hip.

Inflammatory diseases

While inflammation may occur without infection, bacteria or viruses are usually involved. Included under this heading are various infectious diseases, acute and chronic, and a few other conditions. The terms acute and chronic do not refer to severity but signify, respectively, illness of rather sudden onset and usually short duration, against a gradual onset and long duration.

Acute infectious diseases

These are the common infections, nowadays usually of childhood, and of either viral or bacterial origin. Mumps, measles, rubella and chickenpox are common examples. Despite the fact that they are unlikely to leave any impact on the skeleton – unless it be a non-specific line of arrested growth – it can be said with some confidence these problems did not occur in prehistory in the Pacific. They probably have evolved only in the past six thousand years or so with the advent of agriculture and the development of reasonably large aggregations of people. These diseases establish a resistance in the host which tends to be life-long and they thus require largish populations through which they can continue to move to encounter susceptible individuals. They are sometimes called 'crowd diseases'. The requisite minimal populations seem to be quite large, in the order of 50,000 people. In Pacific prehistory such groups hardly existed. For much of prehistoric time the larger islands of Near Oceania seem to have effectively been smaller islands, enclaves of humanity in relatively little contact with neighbours, separated by what has been termed the 'rough quarantine of war'. In Remote Oceania Hawaii is the only island group in which there was in the later prehistoric period a sufficiently large (250,000+) population, relatively mobile within a limited area, to allow evolution of crowd diseases: however, settlement time of perhaps a thousand years and a dense population only towards the time of European contact are insufficient. In other parts of the world the presence of large numbers of other animals, particularly primates, might have provided an evolutionary reservoir, but these are in short supply in Oceania.

One infectious disease presence of which in the Pacific past can be conjectured is scrub typhus, one of the zoonoses. These are infections that are endemic in some animal species but may be transmitted to humans. Typhus is caused by *Rickettsia*, an unusual group that are obligate intracellular bacteria, that is, they cannot live outside host cells. Scrub typhus (tsutsugamushi fever) caused by *Rickettsia tsutsugamushi* is primarily a disease of mites, with rats forming a secondary reservoir. (The related louse-spread strain of typhus, caused by *R. prowazeki*, probably did more damage to Napoleon's Grand Army in Russia than cold and the Cossacks combined.) Clinically there is ulceration at the site of the mite bite, fever and rash and generally widespread symptoms because the linings of the blood vessels through the body are affected. From a long acquaintance as a microbiologist with infectious disease in the Pacific, Miles (in press) has suggested the presence of scrub typhus in the Pacific in prehistoric time. He considers that the illness that afflicted Mendana's crew during their stay in Ndende in the Santa Cruz in 1595 sounds more like scrub typhus than malaria. A journal comment (Quiros 1904: 49) specifically mentions an unusual absence of mosquitos, and only those ashore sickened and died while those who kept to the ship stayed healthy. Miles observes that the ideal camp site, such as the bay chosen by Mendana, is often also the perfect site for an endemic focus of scrub typhus. In 1975 he found this site to be very heavily infected with the local strain of scrub typhus *rickettsiae*. This has been noted as particularly virulent for foreigners, whereas the local people are infected in childhood, when most rickettsial diseases are far less severe than in adults.

Chronic infectious diseases

Two important disease pairs, each with a long relationship with humans, are first considered. These are yaws and syphilis, and tuberculosis and leprosy, They may be paired, because within each pair the causative organisms are closely related.

Yaws and syphilis belong to the treponematoses, diseases caused by related spirochaetes – corkscrew-shaped bacilli. The other two treponematoses, pinta and bejel, are lesser skin diseases, of drier climates and are not Pacific problems. Hudson (1963) suggested that all four conditions were caused by the same organism, and that the differing expressions of disease were due to environmental and cultural factors influencing the mode of infection. DNA analyses support the closeness of the organisms, suggesting two subspecies: *Treponema pallidum pallidum*, causing syphilis and bejel, and *Treponema pallidum pertenue* causing yaws. If there is any distinction, yaws may be the most ancient of the group. It is a disease of warm, moist climates. The spirochaete is spread by contact and perhaps by insects, and enters open skin lesions. These become chronically infected and ulcerated, and may gradually deepen to involve the underlying soft tissues, such as muscle. The soft tissue involvement comprises the primary and seondary phases of the disease. The tertiary stage involves bone; any long bone may be affected but commonly the tibia is first, probably because the shin is a common site of abrasion. Yaws – I assume a subtle distinction of organisms – is not a venereal disease. The origin of syphilis is postulated to date from movement of people to colder climates, where infection was maintained on the warm, moist internal mucosal surfaces of the body. The tertiary stage of syphilis, unlike that of yaws, involves the nervous system, which must speak for some distinction in the organisms. A New World origin for the treponematoses has been claimed (Baker and Armelagos 1988), but the dating of the Pacific record is against it.

The occurrence of the treponematoses in the Pacific has been discussed by Pirie, who comments: 'Yaws is a common and ancient disease in Melanesia, Micronesia and Polynesia, and until the modern and widespread application of antibiotics occurred in all favourable environments' (1971: 189). The spirochaete does have rather fine environmental limits: a mean temperature through the year of 18°C, with an annual rainfall of some 1600mm, and no monthly rainfall of below 65mm – that is, no unduly dry months. This makes occurrence of the disease unlikely in the Tuamotus, the Marquesas, and much of Kiribas, while Hawaii and New Zealand are outside the range (Figure 6.5). Pirie does argue for the presence of yaws on the wetter windward Kona coast in the Hawaiian group on the basis of the varying impact of introduced syphilis. I am sceptical of this – considering the small gene pool and the continuing movement of people within the Hawaiian group, it seems unlikely that one group would possess immunity and another, over the hill, so to speak, vulnerability to the spirochaete, or that the environment over the hill would prove such a barrier. Altitude as well as latitude will influence its distribution, and through Melanesia it is not endemic above about 500 metres.

In the contact record, Cook's observations at Tonga are typical:

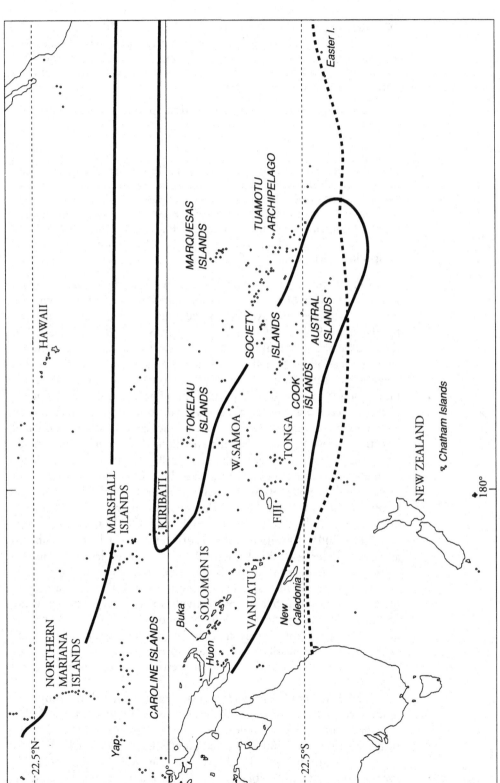

Figure 6.5 Environmental constraints on yaws in the Pacific. The continuous line is the rainfall limit (1650 mm annual rainfall isohyet) and the interrupted line is the temperature limit (minimum winter month mean temperature 18°C). Adapted from Pirie 1971.

We observed several who had ulcers upon different parts of their bodies, some of which had a very virulent appearance, particularly those in the face which were shoking [sic] to look at; on the other hand some seemed to be cured and others in a fair way, but this was not effected without the loss of the nose or the best part of it. (Cook 1967: 170–171)

Later medical workers gave vivid descriptions of the ravages of the disease:

At Boera I got my first real look at a yaws-stricken community. This hideous thing was apparent on the bodies and faces of at least a third of the people, men and women with noses reduced to yawning holes in the middle of a flat scar. Fingers and toes curled like withering twigs. Swarms of flies carried the filth-born germ. I looked into baby faces and saw how the process of healing had drawn their lips together into a featureless surface with an opening so small that you could hardly get a lead pencil through . . . I have since seen many other villages like Boera. (Lambert 1941: 30)

One therefore expects evidence of yaws from the prehistoric tropical Pacific. The bony manifestations of yaws and the differential diagnosis with syphilis have been meticulously discussed and illustrated by Hackett (1976) and Ortner, Tuross and Stix (1992) have added to these observations and distinctions. Periosteal reaction to the common tropical ulcer of the shin also needs to be considered in the differential diagnosis. Despite the likelihood of yaws in the Pacific past, evidence was rather slow in coming, and initially not florid. Stewart and Spoehr (1952) described a 13-year-old individual, from a pre-contact site on Tinian in the Marianas, who showed convincing evidence of treponemal infection of several long bones and the skull. Rather slight tibial manifestations of yaws in Tonga about A.D. 1500, and more florid evidence from about the same era from Eriama in New Guinea have been noted (Pietrusewsky 1969, 1976). However it is in Taumako in the Banks Group, north-eastern outliers of the Solomon Islands, that Lambert's observations (1941) find a reflection in the past. Of 114 individuals from this A.D. 1200–1400 site for whom sufficient material was available to assess the presence of the disease, 49 (43%) showed lesions. This is a conservative figure; of individuals with assessable tibiae, the bone most commonly affected in yaws, 93% showed lesions. This is a far higher incidence than the 1–5% for bony involvement often proposed in the literature (Wilson and Mathis 1930, Steinbock 1976), though Ortner *et al.* (1992) suggest bony involvement in 10–20% of cases. I suppose there exists the remote possibility that this burial mound was reserved for sufferers from yaws, but that is my suggestion, not that of the archaeologists.

The Taumako lesions were assessed against age using a subjective grading from 1 (no lesions) to 4 (gross expansion and distortion of at least one bone). The incidence and severity increased with age: in the under-25-year group 30.8% showed bony lesions, whereas 61.5% of the 35-plus age group showed lesions. The youngest affected was a 13-year-old boy with early tibial lesions (the remains of younger children were too minimal to allow assessment). Cranial involvement, in 18% of those assessable, was higher than noted in other studies.

The impact on human existence on this island must have been devastating, and earlier in the chapter the demographic consequences were suggested. If, at a very conservative estimate, some 40% had skeletal involvement, it is a reasonable assumption that most adults had some soft tissue involvement. Food resources and diet on Taumako seem to have been good, with establishment of a strong bone structure and tall stature. But every child would have seen its fate in the adults around, and secondary infection in the soft tissue lesions must have been a common source of debility and death. How often this picture was repeated in isolated communities in Oceania in prehistory we do not yet know, but the incidence of yaws in this group is unlikely to have been unique.

No such florid evidence is known for the venereal spirochaete. The general view, with which I concur, is that syphilis was not present in the Pacific in prehistory. Amongst the eighteenth-century Maori Newman noted a remarkable absence of syphilis, by contrast with the ravages of gonorrhoea:

Many observers not trained in medicine talk about the frightful effects of the 'awful scourge' syphilis, and say that the Maori population is saturated with it, and that its fearful effects are seen in the sterility of the race and the astonishing mortality existing among the children . . . Several doctors who practice largely among the Maoris assure me that they never saw true syphilis in a Maori. My own experience is that amongst the large numbers of Maoris I have seen I have never been able to detect any evils from this cause . . . I have never seen Maori children with any marks of syphilis. (1881: 469)

Perhaps this may be ascribed to a lingering natural immunity, a legacy of exposure to the related yaws in earlier time in the tropical Pacific. Occasional claims have been made for the presence of syphilis in pre-contact Oceania. However if syphilis were present in confined island populations then one would expect pervasive evidence of it, including the distinctive congenital lesions such as Hutchinson's teeth, and as far as I know these have not yet appeared.

Leprosy and tuberculosis are diseases caused by closely-related rod-shaped bacilli. There is quite a body of writing claiming the presence of leprosy in New Zealand in prehistory, but not in other parts of Remote Oceania. For New Zealand, Buck (1950) even records the names of the canoes said to have brought the disease to the country. Various terms purporting to describe it are noted by Taylor (1855) and Williams (1972). Best wrote that the Maori

assuredly had a form of leprosy, termed *ngerengere*, *tuhawaiki* and *tuwhenua*, in olden times; it was introduced from Polynesia by a canoe since known as *te waka tuwhenua* – the leprosy vessel. This disease was spoken of as a malignant atua . . . The distressing *ngerengere* caused the extremities to drop off joint by joint. (1924: 49)

Despite Best's confidence, there is no material evidence for leprosy in prehistoric New Zealand. Montgomerie (1988) has scrutinized the

New Zealand claims and found them wanting. The picture is confused by the nineteenth-century situation wherein leprosy was brought to New Zealand by trading ships, probably from Asia where it has an ancient history; there is bony evidence of leprosy amongst Chinese goldminers of the 1860s in Otago. Various skin diseases may mimic some of the features of leprosy, and nineteenth-century European medicine, without bacteriological theory, was not itself sophisticated enough to differentiate. Apart from that, leprosy leaves stigmata of its later stages on the facial bones and on hands and feet (Møller-Christensen 1967) and to my knowledge nothing resembling these changes has yet been observed in prehistoric remains from New Zealand or anywhere in Remote Oceania.

A similar caveat applies to reports of tuberculosis in the Pacific in prehistory. In a meticulous report Suzuki (1985) suggests the existence of a pelvic tubercular lesion in a pre-contact Hawaiian. Yet this lesion is not enough. It is not that hallmark of skeletal tuberculosis, Potts disease of the spine, in which an intervertebral disc and adjacent vertebral bodies have collapsed from the infection, with a resulting kyphosis or angular deformity. The back is probably a site of predilection because of load-bearing and exposure to repeated minor trauma. If tuberculosis were present in a closed island population one would expect occasional typical examples of such a lesion. To my knowledge not a single compelling case has yet been described in prehistoric skeletal material from the Pacific.

Other infections

The elimination of so many pathogens from the prehistoric health spectrum should not obscure the fact that there were still plenty around. We are all replete with organisms that usually are properly confined and restrained and useful within their sphere, such as in gut or throat or on the skin, but which may cause problems if they move beyond their limits as a result of a lowering of the body's usual checks and balances. Examples of such Trojan Horse infections range the length of the body to include varieties of pneumonia, urinary infection, meningitis and middle ear disease. The wonder is not so much that such infections occur but rather that they do not occur even more often than they do. Infection, for example, may occur along natural body channels. Bladder and kidney infections are a common complication of pregnancy, probably because the expanding womb obstructs urinary outflow, with stagnation and infection following. From New Zealand comes an example of a large staghorn stone of the kidney typical of that occurring secondary to a urinary infection (Houghton 1975). Chemical analysis confirmed the diagnosis of kidney stone. This was a woman in her mid-twenties. She probably died as a result of the infection and may well have been pregnant – she had previously had children, if the pitting on the pelvis is to be believed. An example such as this, while verging on the isolated oddity category that I do not wish to spend time on, does illustrate the potential of the body's own bacteria (this time of the skin) to cause problems.

Any persisting soft-tissue infection, particularly if deep-seated, may lead to inflammatory involvement of the periosteal surface of the underlying bone, and development of varying degrees of bony reaction. The evidence of this is a reactive roughening and overgrowth of the bone surface. By definition periostitis is superficial and an x-ray will show the underlying bone to be essentially normal. We have an example from the Lapita site of Watom Island in New Britain (Houghton 1989); the surface of the upper end of the shaft of the right femur shows an inflammatory reaction, but the x-ray shows that the deeper cortex was not involved.

Spread of infection more deeply will lead to osteomyelititis. If florid and of long standing this may develop all the classical changes of cloacae, involucrum and sequestrum. The infection leads to the blood supply to a segment of bone being cut off, and this segment (the sequestrum) dies. A cloak of new, rough, superficial bone, the involucrum, is laid down around the sequestrum. Chronic infection in the sequestrum drains to the surface through holes (cloacae) in the involucrum. An example from New Zealand is illustrated in Houghton (1983). In this individual the infection spread through the bloodstream from a compound fracture of the femur, to involve many bones. From the extent of the reaction, death probably occurred about a year after the injury. Osteomyelitis in prehistory was likely eventually to be fatal, though months might supervene between injury and death. By contrast, a periostitis might grumble on for years, neither healing nor causing death.

Parasitic infections

Any organism living on another and doing some harm to it is a parasite, and in that sense bacteria and viruses may be parasites. However the term is usually applied to more substantial organisms, from protozoa to grosser things such as gut worms.

Malaria

The distribution of malaria in the Pacific has been mentioned in Chapter 3 in relation to influences on body growth and development; in Chapter 5 where blood polymorphisms linked with degrees of immunity to the parasite were discussed; and it crops up again in the next chapter. Here there is only a passing comment on the evolution of the disease in Oceania, something on which surprisingly little has been written. Miles (in press) suggests the slow spread of the parasite and its vector, *Anopheles farauti*, through Melanesia as far as the Solomons in prehistoric time. Larvae carried in canoes could only become infected if, when mature, they bit an infected human. It is quite possible that mosquitos had not reached Ndende in the Santa Cruz at the time Europeans arrived. Further west, at Guadalcanal, the illness that afflicted Mendana's expedition in 1568 (Amherst 1901: 358 'a tertian fever') does sound like malaria. East of the Santa Cruz the expedition of Quiros in 1606 stayed nearly two months at Espiritu Santo in Vanuatu, with no illness recorded amongst the crew – an unlikely situation had malaria been present. However at the Santa Cruz in the late 1820s, fever amongst the men of D'Urville's expedition responded to quinine, suggesting that malaria had arrived. This scant recorded pattern in historic time does support Miles' thesis.

Malaria in Melanesia now extends out through the direct chain of major islands and leaps the Near/Remote oceanic boundary to achieve a relatively minor presence in Vanuatu. East of Vanuatu it does not occur, and Fiji and New Caledonia are free of the parasite. There has probably not been significant change in this picture in historic time, although some outliers in Near Oceania such as Ontong Java have been invaded. In New Guinea changing patterns of human settlement and movement have blurred the earlier distinction between Highland populations, largely free of the disease, and lowland populations in whom it was endemic but who showed degrees of natural resistance. There has also been an apparent change from a former dominance of *Plasmodium vivax* to a present ascendancy of *P. falciparum*, again probably due to changing patterns of human congregation and movement, and a wide use of antimalarial drugs (Cattani 1992). Several of the blood polymorphisms said to offer a degree of resistance to the parasite are distinctive in Oceania: the β-thalassaemia is different from that prevalent in South-East Asia while the α-thalassaemia variants also appear to have arisen locally. Neither the sickle-cell trait nor the absence of the Duffy blood groups that elsewhere offer protection to *P. falciparum* are present in Oceania, but ovalocytosis, wherein the red cell membrane appears to be rather inimical to the parasite, is common.

The suggestion arising from these distinctions is that people have lived with the parasite in Oceania for a long time, evolving their own patterns of polymorphisms particularly directed at *P. vivax* resistance. This version of the parasite appears to be of moderate antiquity and is most closely related to the simian *Plasmodian* species. *P. falciparum* appears to be of quite recent evolutionary origin, perhaps only since the origins of agriculture, and derives from avian *Plasmodium* species (Waters, Higgins and McCutchan 1991). Its entry into Oceania may be rather recent. Whatever the antiquity, populations in general through Near Oceania seem reasonably adapted to the parasite and in an earlier chapter it was suggested that variation in physique between groups could not be ascribed to presence or absence of malaria. Interrelationships of host with malaria and other parasites are complex. Prophylaxis against malaria seems to have little effect on the pattern of child growth (McGregor *et al.* 1968), while roundworm infection (ascariasis), malnutrition, and iron-deficiency anaemia all seem, in a perverse way, to offer some protection against the more severe manifestations of malaria (Hendrikse 1987).

Filariasis or elephantiasis

This is caused by the threadlike tissue nematode (roundworm) *Wuchereria bancrofti*, whose habitat is the lymphatic system. The human body is the only host. The mature female worm, which is about twice the size of the male, is about 100 mm long and about 0.25 mm across. The adults live for many years in the lymphatic channels and lymph nodes of the parasitized individual. The fertilized females discharge microfilaria, which are about 0.25 mm long and less than 0.01 mm across, via the lymphatics into the bloodstream. Here they are ingested by the vector, which through most of Oceania is one of the various *Aedes* subspecies of mosquito. Within the mosquito the microfilaria develop over about two weeks into infective larvae and the infection is passed on when the infected mosquito bites another person. Only extended exposure seems to lead to the complications of the disease. These are, most notably, the obstruction of lymphatic drainage of a region, resulting in an often massive swelling of a limb or of the scrotum. Some of the more memorable illustrations from older medical texts are of afflicted people trundling their scrotum along in a wheelbarrow.

A detailed review of the history and probable prehistory of the condition in Polynesia has been given by Laurence (1991). The historical reports begin with Tasman who observed a man with 'St Thomas's arm' near Tongatapu. During Cook's voyages quite detailed observations of elephantiasis were made by J. R. Forster, by Anderson, surgeon on the third voyage, and various others. Curiously, neither the observant and usually scrupulous Cook, nor Joseph Banks, recorded the rather obvious condition, beyond a single mention by Cook at New Caledonia in 1774. Forster's *Resolution* diary notes 'elephant legs' at the Society Islands, Tonga, and New Caledonia, 'men with one or both legs enlarged to a monstrous size' (1778: 487). James King, Second Officer on the third voyage, observed at Tonga 'several people with their Arms and legs swelled the whole length to a prodigious size' (1967: 1366). Despite their deformities the sufferers were generally noted to be active. These and later observations indicate the presence of filariasis from the Tuamotus in the east, to Fiji, and west to New Caledonia. Northward the condition was observed in Kiribas but not in Tuvalu. Hawaii is beyond the range of the mosquitos, as is New Zealand in the far south. The Marquesas also were free at time of European contact, but like Tuvalu and possibly the northern Cook Islands of Manihiki and Penrhyn, they later became parasitized, probably as a result of barrels of drinking water infected with mosquito larvae being brought ashore.

Several of the early observers indicated that the condition was common. Early this century the incidence of filariasis found during medical surveys ranged from 8 to 50% amongst groups in Tahiti, up to

30% in Kiribas, and 6–9% in Samoa, Fiji and the Tokelaus. Before 1950, medical surveys revealed microfilaria in the blood of between 15 and 50% of the populations of the major island groups. There is little reason to suppose that these figures do not reflect the incidence in prehistory – rather, some observers suggested that the incidence abated in the latter part of the nineteenth century. Clearly, filariasis and its attendant problems were significant diseases in the Pacific past. It is something to be alert for in the skeletal record – one suspects that the underlying limb bones would be thickened and hypertrophic.

Gut parasites

Hookworm and roundworm (ascariasis) are common health problems in the tropical Pacific today and presumably were so in the past. Demonstration of this ultimately rests with the extraction from human faeces (coprolites) found in archaeological sites of the characteristic egg cysts of the creatures; but the tropical Pacific environment is unfavourable for coprolite preservation. Hookworm is the greater health problem because of the severe blood loss to the parasite, whereas ascariasis is something that by and large one can live with. New Zealand is too cold to support the hookworm life cycle but ascariasis has been demonstrated in a coprolite from the Wellington region, dated to the protohistoric period but probably valid for prehistory (Andrews 1979). The rather particular Maori attitude towards human excrement and the siting of latrines meant that spread of intestinal parasites would have been discouraged.

Vascular disease

A couple of rather disparate items fall into this category.

Atheroma

Atheroma, hardening of the arteries, is a natural consequence of ageing. The arteriosclerotic plaques have a lot of calcium in them so may survive in the ground. I have seen remains with a clearly-defined atheromatous plaque in the position of the posterior wall of the abdominal aorta, in front of the lumbar spine. This was in a male aged 40–50 years. A microscopic preparation of this specimen was viewed with lack of interest by a meeting of pathologists as just another case of atheroma, until they learned it was 400 years old. Some items of the prehistoric New Zealand diet, such as eel and crayfish (lobster) and some shellfish, have a high cholesterol content (Coster, Hunter and Jackson 1975), though admittedly this is a naive way of looking at the causation of atheroma. Pietrusewsky (1969) has reported atheroma from the 'Ate site in Tonga, about A.D. 1500, in this instance seemingly along the course of some limb arteries, and again in a mature male. The usefulness of these examples is to show that degenerative vascular change is not unique to modern Western populations.

Cribra orbitalia and porotic hyperostosis

Cribra orbitalia – a good descriptive term – refers to a porosity of the roof of the orbit. Porotic hyperostosis is a general thickening of the diploe of the cranial vault and, importantly, a loss of the normal tabular bone structure. This comprises a layer of spongy or cancellous bone, the diploe, sandwiched between inner and outer tables of compact bone. Such a sandwich construction is strong, and effective in dissipating the energy of blows to the head. In porotic hyperostosis the tabular structure disappears so that the entire thickness of the vault appears of cancellous bone. Sometimes porosity akin to that seen in cribra orbitalia appears on the surface of the vault. The emphasis on the loss of normal structure is important, as the normal range of vault thickness is large. All in all, determination as to whether porotic hyperostosis is present is still on the subjective side. Radiology is useful in determining the existence of the condition, but at present we do not have enough comparative material. Although the two lesions, cribra orbitalia and porotic hyperostosis, are believed to have the same aetiology they do not consistently occur together (rather as with lines of arrested growth and enamel hypoplasia).

These conditions are regarded as bony manifestations of anaemia, and as such might be put in the category of general indicators of health, discussed earlier. The best recent analysis of these conditions is by Stuart-Macadam (1987a, b). The basis of the conditions is hypertrophy of the marrow-containing layer of the skull bones when a high rate of red cell replacement is required. They are non-specific – that is, the particular cause of the anaemia cannot be ascertained from the lesions. However it may be suggested from other evidence.

The presence of cribra orbitalia or porotic hyperostosis, or both, in an adult is likely to be the residuum of a childhood situation (Stuart-Macadam 1985, Palkovich 1987). It may be that there is a normal, low incidence of the features in an altogether healthy young population, simply as reflection of actively growing young tissues. The juvenile incidence from various studies ranges from 22.8% to 83%. The low incidence of 22.8% in the juveniles of the Mokapu series from Hawaii (Zaino and Zaino 1975) suggests a relatively good environment for the growing child in prehistoric Hawaii. We do not have enough juvenile data from other parts of Oceania to be able to compare, but the adult incidence in New Zealand is low, 2.3%, compared with 9.9% for the Marianas and 12.5% for Hawaii (Suzuki 1987). One possible reason for the low New Zealand incidence is that the environment is unsuitable for the survival of hookworm, which is a significant cause of anaemia in tropical Pacific populations.

Neoplastic disease

In simple words, neoplastic change in a tissue indicates that some of the cells have decided to go their own way. The terms neoplasm and tumour are synonymous. The behaviour of neoplasms is sometimes described as 'autonomous', and while not really so the term does give an idea of the loss of normal cell controls and constraints. The fundamental division of neoplasms is into benign and malignant varieties.

Benign neoplasms are neither locally invasive nor found to spread distantly through the body. They may still cause problems by continuing growth and pressure, particularly within a confined space such as the cranium. As most benign tumours are of soft tissue there is really no information on them from the past. But it is safe to say they would have occurred – fibroids of the uterus, some growths of thyroid and ovary, and so on – much as today, as part of the fringe unruliness of cells. There are a few bony examples, though it may be argued that these are more hyperplasias than true tumours. Of anthropological interest in this category are the auditory exostoses.

These are rounded outgrowths of dense bone occurring in the auditory canal, the passage leading down to the eardrum. They may enlarge sufficiently to partly block off the canal and diminish hearing and promote infection. Nowadays they are commonly seen in surfies and others involved with water sports – that is, there seems to be an association between their occurrence and prolonged periods in the water. The anthropological interest is their occurrence in groups who are believed to have gained much of their food by swimming and diving. Kennedy (1986) has examined the incidence of auditory exostoses against latitude as an expression of water temperature. She found the highest incidences to occur in middle latitudes (30–45° N and S) among populations who exploit either marine or fresh-water resources. The inference is that while tropical groups may also use such resources, the exostoses are only stimulated to form by colder water. Beyond latitude 45° water is generally too cold for food gathering by prolonged swimming and diving. Frayer (1988) has provided supportive evidence from another European group, while for the Pacific the most comprehensive study is by Katayama (1988a). The Pacific data assembled in these studies are brought together in Table 6.11. The variation in incidence is large, from negligible in the Marianas to more than 20% in the Chatham Islands, Wairau Bar (northern South Island, New Zealand), and Tonga. For most Pacific groups the incidence is low, less than 10%. There are disparities in studies from the same islands, such as Hawaii (Hrdlicka 1935, Snow 1974, Pietrusewsky 1981) and New Zealand (Katayama 1988a). For Hawaii there is concern over the criteria used, as it seems unlikely that within such a relatively small island group the incidence would really vary from nil (Snow) to 24% (Hrdlicka). In New Zealand the localized Moriori and Wairau Bar groups with higher incidences each exploited a large lagoon with food in it accessible by diving. In the simulation studies the effective difference between the climate of Tonga

Table 6.11 Pacific incidence of auditory exostoses. (m) = male series.

Group	% incidence	n	Source
Wairau Bar (NZ)	35.7	30	Katayama 1988a
Chatham Islands	23.1	39	
South Island (NZ)	2.5	40	
North Island (NZ)	6.9	36	
Mangaia (Cook Islands)	2.2	23	
Tongatapu	38.5	27	
Taumako	2.8	59	
Marianas	0	29	
Hawaii (Oahu)	0	1063	Snow 1974
Hawaii	20.3	148	Hrdlicka 1935
Hawaii (Oahu)	13.2	49 (m)	Pietrusewsky 1981
Easter Island	8.6	64 (m)	
Marquesas	2.8	51 (m)	
New Caledonia	2.9	85 (m)	

and the Marquesas, though both lie within the tropics, was brought out. It could be argued that the low Marquesan incidence of auditory exostoses is a reflection of a particularly warm environment, and a lack of fringing reefs to allow much time in the water, whereas Tonga has both reefs in abundance and a relatively cooler climate. On the other hand the southern Cook islanders, whose environment must be similar to that of Tonga, show a low incidence. All in all these Pacific data leave open the cold-water aetiology theory.

Malignant tumours are commonly called cancers, and while the terms are not synonomous the difference does not matter here. Cancers may be primary or secondary. A primary cancer is one arising *de novo* in a tissue, while a secondary cancer is one that has spread from a primary cancer somewhere else, usually by way of the lymphatics or the bloodstream. Primary cancers of bone are rare. Suzuki (1987) has described one in a pre-contact Hawaiian. As stressed earlier, while such a lesion is of passing interest it does not tell us anything about the overall health of a group or even of that individual for most of his or her life. Secondary tumours of bone are common today, but for demographic reasons they also were uncommon in the past. Cancer is a disease of older adults, the incidence rising sharply after the age of 60, and not many people reached that age in the past. I do not recall seeing any lesion in an adult that might be ascribed to secondary cancer. There is a lesser peak of incidence in childhood, and from New Zealand we have a single example of involvement of the spine in a 12-year girl, which possibly is spread from a tumour of the suprarenal gland (Houghton 1980: 145).

For some cancers there is a well-recognized cause. The relationship between smoking and lung cancer is a modern example. We can surmise that some foods and customs in the Pacific past might have led to certain tumours occurring even though there may be no residual evidence of them. In the western Pacific the chewing of betel nut (*Areca catechu*) along with a smear of lime on a leaf is a social custom of some antiquity. It has even been used as an ethnographic marker to distinguish Melanesia from Polynesia. Betel and lime contain carcinogens and there is a recognized association with cancers of the mouth and lips. These doubtless occurred in the past, but are unlikely to leave any skeletal record. In New Zealand bracken root was for many groups at some seasons a major component of diet, and this too is recognized to contain carcinogens. There is a high incidence of gut tumours in animals grazed on bracken, and there is no reason to suppose that humans should not also be susceptible: shikimic acid has been determined to be one of the carcinogenic agents in bracken (Evans and Mason 1965, Evans and Osman 1974).

An unexpected hazard in some Pacific islands may have been natural radioactivity. This has been documented for Niue (Marsden 1960) where the coralline soil has γ-radiation levels some 12 times normal. Root vegetables grown on Niue have very high levels of radioactivity compared with neighbouring islands, as do the bones of semi-domesticated hens and the teeth of both living and prehistoric people. It is possible that human fertility on the island has been affected. Marsden's observations have aroused the interest of mining companies.

Despite these various agents the reality is that cancers were an insignificant health problem in the Pacific past, and paradoxically this is largely for demographic reasons: people simply did not live long enough to be troubled.

Degenerative diseases

Under this heading particularly come various joint problems. Some discussion of terminology is necessary. 'Osteoarthritis' is the term applied to the common wear and tear degeneration of synovial joints. Osteoarthrosis is a better term, without the implication of inflammation, and it is sometimes used, but here I shall stick to the traditional suffix. To ascribe the basis purely as wear and tear is a bit simplistic, but it is a view that will serve. Osteoarthritis applies only to synovial joints – the pathological changes make that point – so in the vertebral column the term is only applicable to the small apophyseal joints linking the articular processes of adjacent vertebrae. Degeneration of the spine centred on the intervertebral discs and adjacent bodies is just that – spinal degeneration. The term 'osteophytosis' is sometimes used in a similar sense, as is the old term, 'spondylosis'. Thus, cervical spondylosis is degeneration involving the intervertebral discs and adjacent vertebral bodies in the neck.

Osteoarthritis

Though a lot of attention is paid to this in skeletal studies the fact is that the synovial joints of the body are rather well suited to taking the forces of a considerable lifetime. These joints have had, through our land-based lineage, a long time to adapt to the demands placed on them. Unless pressed unduly and consistently they do well for us. (Something they did not evolve for was pounding out endless miles on hard modern roads.) In the shorter lifetime of most people in prehistory they gave little trouble unless unduly injured in some incident. Ageing thins the articular cartilage and sharpens the joint perimeter, often with a lipping of bone; this much is normal. Frank outgrowth of osteophytes takes the process into the realms of pathology, and the underlying cancellous bone becomes patchily rarefied. If a joint consistently has undue demands placed on it these changes may appear earlier in life and become more extreme. One of the most extreme changes is eburnation, wherein the bone appears polished and smooth on the articular surfaces. This is evidence that the articular cartilage had disintegrated, allowing the bone ends to bear directly on one another. Eburnation as an occasional finding probably indicates that the particular joint has been rather severely injured at some time, with lasting damage to the articular cartilage.

Of the vigorous life of prehistory, then, the well-adapted limb joints tend to show rather little evidence. I do not know of any Pacific study that compellingly shows a consistent pattern of degeneration in one or other set of joints, consistent with some routine social or cultural activity. (This does not apply to the temporomandibular joint, which is considered with the dentition.) That may simply be because no study has yet been sufficiently focused – differences in patterns of joint degeneration between the people of the more land-based New Zealand environment and that of the warmer, smaller islands may emerge. Neither do I know of significant differences in the pattern of osteoarthritis between the sexes for any group.

Spinal degeneration

If the limb joints have a long adaptive history the vertebral column does not. We have been truly on our feet for perhaps no more than about three million years, and for most of that time we were a lot lighter. The intervertebral discs and the cancellous bone of the vertebral bodies have never come to terms with the new demands placed on them. Spinal degeneration is common and frequently severe in the prehistoric record.

After the age of about 40 years there is significant loss of water content from the intervertebral discs, which shrink very slightly (the reason why we lose height with age). This allows greater movement between adjacent vertebrae, which respond by osteophytic proliferation at the disc margin. These are nature's early attempts at joint stabilization, and the lipping may extend and fuse between adjacent vertebrae, which really does stabilize the joint. The cancellous bone of the vertebrae thins with age, sometimes with collapse and loss of height. Disc contents may protrude into the body of the vertebrae as the classical Schmorl's node. The greater and more consistent the pattern of load-bearing, the earlier these degenerative changes are likely to appear, and this applies as much today as in prehistory. (The activities promoting risk are not always obvious – black-belt judo exponents, for instance, have been shown to have an accelerated rate of spine degeneration.) While from purely skeletal considerations the entire upper body weight appears to be directed through the lower spine, this is not necessarily so in life. A good abdominal musculature makes a hydraulic column of the abdominal contents which takes much of the load. But even in the active past it was not sufficient to prevent early degeneration.

Just where spinal degeneration occurs depends on where the main forces acting on the spine fall. The lumbar spine might seem the obvious place, but a canoe-based existence with much paddling also throws forces onto the cervical and upper thoracic spine. In Table 6.12 the incidence and site of spinal degeneration are compared between some Pacific groups. The skeletal ages at death of the groups are much the same, so differences in incidence cannot be ascribed to different age profiles. The New Zealand Maori do seem to have had a general acceleration of spinal degeneration. Perhaps this is a reflection of a larger, rougher land. Body size does not seem to have much influence, as the group from Nebira are particularly slight but still show significant degeneration.

Table 6.12   Spinal degeneration in Pacific groups. Sample sizes in parentheses.

Group		Cervical	Thoracic	Lumbar
NZ Maori (39)	*combined*	58	50	81
	male	71		
	female	39		
Nebira (18–59)	*combined*	20	30	22
Marquesas (100+)	*combined*	36	21	16
Tonga (100+)	*combined*	24	20	34
	male	28	25	39
	female	17	10	27
Hawaii (164)	*combined*			52
Inuit (38)	*male*	18	17	38
	female	17	20	18

In passing, I would mention a clavicular feature in people who regularly paddle canoes. On the inferior aspect of the medial end of the bone a groove is created where it bears on the first costal cartilage during the extreme downward sweep of the lower paddling arm (Houghton 1980: 116). I am happy to call this a 'first-rib groove', but the term 'costo-clavicular sulcus' is creeping in. This is reminiscent of the problem faced by the neurologist William Gowers who first explained the significance of the knee jerk. According to his son, Ernest Gowers (author of *The Complete Plain Words*, a guide to writing clear English), the father spent much of the rest of his life trying to stop people calling it the 'patellar reflex'.

Other arthritic conditions appear in the skeletal record, or are suggested, but are so uncommon or the diagnosis so unsure as to justify little space. I have seen degeneration of the atlanto-axial region suggestive of rheumatoid arthritis, and ossification of the ligaments embracing the cervical spine fore and aft to give an appearance consistent with ankylosing spondylitis (bamboo spine). The propensity of the modern Polynesian to gout (discussed in the next chapter) means one should be alert for the evidence of gouty arthritis (and uric acid kidney stone) but I have not seen it. The problem may only be manifested under modern conditions.

To summarize these degenerative problems, the limb joints were usually up to the demands placed on them over the average lifetime; the spine was not, and here degeneration beginning in the third decade of life was common, particularly in New Zealand, and often severe.

Iatrogeneic disease

This term today refers to illness caused by treatment. An example would be an allergic skin rash resulting from taking an antibiotic, though unfortunately some examples are much more serious. Here I use the term to refer to some influence on the biology of an individual resulting from a custom of society. The influences seem to fall into two groups, morphological or anatomical, and pharmacological. The morphological variety falls rather into the category of mutilations. The pharmacological variety derives from foods.

Morphological

Trephination implies the making of a hole in the skull of a living person. Trepanation is a synonym. Evidence for it is surprisingly common in prehistory. The purpose was presumably, one way or another, therapeutic. Trephination in the Pacific has been discussed by Margetts (1967) and Heyerdahl (1952a). These authors agree that there is no evidence that trephination was done in Micronesia, but otherwise suggest a pan-Pacific spread of the procedure, from the coast of New Guinea to the Americas. The evidence does not support these claims (Houghton 1977). That trephination was carried out in Melanesia until quite recently is attested by both skulls and eye-witness accounts. There are specimens and reports from New Britain, New Ireland, the Solomons, Vanuatu, New Caledonia, and the Loyalty Islands, and on the other side of the Pacific there is good skeletal evidence from Peru. In between, only a sea-haze of generalizations. There are claims for the fitting of segments of coconut shell to a hole in the head following a wound, with the scalp then healing over it. Apart from the biological unlikelihood of this occurring, the claims are not helped by the additional comments that if part of the brain was lost then it was replaced with a bit of pig brain. These reports belong in the realm of fantasy, and there is at present no worthwhile bony evidence nor convincing ethnographic account to suggest that the procedure was carried out anywhere in Remote Oceania.

Headshaping was performed in various parts of the Pacific: the most accessible comment is provided by Igarashi, Katayama and Takayama (1987). They refer to the reports of the practice from New Guinea, New Britain, Santa Cruz, Vanuatu, Tonga, Hawaii and Tahiti. An account of a traditional method in Hawaii, whereby the infant cranium was compressed between coconut shells padded with bandages, is given by Snow (1974). The effect is more a 'fore-and-aft' deformation that the frontal flattening seen from ancient Egypt. Snow (1974) gives the Hawaiian (Mokapu) incidence as 44%, which suggests an almost routine practice, and notes a similar mode of deformation from Samoa. The effect is seen also in several crania from the 'Ate site in Tonga, dated to about A.D.1500. In an earlier chapter the debate between Pearson and others as to whether the frontal flattening of the Moriori head was the result of artificial deformation was mentioned (Howells 1978). As Pearson (1921) correctly claimed, in this group the shape is natural.

Pharmacological

Through the Pacific there is a considerable catalogue of foods and customs related to ingested items that carry particular risks to humans. One of these, betel, has been discussed earlier, where it was considered as a carcinogenic agent. Some others are problems that people identified and learned to deal with, doubtless through a lot of unfortunate trial and error. Some reef fish are poisonous at certain seasons; ciguatera toxin is produced by some dinoflagellates and accumulates in the larger fish of the food chain. While the effects of the toxin may be severe, and include diarrhoea, skin itching, and a varying degree of paralysis which may persist for months, the poisoning is rarely fatal.

Various food plants contain toxins that must be neutralized by preparation. In the Marianas an association has been identified between certain severe neurological disorders (amyotrophic lateral sclerosis and Parkinsonian dementia) and glycosides contained in the seed of the false sago palm (*Cycas circinalis*) (Kurland 1988). In one form there is a progressive painless weakness and atrophy of limb muscles, in another symptoms more like those of classical Parkinson's disease. Some 10–20 years' exposure to the toxins seems to be necessary before the symptoms appear. As these palm seeds were part of the traditional diet, the diseases are likely to have occurred in prehistory. The risks seem to have been known, for traditionally in preparing the seed for eating it was soaked in several changes of water over a week. The last wash water was fed to chickens, and if they were still alive by the time the nuts had dried then they (the nuts) were deemed safe to eat. With such a lapse of time between the eating of the nut and the appearance of symptoms it is likely that this crude bioassay was not always effective. Perhaps in the ethnographic record there is a hint of these problems in prehistory.

In New Zealand the Maori learned to treat karaka berries by steaming and soaking so as to neutralize a toxin in the kernel. The parts of a plant may vary in toxicity. In New Zealand, the leaves and shoots of the tutu (*Coriaria sarmentosa*) contain a highly poisonous crystalline glyco-side, but from the berries a (somewhat potent) wine has been made.

The name kava refers to the plants *Piper wichmannii* and *Piper methysticum*, and to the drink extracted from the latter. Lebot, Merlin and Lindstrom (1992) in a comprehensive review argue, on biochemical and genetic grounds, that *Piper methysticum* is not a separate species but a group of sterile cultivars selected from somatic mutants of the naturally-occurring and fertile *Piper wichmannii*. If this is so then according to the convention of precedence in taxonomy the name *Piper wichmannii* should disappear. *Piper wichmannii* (if it exists) has a natural distribution from New Guinea to Vanuatu. By contrast *Piper methysticum* occurs as an abundance of cultivars in Vanuatu, extending to Fiji and through all high islands of tropical Polynesia. Islands such as Rapa, Easter Island and New Zealand are too cold. In New Zealand the name was taken over by the related *Macropiper excelsum*, and one of the traditions associated with the drink is maintained in the Maori word *kawa* for ceremonial protocol. In Micronesia kava was known in Pohnape and Kosrae. In Melanesia west of Vanuatu its occurrence and use is only known sporadically, from the Admiralties and coastal New Guinea, where it may be an early historic introduction, and not at all in the Solomons, New Caledonia and the Loyalties. From this distribution and the abundance of sterile cultivars in Vanuatu, Lebot *et al.* (1992) consider Vanuatu to be the origin of selection and cultivation by *Homo sapiens*. Trade and colon-ization carried it elsewhere. If this is so, then its cultivation must have occurred early in human settlement, for it was carried well east.

Preparation of the earthy-chalky drink from the plant root involves pounding or chewing the root and then infusing it in water. The active ingredients in kava are termed kavolactones, complex organic molecules whose structure need not detain us here. The subjective effect of the drink is one of contentment and relaxation without the more belligerent consequences that may arise from alcohol intake. Pharmacologically kava acts to produce an analgesic effect that is weight-for-weight superior to aspirin, a local analgesic effect, an anticonvulsant and muscle-relaxant effect, and apparently it functions also as an antifungal and possibly as an antibacterial agent.

The place of kava in ceremony and in traditional medicine through much of Remote Oceania is well known. Some of the reported uses of kava, such as for dysmenorrhoea and migraine, make sense from a phar-macological viewpoint. In a plea almost as evangelic in tone as the early missionary plaints against kava, Lebot *et al* (1992) make the case for it as a valuable social and possibly medicinal drink and see a promising future for it as a cash crop.

Dental disease

In some ways it is unwise to have a separate category for the dentition. For one thing it switches the classification from one based on cause to a regional or anatomical one. Worse, it tends to perpetuate the attitude that the dentition is something rather apart from and uncoordinated with the rest of the body, a view I have tried to avoid. For all that, it is convenient to discuss the dentition separately, for it has its own particular functions and pathology. It is also likely to be the best preserved material, enamel being the hardest stuff in the body.

From the perspective of dental professionals, anthropological assessments of dental pathology tend to the naive. Thus, on the ready assumption that the extent of alveolar erosion equates with past severity of periodontal disease, Smillie cautions:

Is alveolar bone loss, in the plane in which it is measured, necessarily a consequence of, or at least related to, periodontal disease? As more data are gathered about the nature of periodontal disease, its effect on the hard tissues, and the part played by dental calculus, most of what seemed to be solid ground has proved to be enticing quicksand lurking for those still reciting the dogmas of the past. The destructive value of supra-gingival calculus is very questionable. Worse, recent data throw considerable doubt on the destructive value of sub-gingival calculus, long considered to be a prime factor in the cause of periodontal disease, if not because of its intrinsic qualities, then because of the plaque associated with it. In certain individuals at certain times sub-gingival calculus and plaque may, in fact probably are significant factors in causing periodontal disease. The question then becomes, in which individuals, and when, and why? . . . There may not be any simple and, therefore, from an anthropological point of view, practical way of detecting periodontal disease in other than its severe forms where there will be marked and often irregular loss of alveolar bone visible to the naked eye. (pers. comm.)

And on the relationship between diet and caries, the same authority:

The measurement of the cariogenicity of foodstuffs is still a matter of controversy. Several in vitro methods are used but predictions based on results from these methods do not by any means always agree with the findings of clinical surveys. To declare a diet to be cariogenic because it has a high carbohydrate content is a brave thing to do. Its cariogenicity will depend on the type of carbohydrate, the microflora in the plaques of the population and the presence of trace elements (especially fluoride it seems) in the diet, to cite but a few variables. (*ibid.*)

In a similar vein, Clark (1990) has questioned some of the standard assumptions. With these caveats in mind I shall here discuss patterns of dental pathology but tread particularly lightly in the area of causation.

Tooth wear (or attrition) in its early stages may be as normal as any other process of ageing, only crossing into the realms of pathology when either the tooth or its support have clearly deteriorated. Wear takes away first the cusps and other enamel of the occlusal surface, progressing to exposure of the underlying dentine. This is still very hard and resists abrasion quite well. If with continuing wear the pulp cavity is being threatened, a reactive secondary dentine is formed. This is readily distinguished in archaeological material by its appearance, which is darker than the peripheral primary dentine.

With severe wear even this protective mechanism proves insufficient, and the pulp cavity is breached. The ensuing infection manifests as an abscess related to the roots of the tooth, evidenced in archaeological material as a residual excavation or rarefaction of alveolar bone. Sometimes an osteomyelitis has been set up in the bone.

The cusps of the teeth have been ascribed importance in the process of mastication, but it is doubtful if this really is so. Some wear of the teeth with loss of cusps, right through to the stage of early formation of secondary dentine, creates an occlusal surface with a very sharp peripheral edge of enamel, which is still extremely efficient in breaking down food during the shearing phase of mastication. The cusps may play some part during growth of the masticatory system, providing neural feedback as to the precise positioning of the teeth.

In New Zealand over the time of prehistoric settlement a change in the pattern of tooth wear is observable (Houghton 1978). The earlier settlers, before about A.D. 1500 and exemplified in the people from Wairau Bar, Sarahs Gully and Castlepoint – and particularly the latter – show rather little wear. Using Molnar's (1971) grading for wear, individuals at about age 20 years in the earlier prehistoric barely show grade 4 wear anywhere on the dentition, this progressing to grades 6–7 at age 35 years. In contrast in the later prehistoric, wear intensifies to such an extent that an individual in the early 20s often shows the terminal stages in the first molar, that is, grade 7 or 8, with evidence of related infection in the supporting bone. There is a change in the position of maximum wear on the dental arcade: in the early prehistoric, wear appears earliest and most markedly on the incisors, whereas in the late prehistoric wear is early and severe on the molars and the incisors are relatively spared. By age 40 years in the later prehistoric, substantial loss of teeth occurred, particularly in the posterior part of the arcade, with a few anterior teeth often being spared. Salivary calculus is not marked in the early prehistoric period and is almost absent in the individuals of the later prehistoric. From any era some loss of contour of the alveolus, suggestive of slight periodontal disease, may be evident, but in the later prehistoric particularly the alveolus is generally healthy.

In the light of evidence of climatic change in New Zealand over the period, this changing pattern of dental wear has been interpreted as indicating a shift from a diet with relatively soft vegetable and meat components, such as kumara and birds and sea-mammals, to a more abrasive diet that included gritty shellfish and fern root (Houghton 1978b).

Wear also occurs between the teeth as a consequence of their slight movement against one another, and the extent to which this wear occurs is proportional to the occlusal wear. A moderate amount of such interproximal wear will shave a good millimetre off the length of each tooth. Many years ago Begg (1954) in a series of superb papers drew attention to this natural orthodontics, whereby ample room is created for teeth yet to erupt.

A distinctive pattern of wear in some individuals in New Zealand is the so-called 'fern-root plane'. Here the occlusal wear spills over on to the buccal surface of posterior teeth, particularly first molars. With this may be associated a gradual lateral dislocation of the roots of the tooth through the alveolus, the whole tooth being eventually turned out of its socket. Exposure of the pulp cavity may occur, as in any case of continuing wear, but sometimes with the remarkable appearance of a molar lying almost on its side with the pulp cavity exposed from apex of the root through to the crown. Taylor (1963) has stressed the distinct and separate nature of the 'fern-root plane' and the occurrence of dislocation. While strictly this may be correct, Simpson (1981) has shown that on the material available the two features are indeed commonly associated and are largely confined to the regions of the country where fern root was a substantial item of diet – the northern part of the North Island, and the Chathams. That is, the basis for these features found in New Zealand does seem to lie with the use of fern root. The term 'fern-root plane' seems to have been coined by Buck (1925) who attributed it to the regular chewing at a rhizome. Rather similar patterns have been reported from other parts of the world, so other dietary items may create the same pattern.

Most other Pacific habitats lie within the tropics where foods tend to be softer, and I know of no other human group in the region in whom tooth wear consistently leads on to significant pathology. For example in Taumako, individuals at about age 30 generally have reached only grade 2 wear on the molars. For the Marianas, Roy (1989) noted a mean level of adult tooth wear (all ages) of less than 3 on Molnar's scale – that is, on most teeth in most individuals the enamel was not breached. Rather, much dental disease in the tropical Pacific centres around the state of the alveolus.

On the teeth from the fifteenth-century 'Ate site in Tonga, Taylor (1971) observed moderate to marked supragingival calculus in 40% of 45 dentitions, but the likelihood of subgingival calculus in only 20%. General alveolar bone resorption suggestive of chronic periodontal disease was evident, and sometimes an extreme resorption in relation to individual teeth. In a probable late-prehistoric series from Rota in the Marianas, Hanson (1987) recorded calculus, but the condition of the alveolus was not given. From the San Antonio site and others on Saipan in the Marianas, Roy (1989) recorded a pattern of moderate calculus and slight resorption of alveolar bone. In Micronesia the dentitions frequently show betel staining, and here and in Melanesia the impact of a soft diet on the health of the gingiva and alveolus may be accentuated by the habit of betel chewing, along with its smear of lime. The carcinogenic potential of this mix has already been mentioned. The habit may also lead to an astonishing accumulation of calculus around the posterior teeth such as to encase them in a plaster-like cast with only the occlusal surfaces showing through. In the sixteenth-century population from Taumako such accumulation of calculus is gross, and it is hard to resist the conclusion that the effects on the alveolus have been severe. In this population the alveolar bone is resorbed and support for the teeth is progressively lost – presumably as a reflection of a severe overlying infection – so that an individual at 40 years was likely to be edentulous. At this age the teeth themselves may be still pristine, unworn and uncarious. The problem is that there is no bone left to hold them in place. We have a situation wherein a pattern of pathology totally different from that evident in late prehistoric New Zealand has the same consequence at much the same age: the edentulous state. Generally, however, the tropical Pacific pattern of alveolar destruction is less florid, most groups evidencing slight wear, slight calculus, and a slight-to-moderate degree of alveolar erosion.

The incidence of dental caries also varies through Oceania. In New Zealand the early Wairau Bar population shows some minor lesions, but in the later prehistoric it is rare. In tropical Polynesia the incidence is sharply higher. For Easter Island (but scoring on the number of teeth involved rather than by individual) Owsley, Miles and Gill (1985) recorded a 27% incidence in adults. From prehistoric Hawaii Keene (1986) noted an adult incidence increasing from 9% in the 18–25-year range, to 49% in the 26–40-year range, and 68% in the 40-plus age range. The lesions were overwhelmingly in the root and cervical region of the teeth. Taylor (1971) noted caries in 22% of 45 adult dentitions from the Tongan 'Ate series: the lesions were distributed quite evenly through both upper and lower dentitions, and again were nearly all root or cervical in location. For the Sigatoka population Visser (1994) noted almost 75% of the adults as having caries, with the incidence rising sharply with age.

In contrast to these records from tropical Polynesia, a review of dental caries in skeletal samples from the Marianas (Hanson 1993) reports a range from 0% to no more than 11%. In the large Taumako series the incidence was 1.5%. A 'strong inverse relationship between the prevalence of dental caries and the intensity of betel chewing' was noted by Moller, Pindborg and Effendi (1977: 64), who suggested that it may relate to the film established on the teeth, or to high fluoride content of the betel quid. Drinking water in some areas in Micronesia also is high in fluoride.

These varieties of infection in and around the teeth – caries and alveolar bone involvement for whatever reason – were probably significant in determining the general health of groups in the past. (The betel habit is good for the enamel but bad for the gums.) A grumbling nidus of infection could provide a source for bloodstream spread if for any other reason – some other illness, or a seasonal shortage of food – an individual was in less than rude health. While it is a long way from the Pacific, a good illustration of the potential of the body's own organisms to be lethal is revealed in the *Gentleman's Magazine*. This was an annual publication in England during much of the eighteenth century, and usually contained a 'Yearly Bill of Mortality for London', an analysis of the records of deaths in the city for the previous year. That for 1757 is typical: when deaths from the common infectious diseases are stripped out along with what is classed as 'old age' and heart failure (some of which would have had an infective basis) then 'teeth' emerge as the major killer, far surpassing any other pathology.

Problems of the temporomandibular joint are usually considered in tandem with those of the dentition. It seems a reasonable postulate that groups showing much tooth wear are likely to have plenty of degeneration at the joint. In Oceania, therefore, groups from the tropics are likely to have few joint problems while those from the later prehistoric in New Zealand are likely to have many. Surprisingly, such an association is uncommon: certainly the tropical incidence of temporomandibular pathology is low, but so is it also in New Zealand, even in the presence of severe tooth wear. This was the clear finding of a New Zealand study by Stokes and Fitzharris (1991), and of another in far-away Nubia (Sheridan *et al.* 1991). In this latter study, temporomandibular joint degeneration was associated with long-standing loss of posterior teeth, but this situation was not assessed in the New Zealand study.

However from the tropical Pacific do come some remarkable examples of gross change at the temporomandibular joint. In the Sigatoka population of about 1900 b.p. most of the males showed a degenerative remodelling of the articular eminence, sometimes to the stage of its effective flattening and disappearance, with reactive bone formation in the condylar fossa. In this way the superior joint surface was converted from one of fossa and eminence to almost a plane surface. The condyles showed flattening and peripheral lipping, and a skewing such that the medial pole was turned more posteriorly. Eburnation was not present, indicating that soft tissues remained interposed between the bone surfaces.

In contrast to the studies cited earlier, the temporomandibular joint changes showed a strong positive correlation with the extent of wear on the teeth, particularly the molars. As Visser comments: 'the causal relationship between tooth loss, tooth wear, and temporomandibular joint progressive remodelling is not secure' (1995: 115). He ascribes the remarkable pathology to the custom of preparing kava root for infusion by chewing. There are good ethnographic accounts of this being done by young men; in the Sigatoka group the women showed a low incidence of such joint degeneration. The neurological effects of kava lend themselves to these changes. There is likely to be a loss of sensory feedback from the joint, with a deterioration of the usual precise protection of the joint by the muscles, and at the same time a suppression of painful stimuli resulting from the loss of such protection from damage. The joint in fact comes rather to the plane-surface form, with minimal articular eminence, characteristic of herbivores with their prolonged periods of leisurely mastication.

LEGACY

7

Following European contact, and exposure to diseases to which they had no natural immunity, the people of the Pacific suffered the devastation of epidemics that has been the melancholy fate of every newly encountered and previously isolated group. In the Americas the astonishing ease with which a few Spaniards came to dominate great indigenous civilizations is ascribable more to the allies of disease, such as smallpox, than to any military superiority. The near-extinction of the New England Indians has been related to initial epidemics of plague and smallpox and a subsequent high incidence of tuberculosis and dysentery (Cook 1973).

On small Pacific islands distance and isolation were for a while some protection, but the vulnerability was there and the record is clear and depressing (McArthur 1968). It is said that 25% of the population died when an epidemic of measles swept through Fiji in 1875. In Tonga a succession of introduced epidemic diseases seems to have reduced the population through the nineteenth century, but no clear major episode like that of measles in Fiji is recorded. Through the same period, influenza, dysentery and measles decimated the populations of Samoa, the Cook Islands and what is now termed French Polynesia. A smallpox epidemic spread through the Society Islands in the 1840s. And so on. Culpability for these episodes in human history is sometimes taken as one of those hair shirts that people of European origin should put on. Yet it was an inevitable biological consequence of a shrinking world, and for better or worse, no group was ever going to remain in pristine isolation. I do not intend to discuss further this historic stage of the Pacific past. Rather, the purpose of this last brief chapter is to look at some aspects of health and health problems of contemporary Pacific people that clearly, possibly, or speculatively may be linked to their origins – that is, consequent on their selective adaptation to the distinctive oceanic environment, along with the operation of genetic drift and the founder effect in this small-island world.

Normal growth and development

It is the common observation of those involved with school sports teams in Pacific communities that Polynesian children mature earlier than those of European ancestry. From about the age of 12 years the Polynesian children are ahead in stature, and particularly in muscularity. Figure 7.1, adapted from Tonkin (1970), illustrates this for Maori and

Figure 7.1
Growth curves for height and weight for Maori and European males in New Zealand. Adapted from Tonkin 1970.

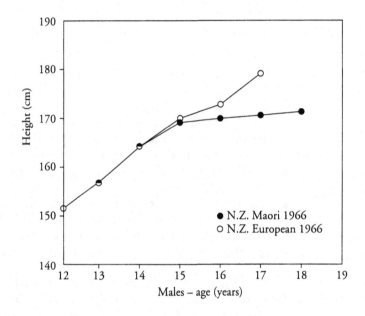

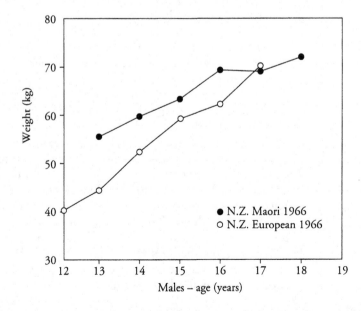

European males in New Zealand. By about 17 years those of European stock have often caught up in stature but generally remain behind in muscle mass. The muscularity remains impressively obvious in adult Polynesian sportsmen and women, and the contrast with European physique is well seen on the rugby field or netball court.

Dental development is a useful indicator of the stage of growth and maturation of the body, though the correlation with skeletal maturation may differ between groups. On limited numbers Fry (1976) has shown Polynesian children to have some teeth erupting up to two years earlier than with children of European stock.

Such a pattern in Polynesians could be regarded as just another example of the generally earlier maturation of tropical peoples (Eveleth and Tanner 1976). However an alternative view – for these are not really tropical peoples – is that such accelerated maturation conferred in the past a selective advantage. Children and adolescents would have had better chances of survival in the oceanic environment. But probably more significant is the advantage early maturation would confer on small groups struggling to avoid extinction on small islands. In the previous chapter, evidence was given for the early age of menarche in Remote Oceania – 11–12 years in Tahiti and in Fiji. For the New Zealand Maori a more recent figure of 12.7 years is cited (Eveleth and Tanner 1976). Such earlier fecundity meant a greater chance of survival for a small founding population.

Health problems

Congenital defects

The period of settlement of the Pacific must have seen the founder effect enacted again and again, often involving very small groups. While from this limited gene pool any more deleterious homozygotes would have been quickly eliminated, the situation favoured the emergence of distinctive incidences of congenital defects, and there are data for several. The Maori incidence of cleft palate is 1.867 per 1000 compared with the New Zealand European incidence of 0.643 per 1000. By contrast, the Maori incidence of cleft lip, with or without cleft palate (this having a different genetic basis to cleft palate alone), is 0.397 per 1000, against 1.195 for Europeans (Chapman 1983). Comparable figures are cited for Hawaii (Chung *et al.* 1987). That is, Polynesians are about three times as likely as Europeans to be born with cleft palate, but Europeans are three times as likely to be born with cleft lip.

Club foot (talipes equino-varus) shows a similarly distinctive pattern. In the contemporary Maori the figures are 6.5–7 per 1000 births, against 1–2 per 1000 births for New Zealanders of European origin

(Cartlidge 1983). In Hawaii, Chung *et al.* (1969) determined an incidence of 0.567 per 1000 births for Asians, 1.121 for Europeans, and 6.812 for Hawaiians.

Neural tube defects (anencephalus and spina bifida) have a genetic component (Elwood and Elwood 1980). The risk to people of Polynesian origin is significantly lower than for other groups. For spina bifida the New Zealand Maori rate of 0.58 per 1000 births is lower than for Europeans (1.0 per 1000), but for those from the tropical Pacific the rate (0.14 per 1000) is markedly lower still (Borman and Cryer 1993). A similar pattern was apparent for anencephalus. This cline presumably mirrors the greater admixture in the Maori of European genes.

A dental congenital defect in people of Polynesian ancestry is a distinctive patchy defect of mineralization of tooth enamel, a variety of *amelogenesis congenita imperfecta*. The enamel is yellow to brown, chips readily, and decays very rapidly. It has been identified in the Maori, in Cook Islanders, and Marquesans. Some 2% of Maori and Cook Islanders show the defect. The pedigrees obtained are consistent with the abnormality being inherited as an autosomal dominant, the gene having reduced penetrance (Smillie, Rodda and Kawasaki 1986).

Tissue polymorphisms

Distinctive Pacific frequencies (compared with Europeans) have been documented for several tissue genetic systems. For the HLA system, many alleles, such as HLA-A1 and A3, and HLA-B5, B8, B12, B17 and B27, are either absent or occur in low frequencies in Polynesians. There is a higher frequency of A2, A9, BW22 and B40 in Polynesians. The processes of drift and possibly, though less clearly, selection, have shaped these distinctive patterns. Because of the central place of these systems in some body responses to disease there may be clinical implications, but this aspect is still largely unexplored. Polynesians have a much higher risk than Europeans of developing chronic renal disease, and it has been suggested that this susceptibility lies in the histocompatibility or immune response gene frequencies (Neale and Bailey 1990). There may be a similar basis to the incidences of systemic lupus erythematosus and rheumatic fever, which are more common in the Maori compared with the European, and rheumatoid arthritis, Paget's disease and ankylosing spondylitis which are much less common in the Maori (Caughey and Woodfield 1983).

Several blood polymorphisms have been discussed in Chapter 5, though from a rather different perspective. In the ABO system Polynesians show low or absent B and A2, and in the Rh system a high frequency of cDE and a low frequency of CDE. Extensive references to the literature on blood polymorphisms in the Pacific are provided by Woodfield *et al.* (1987). Amongst the matters of clinical interest emerging from DNA studies is the realization that Polynesian people have an appreciable incidence of alpha-thalassaemia: Mickleson *et al.* (1985)

suggest a figure of about 16%. Dacie writes that 'where thalassaemia is present, there is now, or has been in the past, much malaria' (1988: 398). However, although the evidence points to the condition being a balanced polymorphism that provides, in a manner yet uncertain, a degree of resistance to malaria, the parasite is not found in Polynesia. Rather, the Polynesian incidence of thalassaemia may be viewed as a legacy of an ancestral sojourn within the malarial regions of Near Oceania or Island South-East Asia. Carried on in the genes, it has not been sufficiently deleterious to have been selected out.

There is a well-documented problem of anaemia in the young Maori, and in other Polynesian children (Tonkin 1960, Akel *et al.* 1963, Neave, Prior and Toms 1963, Cantwell 1973, Wood and Gans 1984, Moyes, O'Hagen and Armstrong 1990). Some of these clinical studies go back much more than a decade, and while concluding that the problem was multifactorial – inadequate dietary iron, infections, gastrointestinal blood loss, even growth patterns – the impression is that the clinicians have been a little perplexed as to just why the problem of anaemia should be so persistent. Mickleson *et al.* (1985) have pointed out the necessity to bear thalassaemia in mind when investigating anaemia in the Polynesian infant and child. In a series investigated in Vanuatu most hypochromic anaemia could be attributed to alpha-thalassaemia of the single-gene-deficiency type, rather than to iron-deficiency or parasitism (Bowden *et al.* 1985). So while the aetiology of anaemia in Polynesian infants and children may well be dominated by iron-deficiency and socioeconomic factors, there are other influences in the evolutionary backdrop. Iron-deficiency may also influence the alpha/beta globin chain ratio, lowering alpha-synthesis and exacerbating any deficiency.

Orthopaedic problems: fractures and dislocations

The distinctive bowing of Polynesian long bones in both upper and lower limbs has been discussed in Chapter 4. In part this bowing is a necessary requirement for accommodation of the substantial musculature, but it carries the associated advantage that an appropriate bowing of the long bones alleviates the bending stresses of movement. The particularly distinctive morphology of the Polynesian femur has been described. Apart from possibly causing perplexity during internal fixation of fractures, such morphology may matter clinically. Despite the compensating reorientation of the acetabulum, the Polynesian hip may be rather more susceptible to dislocation, though I know of no data on this. On the other hand, Stott and Gray (1980) showed that Maori women had a significantly lower incidence of femoral neck fractures than women of European stock. The incidence for Maori males was similar to that for European males, which suggests that it is not the morphological (shape) element that contributes to the differing female incidence. Polynesians have been shown to have bone density, measured in the forearm, about 20% greater than that of Europeans (Reid, Mackie and Ibbertson 1986). This difference does not seem to have an endocrine basis (Reid *et al.* 1990) and is unlikely to have a dietary explanation. It presumably relates to the inherently greater muscularity of the Polynesian, for, as might be expected, there is a strong correlation between bone density, and muscle mass and function (Ellis and Cohn 1975, Menkes *et al.* 1993). This greater bone density might reduce susceptibility to fracture in Polynesian women because of a lesser liability to osteoporosis from middle-age onward.

Middle ear disease

Middle ear disease in Polynesian children is a major health problem (Stanhope *et al.* 1976, 1978). An incidence of 10–15% chronic disease has been recorded in Maori primary school children, with 20–25% showing some hearing loss (Tonkin 1974). Immigrant Tokelau children to New Zealand have shown a startling 26% incidence of middle ear disease (Tonkin 1975). The question is whether there is some predisposing physical basis, or whether it is a reflection of living standards. There is a clinical impression that the Polynesian auditory tube is relatively short and the opening patulous or spread open, and the studies of the Polynesian cranial base suggesting the nasopharynx to be some 30% more capacious than the European one might fit with that impression. Whether this predisposes them to infection is another matter. Some clinicians are adamant that 'The problem of running ears is basically a problem of socioeconomic status' (Roydhouse 1977: 25). This view is supported by a study that compared the incidence of otitis media (among other conditions) in a group of children of predominantly Maori origin, with that in a group of European origin (Hood and Elliot 1975). The groups were selected for equal (in fact they were of rather high) socioeconomic status. The interesting result was that the European children had a significantly greater incidence of middle ear disease than the Maori group. The alarming incidence of otitis media in immigrant Tokelauans is not matched back in the islands, where an incidence of a little over 1% has been noted (Tonkin 1975). Again, this suggests that socioeconomic factors or some newly-encountered environmental agent may dominate.

We investigated this matter further by examining radiologically a large series (n >100) of prehistoric New Zealand crania for evidence of past middle ear disease. Comprehensive views were taken (Townes, basal and rotated lateral) with entirely negative results. That is, no individual showed changes suggestive of past chronic middle ear infection. This contrasts with studies on North American prehistoric material which indicate infection rates of from 9 to 50% (Gregg *et al.* 1965, Homoe and Lynnerup 1991). The New Zealand finding is against any innate predisposition towards middle ear infection in Polynesians and supports Roydhouse's contention that socioeconomic factors are what matter.

Lung disease

Perhaps surprisingly, the capaciousness of the Polynesian respiratory tract in its upper part does not continue down. There are several studies of Polynesian groups as widely separated as the Maori and the Marquesans that show vital capacity, corrected for body size, to be some 8–9% less than in Europeans (de Hamel and Welford 1983, Asher *et al.* 1987, Neukirch *et al.* 1988). This is not a problem in the healthy individual, for the crucial functional parameter is the ratio of forced expiratory volume in one second to vital capacity (FeV1/VC). This ratio in fact is slightly higher, or better, in the Polynesian than in the European and active, healthy Polynesians show levels of maximum oxygen consumption well above those of the average European, a reflection of muscularity. The capacious upper respiratory tract of the Polynesian should ensure good airflow, and the lesser vital capacity may be at ease in the oceanic environment of high humidity. The evolutionary thesis also suggests that the substantial musculature has evolved principally for heat production rather than locomotion: the former function requires sustained activity at a rather low level (about 15% of maximum voluntary effort) rather than a demand consistently approaching the maximum. However fatter, westernized Polynesians may show a much impaired maximum oxygen consumption, and any deterioration of pulmonary function in such individuals, if they have a relatively small lung size against total body size, is likely to have compounding consequences.

Metabolic disorders

A group of 'interrelated metabolic problems that are contributing to the health risk among Polynesians . . . includes obesity, hypertension, hyperglycemia, hyperuricemia, hypertriglyceridemia, and the clinical disorders of diabetes, gout, and coronary artery disease' (Prior *et al.* 1978: 257).

The clinical literature on this complex array of disorders is vast, and the background to any one disorder is likely to differ somewhat from group to group. Some suggestions are made here, linking the evolutionary background of the Polynesian people to their tendency to disorders of lipid metabolism, to run to fat, and problems of non-insulin-dependent diabetes and gout.

The substantial literature on muscle fibre types in humans has been reviewed by Saltin and Gollnick (1983). They conclude that while particular muscles have a predominance of one or other fibre type, most muscles have a mean fibre composition of about 50% of each type, and with a reasonably even distribution of fibre types through the substance of any muscle. While it is recognized that some modification of fibre type is achievable by specific physical training, 'The large range of the percentages of ST [Type 1] and FT [Type II] fibers observed in the population in these studies implies that for a certain percent of the population of males and females, their muscles are predominantly of one or the other fiber type' (Saltin and Gollnick 1983: 573). This inter-individual variation appears to be genetically determined; for example, studies with mono- and dizygotic twins have shown almost identical fibre composition of the vastus lateralis muscle in the monozygotic but not in the dizygotic twins.

The adaptation by people to the oceanic environment saw the emphasis on individuals with a large muscle mass that was particularly directed towards the maintenance of body temperature: this much has been discussed. Shivering is an isometric contraction, efficient in heat production, and at about 15% of maximum voluntary effort may be maintained for long periods (Lind 1980). Type II fibres are implicated in shivering (Parker and George 1974, Martineau and Jacobs 1988, Walters and Constable 1993). Type II fibres, particularly Type IIb, use an anaerobic mode of respiration, by glycogenolysis, utilizing the internal glycogen stores of the fibres, and glucogenesis from circulating blood sugar. (This contrasts with Type I fibres which have an aerobic mode of respiration. The basis of this involvement of Type II fibres has been suggested to be the vasoconstriction and relatively poor oxygenation within the muscles at such times.) A significant correlation between individual obesity and a preponderance of Type II muscle fibres has been established (Staron *et al.* 1984, Wade, Marbut and Round 1990). Type II fibres, particularly IIb, are essentially unable to metabolize free fatty acids, which are therefore likely to be deposited as fat in the body. Environmental selection for muscle fibre type suited to shivering – and lots of such fibres – may thus be a factor that renders people of Polynesian ancestry susceptible to the development of obesity, because of this inability to readily metabolize free fatty acids. The same process might contribute to the higher levels of triglycerides found even in Polynesian adolescents, and be implicated in other problems of lipid metabolism. Muscle biopsies on individuals of Polynesian ancestry, where a preponderance of Type II fibres is predicted, will be interesting.

The incidence of non-insulin-dependent diabetes mellitus (NIDDM) (and prediabetic states) in some Pacific populations is high, and rises with urbanization and the change to a Western diet: a selection of data is given in Table 7.1. Obesity is sometimes seen as being linked to the development of such diabetes, to the extent that the term 'diabesity' has been coined. Of this association Taylor comments: 'Obesity has been ascribed great significance in the aetiology of NIDDM by many writers. However there is much evidence that obesity is irrelevant and little or no direct evidence that it plays an aetiological role' (1989: 77–78). In the Pacific arena not many clinicians might agree with this. It may be that in these considerations global generalizations are unlikely to be valid and each group has to be considered as a particular case. Lean NIDDM individuals do appear to differ from obese NIDDM in production, and peripheral action, of insulin (Kelley, Moran and Mandarino 1993). Lean NIDDM individuals have muscle insulin sensitivity equivalent to that of age-matched lean non-diabetic subjects, whereas obese NIDDM indi-

Table 7.1 Age-standardized incidence (percentage of adult population) of , diabetes in some Pacific populations. Data from Prior *et al.* 1978, Wessen 1992, and Dowse *et al.* 1994.

Group		Incidence of diabetes
Wanigela (New Guinea urban)	*male*	27.5
	female	33.0
Wanigela (New Guinea rural)	*male*	11.7
	female	17.0
Nauru	*male*	33.4
	female	32.1
Fiji (rural)	*male*	2.1
	female	2.1
Fiji (urban)	*male*	5.9
	female	10.3
Tonga	*male*	4.4
	female	9.4
Tokelau	*male*	2.3
	female	7.5
Tokelau (NZ)	*male*	4.6
	female	12.9
Pukapuka	*male*	0.4
	female	1.6
Niue	*male*	5.6
	female	8.9
Rarotonga	*male*	5.5
	female	8.5
NZ Maori	*male*	11.7
	female	14.9
NZ European	*male*	1.5
	female	3.1

viduals tend to be insulin-resistant. For the frequently-obese Polynesian the link with the diabetic condition may again involve skeletal muscle. The impaired ability of insulin to stimulate glucose uptake in skeletal muscle (insulin resistance) has been shown to be localized to the non-oxidative pathway (Beck-Nielsen *et al.* 1992), which implicates Type II fibres.

For a generation, the memorable epithet in any consideration of the aetiology of non-insulin-dependent diabetes has been the 'thrifty geno-type' of J. V. Neel. In 1962 Neel proposed that the persistence of such a diabetic tendency must mean that the diabetic genotype had some advantage. This was held to be the ability to respond with rapid insulin production so that calories could be stored more efficiently in times of plenty. The foods of westernization were argued to overstimulate the mechanism leading to insulin antagonists and pancreatic exhaustion. The concept of insulin antagonists has not survived, but in 1976 Neel proposed that the thrifty genotype concept might be a good example of the right idea based on the wrong data. In the revised version the primary event is insulin resistance, partly inherited and partly acquired through a sedentary, overfed modern existence. Those with a rapid insulin response secrete more insulin, which exhausts the receptors, makes them fatter as individuals, and eventually exhausts the pancreatic cells. Hyper-insulinaemia is the central factor. Those groups often drawn under the thrifty umbrella are as disparate in environment as the people of Oceania, Australian Aborigines, North American Indians, Asian Indians, and Chinese.

The thrifty genotype epithet is memorable, but it hardly serves to elucidate the complex and varying metabolic patterns and problems underlying the diabetic state. At the very least it is arguable whether the ability to suddenly pour out quantities of insulin has any relevance to seasonal periods of famine and plenty. It is sometimes claimed that the people of Remote Oceania would be particularly advantaged by possession of a thrifty genotype, yet in Fiji today the incidence of non-insulin-dependent diabetes is much higher in Indians of Asian origin than in indigenous Fijians (Wessen 1992). The low incidence of non-insulin-dependent diabetes in people of European origin, against which others seem to be judged, has been achieved in the face of a rather good historical record of some disastrous dietary times. It is doubtful whether the Chinese historical record portrays worse episodes. It may be that the thrifty genotype has become one of those pithy tags, readily taken as an explanation and initially helpful, but which eventually come to stand in the way of clarification of a complex spectrum of problems. There are other examples of such ultimately obscuring catchphrases in human biology; Bjork's (1947) 'growth rotations' of the mandible, and the 'functional matrix' of Moss (1983) are a couple.

Abnormalities of uric acid metabolism are common among Polynesians. Hyperuricaemia, defined as a level greater than 7mg% (0.42 mmol/litre), is evident in 49% of Maori males and 42% of females (Brauer and Prior 1978). Significantly, raised levels of uric acid in the blood are already found amongst adolescent Maori (Stanhope and Prior 1975). The problem extends through the South Pacific Polynesian domain, with rather similar levels of hyperuricaemia being recorded in Samoa, the Cook Islands, the Tokelaus, and Tonga (Prior and Rose 1966, Prior and Stanhope 1977, Prior et al. 1987). With this clear tendency of the South Pacific Polynesian towards elevated serum uric acid levels, it is surprising that a study on native Hawaiians (Healey et al. 1966) found serum uric acid significantly lower than in the Maori, though still higher than European levels. Further studies on Hawaiian groups would be of interest.

Gout, the gross manifestation of uric acid disorders, varies in incidence in men from 2.4% in Samoa and Pukapuka, to 5.5% in Rarotonga and 8.8% in the New Zealand Maori. Women generally show incidences of less than 1%, being protected by hormonal factors.

The body pool of uric acid derives from nucleotides and nucleoproteins, breakdown products of meat. Excretion is predominantly through the kidneys. Raised blood uric acid may be a consequence of a diet rich in precursors, excess production by the body, or of deficient excretion by gut and kidney. On the excretory side there is evidence that in the New Zealand Maori there is reduced renal urate clearance compared to Europeans (Gibson *et al.* 1984), and this has been related to reduced proximal tubular sodium clearance (Maling *et al.* in press). (In passing it may be noted that Polynesians may have significantly larger kidneys than Europeans: Reid *et al.* 1990.) On the input side, usually about one-third of nucleoprotein is derived from the diet, and about two-thirds is derived from within the body by breakdown of tissues in the natural course of turnover and replacement and repair (Dieppe and Calvert 1983). The tissue contributing most to endogenous purine, and thus to uric acid, is skeletal muscle. Meat typically contains about 50 mg of purine nitrogen (equivalent to 170 mg of uric acid) in every 100 grams. Levels in other foods are much lower. Subjects fed the equivalent of 1.2 kg of meat per day double their serum uric acids. The skeletal muscle mass of a lean individual of 75 kg is about 30–33 kg, and turnover of myosin, the dominant muscle protein, is about 30 days or a little less. That is, a muscular male is making quite a meal of himself each day.

There is support for the view that an individual's muscle mass correlates with blood uric acid level. Brauer and Prior noted 'A previously unreported relationship linking muscle size to the incidence of gout in males' (1978: 466) as a major finding of their study of gout in the New Zealand Maori. While their paper was in press, an American study was published (Fessel and Barr 1978), in which blood uric acid levels of some 69,000 individuals were examined by multivariate analysis against 77 other body variables. In this study the dominant association of raised serum uric acid levels again was with lean body mass. The authors concluded that, with regard to raised levels of uric acid in the blood, it seemed reasonable to infer that muscle bulk played an influential role.

The evidence, then, is that the muscular Polynesian has both an input and an output problem with uric acid. Factors incidental to a modern westernized pattern of existence, such as the intake of purine-rich beer, may readily tip people of this genetic predisposition into the clinical condition.

RETROSPECT

The view taken in this book has been that the people of the Pacific, through selective adaptation and genetic drift, have to a large extent been physically shaped by their environment. This view contrasts with those that interpret the variety in terms of ancestral strains and racial types and the mixing of these. In the light of more than a century of evolutionary biology the persistence of these latter views may seem, to some, unnecessary.

In Near Oceania the variety of human biology, whether expressed openly in the phenotype or covertly in the genome, is well known. There should be nothing surprising in this finding. No other part of the world so lends itself to the prospect of human variation as does the expanse of this microcosm of small and large islands. Nowhere else on the globe have the processes of adaptation and drift had such opportunity. Here also are the small populations that provide 'great opportunities for a genetic revolution . . . It would not surprise me if under these circumstances new species could arise in a period measured only in thousands or even hundreds of years' (Mayr 1970: 349). What is being suggested here of course is nothing like speciation, but only significant phenotypic change within a species.

For *Homo sapiens*, this region, where land erratically meets water, is effectively a species boundary, and at such boundaries organisms are placed under environmental stress. Here phenotypic and genotypic variability of organisms tends to be high, and 'from the molecular to the biogeographic levels of organization, stressful environmental conditions underlie much evolutionary change' (Parsons 1991: 1). Such variability may be seen, in a rather teleological sense, as an environmental probe – species groping for the right phenotype to enable them to move on.

Evolutionary change may be most likely to occur in populations from habitats not quite at the environmental extreme, where there is a certain degree of stress associated with sufficient metabolic energy to accommodate some adaptive change. Such populations may show considerable phenotypic plasticity, and this is probably the situation with *Homo sapiens* in Near Oceania. Extreme selective pressure may, however, reduce the plasticity that can be expressed by a selected genotype, and in this regard there does not seem to be much plasticity in the Polynesian phenotype. Nonetheless the peoples of Remote Oceania do show the broad climatic tolerance characteristic of high-latitude, colder-climate individuals, compared with most of the peoples of Near Oceania, in typical low-latitude environments of limited range.

That the Pacific should have been the last area of the world to be settled is unsurprising, for here *Homo sapiens*, a terrestrial animal, faced a foreign environment. Other organisms, animals and plants, had colonized it, but randomly, by accident of drift on wind and current. The human settlement is set apart in that it was deliberate and controlled, and thus a matter of intellect. In that sense, if it is possible to rank such matters, the exploration and colonization of the Pacific may, in its generations, be regarded as the supreme collective human achievement.

REFERENCES

Abbie, A. A. 1957. Metrical characters of a central Australian tribe. *Oceania* 27: 220–243.

—— 1966. Physical characteristics. In *Aboriginal man in South and Central Australia*, ed. B. C. Cotton, pp. 9–45. Adelaide: University of Adelaide.

Adams, W. Y., Van Gerven, P. and Levy, G. 1978. The retreat from migrationism. *Annual Review of Anthropology* 7: 483–532.

Aguera y Infanzon, F. A. de. 1908. *Journal of the Principal Occurrences during the Voyage of the Frigate Santa Rosalia in the year 1770*. Cambridge: Hakluyt Society.

Aiello, L. and Dean, C. 1990. *An Introduction to Human Evolutionary Anatomy*. London: Academic Press.

Akel, R. N., Frankish, J. D., Powles, C. P., Tyler, H. R., Watt, J. M., Weston, H. J. and Prior, I. A. M. 1963. Anaemia in Maori and European infants and children on admission to hospital. *New Zealand Medical Journal* 62: 28–33

Alexander, R. D., Hoogland, J. L., Howard, R. D., Noonan, K. M. and Sherman, P. W. 1979. Sexual dimorphism and breeding systems in Pinnipeds, Ungulates, Primates and Humans. In *Evolutionary Biology and Human Social Behaviour*, ed. N. A. Chagnon and W. Irons, pp. 402–435. North Scituate: Duxbury Press.

Alkire, W. M. 1978. *Coral Islanders*. Arlington Heights: AHM Publishing.

Allbrook, D. 1961. The estimation of stature in British and East African males. *Journal of Forensic Medicine* 8: 15–28.

Allen, J. A. 1877. The influences of physical conditions on the genesis of species. *Radical Review* 1: 108–140.

Allen, J., Gosden, C., Jones, R. and White, J. P. 1988. Pleistocene dates for the human occupation of New Ireland, northern Melanesia. *Nature* 33: 707–709.

Allen, J., Gosden, C. and White, J. P. 1989. Human Pleistocene adaptations in the tropical island Pacific; recent evidence from New Ireland, a Greater Australian outlier. *Antiquity* 63: 548–561.

Alvesalo, L. and Tigerstedt, P. M. A. 1974. Heritabilities of human tooth dimensions. *Hereditas* 77: 311–318.

Amherst of Hackney and Thomson, B. 1901. *The Discovery of the Solomon Islands*. London: Hakluyt Society.

Anderson, A. J. 1991. The chronology of colonization in New Zealand. *Antiquity* 65: 767–795.

Anderson, D.M. and Webb, R.S. 1994. Ice age tropics revisited. *Nature* 367: 23–24.

Anderson, W. 1967. A journal of a voyage made in His Majesty's sloop *Resolution*. In *The Journals of Captain James Cook on his Voyages of Discovery. The Voyage of the Resolution and Discovery 1776–1780*, ed. J. C. Beaglehole, pp. 723–986. Cambridge University Press for Hakluyt Society.

Andrews, J. R. H. 1979. Ascaris egg in coprolite material. *New Zealand Medical Journal* 89: 274.

Asher, M. I., Douglas, C., Stewart, A. W., Quinn, J. P. and Hill, P. M. 1987. Lung volumes in Polynesian children. *American Review of Respiratory Diseases* 136: 1360–1365.

Ashworth, A. 1968. An investigation of very low calorie intakes reported in Jamaica. *British Journal of Nutrition* 22: 341–355.

Baker, B. J. and Armelagos, G. J. 1988. The origin and antiquity of syphilis. *Current Anthropology* 29: 703–720.

Baker, P. T. 1984. Migrations, genetics and the degenerative diseases of South Pacific islanders. In *Migration and Mobility*, ed. A. J. Boyce, pp. 209–239. London: Taylor and Francis.

Baker, P. T., Hanna, J. M. and Baker, T. S. (eds). 1986. *The Changing Samoans: behaviour and health in transition.* Oxford: Clarendon Press.

Baker, R. 1975. The Maori Pelvis. Unpublished MA thesis, University of Auckland.

Balliet, G., Rothhammer, F., Carnese, F. R., Bravi, C. M. and Bianchi, N. O. 1994. Founder mitochondrial haplotypes in Amerindian populations. *American Journal of Human Genetics* 54: 27–33.

Ballinger, S. W., Schurr, T. G., Torrini, A., Gan, Y-Y., Hodge, J. A., Hassan, K., Chen, K. H. S. and Wallace, D. C. 1992. Southeast Asian mitochondrial DNA analysis reveals genetic continuity of ancient Mongoloid migrations. *Genetics* 130: 139–152.

Barnett, C. 1983. The Maori Spine. Unpublished BA thesis. University of Otago.

Barnett, E. and Nordin, B. E. C. 1961. The radiological diagnosis of osteoporosis: a new approach. *Clinical Radiology* 11: 166–174.

Barnett, T., Graham, N., Cane, M., Zebiak, S., Dolan, S., O'Brien, J. and Legler, D. 1988. On the prediction of the *El Nino* of 1986–1987. *Science* 241: 194–196.

Beals, K. L. 1974. Headform and climatic stress. *American Journal of Physical Anthropology* 37: 85–92.

Beals, K. L., Smith, C. L. and Dodd, S. M. 1983. Climate and the evolution of brachycephalization. *American Journal of Physical Anthropology* 62: 425–438.

Beals, K. L., Smith, C. L. and Kelso, A. J. 1992. ABO phenotype and morphology. *Current Anthropology* 33: 221–224.

Beaven, D. W. 1977. Maori health. *New Zealand Medical Journal* 85: 540.

Beck-Nielsen, H., Vaag, A., Damsbo, P., Handberg, A., Nielsen, O. H., Henriksen, J. E. and Thyeronn, P. 1992. Insulin resistance in skeletal muscles in patients with NIDDM. *Diabetes Care* 15: 418–429.

Begg, P.R. 1954. Stone age man's dentition. *American Journal of Orthodontics* 40: 298–312; 373–383; 462–475; 517–531.

Behnke, A. R. 1959. The estimation of lean body weight from 'skeletal' measurements. *Human Biology* 31: 295–315.

Bell, D.G., Tikuisis, P. and Jacobs, I. 1992. Relative intensity of muscular contraction during shivering. *Journal of Applied Physiology* 72: 2336–2342.

Bellingshausen, T. 1945. *The Voyage of Captain Bellingshausen to the Antarctic Seas 1819–1821*, translated and edited by F. Debenham. Cambridge: Hakluyt Society.

Bellwood, P. S. 1985a. On Polynesians and Melanesians. *Journal of the Polynesian Society* 95: 131–134.

—— 1985b *Prehistory of the Indo-Malaysian Archipelago.* Sydney: Academic Press.

—— 1989. The colonization of the Pacific: some current hypotheses. In *The Colonization of the Pacific. A Genetic Trail,* ed. A. V. S. Hill and S. W. Serjeantson, pp. 1–59. Oxford: Clarendon Press.

Bennett, K. A. 1972. Lumbo-sacral malformations and spina bifida occulta in a group of proto-historic Modoc Indians. *American Journal of Physical Anthropology* 36: 435–440.

Berglund, O. 1963. The Bony Nasopharynx. *Acta Odontologica Scandinavica* 21 (suppl. 35): 1–137.

Bergman, R. A. M. and The, T. H. 1955. The length of the body and long bones of the Javanese. *Documenta de Medicina et Tropica* 7: 197–214.

Bergmann, C. 1847. Uber die verhaltuisseder Warmeokonomie der Thiere zu ihrer Grosse. *Göttinger Studien* No. 8.

Berry, A. C. and Berry, R. J. 1967. Epigenetic variation in the human cranium. *Journal of Anatomy* 101: 361–379

Best, E. 1924. *The Maori.* Wellington: Tombs.

Biggs, B. 1974. A drift voyage from Futuna to Cikobia. *Journal of the Polynesian Society* 83: 361–365.

Bindon, J. R. and Baker, P. T. 1985. Modernization, migration and obesity among Samoan adults. *Annals of Human Biology* 12: 67–76.

Bjork, A. 1947. *The Face in Profile.* Copenhagen: Odontologisk Boghandels.

Blackwood, B. 1931–32. Report on fieldwork in Buka and Bougainville. *Oceania* 2: 199–219.

—— 1935. *Both Sides of Buka Passage.* Oxford: Clarendon Press.

Blair, E. and Robertson, J. A. 1903. *The Philippine Islands 1493–1898.* Cleveland: Arthur H. Clark.

Blake, N. M., Hawkins, B. R., Kirk, R. L., Bhatia, K., Brown, P., Garruto, R. M. and Gajdusek, D. C. 1983. A population and genetic study of the Banks and Torres Islands Vanuatu and of the Santa Cruz Islands and Polynesian Outliers, Solomon Islands. *American Journal of Physical Anthropology* 62: 343–361.

Blanco, R. A., Acheson, R. M., Canosa, C. and Salomon, J. B. 1974. Weight, height and lines of arrested growth in young Guatemalan children. *American Journal of Physical Anthropology* 40: 39–48.

Bligh, W. 1961. *The Mutiny on Board H. M. S. Bounty.* New York: Signet.

Bocquet-Appel, J. and Masset, C. 1982. Farewell to paleodemography. *Journal of Human Evolution* 11: 321–333.

Bogin, B. 1988. *Patterns of Human Growth.* Cambridge University Press.

Borman, B. and Cryer, C. 1993. The presence of anencephalus and spina bifida in New Zealand. *Journal of Paediatrics and Child Health* 29: 282–288.

Bougainville, L. A. de. 1772. *A Voyage Round the World.* London: J. Nourse.

Bouvier, M. and Hylander, W. L. 1981. Effect of bone strain on cortical bone structure in macaques *Maca mulatta. Journal of Morphology* 167: 1–12.

Bowden, D. K., Hill, A. V. S., Higgs, D. R., Weatherall, D. J. and Clegg, J. B. 1985. The relative roles of genetic factors, dietary deficiency and infection in anaemia in Vanuatu, Southwest Pacific. *Lancet* 2: 1025–1028.

Boyce, A. J., Harrison, G. A., Platt, C. M., Hornabrook, R. W., Serjeantson, S., Kirk, R. L. and Booth, P. B. 1978. Migration and genetic diversity in an island population: Karkar, Papua New Guinea. *Proceedings of the Royal Society* B 202: 269–295.

Brace, C. L. and Hinton, R. J. 1981. Oceanic tooth-size variation as a reflection of biological and cultural mixing. *Current Anthropology* 225: 549–569.

Brace, C. L. and Hunt, K. D. 1990. A nonracial craniofacial perspective on human variation: A(ustralia) to Z(uni). *American Journal of Physical Anthropology* 82: 341–360.

Brace, C. L. and Mahler, P. E. 1971. Post-pleistocene changes in the human dentition. *American Journal of Physical Anthropology* 34: 191–204.

Brauer, G. W. and Prior, I. A. M. 1978. A prospective study of gout in New Zealand Maoris. *Annals of Rheumatic Diseases* 37: 466–472.

Broadbent, B. H., Broadbent, B. H. Jr. and Golden, W. H. 1975. *Bolton Standards of Dentofacial Developmental Growth.* St Louis: C. V. Mosby Co.

Bromage, T. G. 1980. A brief review of cartilage and controlling factors in chondrocranial morphogenesis. *Acta Morphologica Neerlando-Scandinavica* 18: 317–322.

Brown, T. 1973. *Morphology of the Australian Skull.* Australian Aboriginal Studies no. 49. Canberra: Australian Institute of Aboriginal Studies.

Brown, T. and Barrett, M. J. 1964. A roentgenographic study of facial morphology in a tribe of Central Australian Aborigines. *American Journal of Physical Anthropology* 22: 33–42.

Buck, P. H. 1922–3. Maori somatology. *Journal of the Polynesian Society* 31: 27–44, 145–153, 159–170, 32: 21–28.

—— 1925. The pre-European diet of the Maori. *New Zealand Dental Journal* 20: 203–217.

—— 1938. *Vikings of the Sunrise.* New York: Frederick A. Stokes.

—— 1950a. *Material Culture of Kapingamarangi.* Bernice P. Bishop Museum. Bulletin 200.

—— 1950b. *The Coming of the Maori.* Wellington: Government Printer.

Budd, G. M. 1965. Effects of cold exposure and exercise in a wet, cold Antarctic climate. *Journal of Applied Physiology* 20: 417–422.

Budd, G. M., Fox, R. H., Hendrie, A. L. and Hicks, K. E. 1974. A field study of thermal stress in New Guinea villagers. *Philosophical Transactions of the Royal Society* B 268: 393–400.

Budd, G. M., Brotherhood, J. R., Hendrie, A. l. and Jeffrey, S. C. 1991. Effects of fitness, fatness, and age on men's responses to whole body cooling in air. *Journal of Applied Physiology* 71: 2387–2393.

Buikstra, J. E., Frankenberg, S. R. and Konigsberg, L. W. 1990. Skeletal biological distance studies in American physical anthropology. *American Journal of Physical Anthropology* 82: 1–7.

Bulmer, S. 1979. Prehistoric ecology and economy in the Port Moresby region. *New Zealand Journal of Archaeology* 1: 5–27.

Buranarugsa, M. and Houghton, P. 1981. Polynesian head form: an interpretation of a factor analysis of Cartesian co-ordinate data. *Journal of Anatomy* 133: 333–350.

Burton, R. P. and Edholm, O. G. 1955. *Man in a Cold Environment.* London: Edward Arnold.

Butler, P. M. 1939. Studies in the mammalian dentition – and of the differentiation of the postcanine dentition. *Proceedings of the Zoological Society, London* B 109: 1–36.

Byron, J. 1964. *Byron's Journal of his Circumnavigation 1764–1766*, ed. R. E. Gallagher. Cambridge: Hakluyt Society.

Cachel, S. M. 1979. A functional analysis of the primate masticatory system and the origin of the anthropoid post-orbital septum. *American Journal of Physical Anthropology* 50:1–18.

Calder, W. A. 1984. *Size, Function, and Life History*. Cambridge, Mass.: Harvard University Press.

Cameron, N. 1991. Measurement issues related to the anthropometric assessment of nutritional status. In *Anthropometric Assessment of Nutritional Status*, ed. J. H. Himes, pp 347–364. New York: Wiley-Liss.

Cann, R. L., Stoneking, M. and Wilson, A. C. 1987. Mitochondrial DNA and human evolution. *Nature* 325: 31–36.

Cann, R. L. and Wilson, A. C. 1983. Length mutations on human mitochondrial DNA. *Genetics* 104: 698–711.

Cantwell, R. J. 1973. A prospective study of Maori infant health and the problem of nutritional anaemia. *New Zealand Medical Journal* 78: 61–65.

Carano, P. and Sanchez, P. C. 1964. *A Complete History of Guam*. Tokyo: Tuttle.

Carey, J. W. and Steegman, A. T. 1981. Human nasal protrusion, latitude and climate. *American Journal of Physical Anthropology* 56:313–319.

Carlsson, D. S. and Van Gerven, D. P. 1977. Masticatory function and post-Pleistocene evolution in Nubia. *American Journal of Physical Anthropology* 46: 495–506.

Carteret, P. 1965. *Philip Carteret's Voyage Round the World 1766–1769*, ed. H. M. Wallis. Cambridge: Hakluyt.

Cartlidge, I. J. 1983. Club foot in the Polynesian: an epidemiological survey. *New Zealand Medical Journal* 96: 515–517.

Cassidy, C. M. 1984. Skeletal evidence for prehistoric subsistence adaptation in the Central Ohio river valley. In *Paleopathology at the Origins of Agriculture*, ed. M. N. Cohen and G. J. Armelagos, pp. 307–346. Orlando: Academic Press.

Cattani, J. A. 1992. The epidemiology of malaria in Papua New Guinea. In *Human Biology in Papua New Guinea*, ed. R. D. Attenborough and M. P. Alpers, pp. 302–312. Oxford: Clarendon Press

Caughey, D. E. and Woodfield, D. G. 1983. HLA and rheumatic disease associations in Polynesians and Europeans. *Australian and New Zealand Journal of Medicine* 13: 218–9

Cavalli-Sforza, L. L. and Bodmer 1971. *The Genetics of Human Populations*. San Francisco: Freeman.

Cavalli-Sforza, L. L., Piazza, A., Menozzi, P. and Mountain, J. 1988. Reconstruction of human evolution: bringing together genetic, archaeological and linguistic data. *Proceedings of the National Academy of Science* 85: 6002–6006.

Chai, C. K. 1967. *Taiwan Aborigines*. Cambridge, Mass.: Harvard University Press.

Chamla, M-C. 1964. The increase in stature in France from 1880 to 1960: A comparison with the countries of Western Europe. *Yearbook of Physical Anthropology* 12: 146–183.

Champness, L. T., Bradley, M. A. and Walsh, R. J. 1963. A study of the Tolai in New Britain. *Oceania* 34: 66–75.

Chapman, C. J. 1983. Ethnic differences in the incidence of cleft lip and/or cleft palate in Auckland 1960–1976. *New Zealand Medical Journal* 96: 327–329.

Chen, K. H., Cann, H., Chen, T. C., Van West, B. and Cavalli-Sforza, L. 1985. Genetic markers of an aboriginal Taiwanese population. *American Journal of Physical Anthropology* 66: 327–337.

Cheverud, J. M. and Buikstra, J. E. 1981. Quantitative genetics of skeletal non-metrics in the Rhesus macaques on Cayo Santiago. *American Journal of Physical Anthropology* 54: 43–49.

Christiansen, S. 1975. *Subsistence on Bellona Island: language and culture on Rennell and Bellona Islands.* Volume 5. Copenhagen: National Museum of Denmark.

Chung, C. S., Mi, M. P. and Beechert, A. M. 1987. Genetic epidemiology of cleft lip with or without cleft palate in the population of Hawaii. *Genetics and Epidemiology* 4: 415–423.

Chung, C. S., Nemechek, R. W., Larsen, F. J. and Ching, G. H. S. 1969. Genetic and epidemiological studies of club foot in Hawaii. *Human Heredity* 19: 321–342.

Clark, N. G. 1990. Periodontal defects of pulpal origin: evidence in early man. *American Journal of Physical Anthropology* 82: 371–376.

Clark, R. P. and Edholm, O. G. 1985. *Man and his Thermal Environment.* London: Edward Arnold.

Clark, R. S., Hellon, R. F. and Lind, A. R. 1958. Duration of sustained contractions of the human forearm at different muscle temperatures. *Journal of Physiology* 143: 454–462.

Clark, S. K. 1982. The association of early childhood enamel hypoplasias and radiopaque transverse lines in a culturally diverse skeletal sample. *Human Biology* 54: 77–84.

Clerke, C. 1961. Clerke's Log. *In The Journals of Captain James Cook on his Voyages of Discovery. The Voyage of the Resolution and Adventure 1772–1775,* ed. J. C. Beaglehole, pp. 753–766. Cambridge University Press for Hakluyt Society.

—— 1964. In *Byron's Journal of his Circumnavigation,* ed. R. E. Gallagher, pp. 210–213. Cambridge: Hakluyt Society.

Coale, AJ. and Demeny, J. P. 1983. *Regional Model Life Tables and Stable Populations.* New York: Academic Press.

Cook, J. 1955. *The Journals of Captain James Cook on his Voyages of Discovery. The voyage of the Endeavour 1768–1771,* ed. J. C. Beaglehole. Cambridge University Press for Hakluyt Society.

—— 1961. *The Journals of Captain James Cook on his Voyages of Discovery. The Voyage of the Resolution and Adventure 1772–1775,* ed. J. C. Beaglehole. Cambridge University Press for Hakluyt Society.

—— 1967. *The Journals of Captain James Cook on his Voyages of Discovery. The voyage of the Resolution and Discovery 1776–1780,* ed. J. C. Beaglehole. Cambridge University Press for Hakluyt Society.

Cook, S. F. 1973. The significance of disease in the extinction of the New England Indians. *Human Biology* 45: 485–508.

Coon, C. S., 1962. *The Origin of Races.* New York: Knopf

—— 1965. *The Living Races of Man.* London: Jonathan Cape.

Cordy, R. 1983. Social stratification in the Mariana Islands. *Oceania* 53: 272–276.

Corney, B. G. 1904. *The Voyages of Pedro Fernandez de Quiros,* ed. C. Markham. Cambridge: Hakluyt Society.

Corruccini, R. S. 1975. Multivariate analysis in biological anthropology: some considerations. *Journal of Human Evolution* 4: 1–19.

Coster, G. D., Hunter, J. D. and Jackson, D. 1975. Fat analysis of some New Zealand foods. *Journal of the New Zealand Dietetic Association* 29: 1–4.

Cotes, J. E. 1979. *Lung function: Assessment and Application in Medicine.* 4th ed. Oxford: Blackwell.

Cotes, J. E., Reed, J. W. and Mortimore, I. L. 1982. Determinants of capacity for physical work. *Symposia for the Society for the Study of Human Biology* 22: 39–64.

Court, A., 1948. Wind chill. *Bulletin of the American Meteorological Society* 29: 487–493.

Cox, M. and Scott, A. 1992 Evaluation of the obstetric significance of some pelvic characters in an eighteenth century British sample of known parity status. *American Journal of Physical Anthropology* 89: 431–440.

Crittingden, C. and Baines, J. 1989. On assessing nutritional status in the Papua New Guinea Highlands. *Current Anthropology* 30: 208–209.

Cronk, C. E. and Roche, A. F. 1982. Race- and sex-specific reference data for triceps and subscapular skinfolds and weight/stature. *American Journal of Clinical Nutrition* 35: 347–354.

Crozet, M. 1985. In *Early Eyewitness Accounts of Maori Life 2. Extracts from Journals Relating to the Visit to New Zealand in May–July 1772 of the French Ships Mascarin and Marquis de Castries under the Command of M.-J, Marion du Fresne,* transcription and translation by Isabel Ollivier, pp. 5–17. Wellington: Alexander Turnbull Library Endowment Trust with Indosuez New Zealand.

Currey, J. 1984. *The Mechanical Adaptations of Bones.* Princeton: Princeton University Press.

Dacie, J. 1988. *The Haemolytic Anaemias* Vol II. Edinburgh: Churchill Livingstone

Daiger, S. P., Reed, L., Huang, S-S., Zeng, Y-T., Wang, T., Lo, W. H. Y., Okano, Y., Hase, Y., Fukuda, Y., Oura, T., Tada, K. and Woo, S. L. C. 1989. Polymorphic DNA haplotypes at the phenylalanine hydroxylase (PAH) locus in Asian families with phenylketonuria. *American Journal of Human Genetics* 45: 319–324.

Damon, A. 1977. *Human Biology and Ecology.* New York: Norton.

Damon, A. and Goldman, R. F. 1964. Predicting fat from body measurements. *Human Biology* 36: 32–44.

Daniels, F. and Baker, P.T. 1961. Relationship between body fat and shivering in air at 15°C. *Journal of Applied Physiology* 16: 421–425.

Davenport, J. 1992. *Animal Life at Low Temperature.* London: Chapman and Hall.

Davidson, J. M. 1969. Archaeological excavations in two burial mounds at 'Atele, Tongatapu. *Records of the Auckland Institute and Museum* 6: 251–286.

——— 1984. *The Prehistory of New Zealand.* Auckland: Longman Paul.

Davis, P. J. and Hersh, R. 1986. *Descartes' Dream: the World According to Mathematics.* London: Pelican.

De Hamel, F. A. and Welford, B. 1983. Lung function in Maoris and Samoans working in New Zealand. *New Zealand Medical Journal* 96: 560–562.

De Souza, P. and Houghton, P. 1977. The mean measure of divergence and the use of non-metric data in the estimation of biological distances. *Journal of Archaeological Science* 4: 163–169.

DeLattre, A. and Fenart, R. 1960. *L'Hominisation du Crâne.* Paris: Editions du Centre National de la Recherche Scientifique.

Dening, G. M. 1972. The geographical knowledge of the Polynesians and the nature of inter-island contact. In *Polynesian Navigation,* ed. J. Golson, pp. 102–131. Wellington: Polynesian Society.

—— 1980. *Islands and Beaches, Discourse on a Silent Land: Marquesas 1774–1880.* Melbourne: University Press of Melbourne.

Dennett, G. and Connell, J. 1988. Acculturation and health in the Highlands of Papua New Guinea. *Current Anthropology* 29: 273–295.

Dennison, K. J. 1992. *The human skeletal material from the marae Te Tahata, Tepoto, northern Tuamotu archipelago.* Dunedin: Otago University, Department of Anatomy.

Dez, le. 1985. Summary of a new voyage to Australasia. In *Early Eyewitness Accounts of Maori Life 2. Extracts from Journals Relating to the Visit to New Zealand in May–July 1772 of the French Ships Mascarin and Marquis de Castries under the Command of M.-J. Marion du Fresne,* transcription and translation by Isabel Ollivier, pp. 255–361. Wellington: Alexander Turnbull Library Endowment Trust with Indosuez New Zealand.

Diamond, J. 1988. Express train to Polynesia. *Nature* 336: 307–308.

Dieppe, P. and Calvert, P. 1983. *Crystals and Joint Disease.* London: Chapman and Hall.

Dobbing, J. 1982. Maternal nutrition, breast feeding, and contraception. *British Medical Journal* 284: 1725–1726.

Domett, K. 1994. *Femoral and Acetabular Relationships in Polynesian Skeletal Material.* Unpublished MSc (Hons) study, Department of Anatomy and Structural Biology, University of Otago.

Doran, E. 1972. Wa, vinta, and trimaran. *Journal of the Polynesian Society* 81: 144–159.

Douglas, M. T. 1991. Wryneck in the ancient Hawaiians. *American Journal of Physical Anthropology* 84: 261–272.

Dowse, G. and Zimmet, P. 1993. The thrifty genotype in non-insulin-dependent diabetes. *British Medical Journal* 306: 532–533.

Dowse, G. K., Spark, R. A., Mavo, B., Hodge, A. M., Erasmus, R. T., Gwalimu, M., Knight, L. T., Koki, G. and Zimmet, P. Z. 1994. Extraordinary prevalence of non-insulin-dependent diabetes mellitus and bimodal plasma glucose distribution in the Wanigela people of Papua New Guinea. *Medical Journal of Australia* 160: 767–774.

Du Clesmeur, M. 1914. Journal of the Marquis de Castries. In *Historical Records of New Zealand,* ed. R. McNab, pp. 445–481. Wellington: Government Printer.

DuBois, D. and DuBois, W. F. 1915. The measurement of the surface area of man. *Archives of Internal Medicine* 15: 858–881.

Duckworth, W. L. H. 1900. On a collection of crania, with two skeletons of the Moriori, or aborigines of the Chatham Island. *Journal of the Anthropological Institute* 30: 141–152.

Ducros, A. and Ducros, J. 1979. Recent anthropological data on the Ammassalimiut Eskimo of East Greenland: comparative studies. In *Physiological and Morphological Adaptation and Evolution,* ed. W. A. Stini, pp. 55–67. The Hague: Mouton.

Ducros, J. and Ducros, A. 1987. Age at menarche in Tahiti. *Annals of Human Biology* 146: 559–62

Dunn, L. C. and Tozer, A. M. 1928. An anthropometric study of Hawaiians of pure and mixed blood. *Papers of the Peabody Museum* 9: 89–211.

Durnin, J. V. G. A. and Womersley, J. 1974. Body fat assessed from total body density and its estimation from skinfold thicknesses: measurements in 481 men and women aged from 16 to 72 years. *British Journal of Nutrition* 32: 77–97.

Dyen, I. 1965. A lexicostatistical classification of the Austronesian languages. *International Journal of American Linguistics, Memoir 19.*

Eberhart, R. C. 1985. Thermal models of single organs. In *Heat Transfer in Medicine and Biology,* ed. A. Shitzer and R. C. Eberhart, pp. 261–324. New York: Plenum Press.

Edmundson, W. 1980. Adaptation to undernutrition: how much food does man need? *Social Science and Medicine* 14D: 119–126.

Ellis, K. J. and Cohn, S. H. 1975. Correlation between skeletal calcium mass and muscle mass in man. *Journal of Applied Physiology* 38: 455–460.

Ellis, W. 1827. *A Narrative of a Tour through Hawaii.* London: Fisher and Jackson.

Elsner, R. W. 1963. Skinfold thickness in primitive peoples native to cold climates. *Annals of the New York Academy of Sciences* 110: 503–514.

Elwood, J. M. and Elwood, J. H. 1980. *Epidemiology of Anencephalus and Spina Bifida.* Oxford: Oxford University Press.

Enlow, D. H. 1982. *Handbook of Facial Growth.* Philadelphia: W. B. Saunders.

Erskine, J. E. 1967. *Journal of a Cruise Among the Islands of the Western Pacific.* London: Dawson.

Evans, I. A. and Mason J. 1965. Carcinogenic activity of bracken. *Nature* 208: 913–914.

Evans, I. A. and Osman, M. A. 1974. Carcinogenicity of bracken and shikinic acid. *Nature* 250: 348–349.

Eveleth, P. B. 1975. Differences between ethnic groups in sex dimorphism of adult height. *Annals of Human Biology* 2: 35–39.

Eveleth, P. B. and Tanner, J. M. 1976. *Worldwide Variation in Human Growth.* Cambridge: Cambridge University Press.

—— 1990. *Worldwide Variation in Human Growth,* 2nd ed. Cambridge: Cambridge University Press.

Feinberg, R. 1988. *Polynesian Seafaring and Navigation; Ocean Travel in Anutan Culture and Society.* Kent: Kent State University Press.

Feldesman, M. R., Kleckner, J. G. and Lundy, J. K. 1990. Femur/stature ratios and estimates of stature in mid- and late-Pleistocene fossil hominids. *American Journal of Physical Anthropology* 83: 359–372.

Ferris, B. G. and Smith, C. W. 1953. Maximum breathing capacity and vital capacity in female children and adolescents. *Pediatrics,* 12: 341–352.

Ferris, B. G., Whittenberger, J. L. and Gallagher, J. R. 1952. Maximum breathing capacity and vital capacity of male children and adolescents. *Pediatrics,* 9: 659–670.

Ferro-Luzzi, A., Norgan, N. G. and Durnin, J. V. G. A. 1978. The nutritional status of some New Guinea children as assessed by anthropometric, biochemical and other indices. *Ecology of Food and Nutrition* 7: 115–128

Fessel, W. J. and Barr, G. D. 1978. Uric acid, lean body weight, and creatinine interactions: results from regression analysis of 78 variables. *Seminars in Arthritis and Rheumatism* 7: 115–121

Finau, S. A., Stanhope, J. M., Prior, I. A. M., Joseph, J. G., Puloka, S. T. and Leslie, P. N. 1983. The Tonga cardiovascular and metabolic study. *Community Health Studies* 7: 57–77.

Finney, B. R., 1979. *Hokule'a: the Way to Tahiti.* New York: Dodd, Mead.

Finney, B. R., Frost, P., Rhodes, R. and Thompson, N. 1989. Wait for the west wind. *Journal of the Polynesian Society* 98: 261–302.

Finney, B. R., Kilonsky, B. J., Somsen, S. and Stroup, E. D. 1986. Relearning a vanishing art. *Journal of the Polynesian Society* 95: 41–90.

Firth, R., 1963. *We, the Tikopia.* Boston: Beacon Press.

Fisher, R. A. 1936. 'The Coefficient of Racial Likeness' and the future of craniometry. *Journal of the Royal Anthropological Institute* 66: 57–63.

Flower, W. H., 1896. Moriori crania. *Journal of the Anthropological Institute* 26: 295–296.

Food and Agricultural Organisation/World Health Organisation. 1973. *Energy and Protein Requirements.* WHO Technical Report Series No. 522. Geneva: Food and Agricultural Organisation/World Health Organisation.

—— 1985. *Energy and Protein Requirements.* Technical Report Series No. 724. Geneva: Food and Agricultural Organisation/World Health Organisation.

Forster, J. R. 1778. *Observations Made During a Voyage Round the World with Captain Cook.* London: Robinson.

—— 1982. *The Resolution Journal of Johann Reinhold Forster 1772–1775.* London: Hakluyt Society.

Fox, A. 1983. Pa and people in New Zealand: an archaeological estimate of population. *New Zealand Journal of Archaeology* 5: 5–18.

Fox, R. H., Budd, G. M., Woodward, P. M., Hackett, A. J. and Hendrie, A. L. 1974. A study of temperature regulation in New Guinea people. *Philosophical Transactions of the Royal Society* B. 268: 375–391.

Frayer, D. W. 1988. Auditory exostoses and evidence for fishing at Vlasac. *Current Anthropology* 29: 346–348

Freycinet, L. C. de S. 1978. Hawaii in 1819. *Pacific Anthropological Records* No. 26.

Friederici, G. 1913. Wissenschaftliche Ergebnisse einer amtliche Forschungsreise nach dem Bismarck-Archipel im Jahre 1908. II. Beitrage zur Volkerkunde Sprachenkunde von Deutsch-Neuguinea. 1912. III. Untersuchungen über eine melanesische Wundertsrasse, 1913. Erganzungsheft nos. 5, 7 der Mitteilungen aus den Deutschen Schutzgebieten: Berlin

Friedlaender, J. S. 1975. *Patterns of Human Variation.* Cambridge, Mass.: Harvard University Press.

—— 1987a. *The Solomon Islands Project.* Oxford: Clarendon.

—— 1987b. Conclusion. In *The Solomon Islands Project,* ed. J. S. Friedlaender, pp. 351–362. Oxford: Clarendon.

Friedlaender, J. S. and Page, L. B. 1987. Epidemiology. In *The Solomon Islands Project,* ed. J. S. Friedlaender, pp. 89–124. Oxford: Clarendon.

Friedlaender, J. S. and Rhoads, J. G. 1987. Longitudinal anthropometric changes in adolescents and adults. In *The Solomon Islands Project,* ed. J. S. Friedlaender, pp. 283–306. Oxford: Clarendon.

Frisancho, A. R. 1974. Triceps skin fold and upper arm muscle size norms for assessment of nutritional status. *American Journal of Clinical Nutrition* 27: 1052–1058.

Frisancho, A. R. and Flegel, P. N. 1983. Elbow breadth as a measure of frame size for US males and females. *American Journal of Clinical Nutrition* 37: 311–314.

Froehlich, J. W. 1970. Migration and plasticity of physique in the Japanese-Hawaiians in America. *American Journal of Physical Anthropology* 32: 429–442.

—— 1987. Fingerprints as phylogenetic markers in the Solomon Islands. In *The Solomon Islands Project*, ed. J. S. Friedlaender, pp. 175–214. Oxford: Clarendon.

Froehlich, J. W. and Giles, E. 1981a. A multivariate approach to fingerprint variation in Papua New Guinea: the implications for prehistory. *American Journal of Physical Anthropology* 54: 73–91.

—— 1981b. A multivariate approach to fingerprint variation in Papua New Guinea: perspectives on the evolutionary stability of dermatoglyphic markers. *American Journal of Physical Anthropology* 54: 93–106.

Froese, G. and Burton, A. C. 1957. Heat losses from the human head. *Journal of Applied Physiology* 10: 235–241.

Fry, E. I. 1976. Dental development in Cook Island children. In *The Measures of Man*, ed. E. Giles and J. S. Friedlander, pp. 164–180. Cambridge, Mass.: Peabody Museum Press.

Fuji, A. 1960. On the relation of long bone lengths of limb to stature. *Bulletin of the School of Physical Education, Juntendo University* 3: 49–61.

Gabel, N. E. 1958. *A Racial Study of the Fijians.* University of California Anthropological Records Vol 20.

Galdikas, B. M. F. and Wood, J. W. 1990. Birth spacing patterns in humans and apes. *American Journal of Physical Anthropology* 83: 185–191

Galton, F. 1892. *Fingerprints.* London: Macmillan.

Garcia, F. 1937. *Life and Martyrdom of the Venerable Father Diego Luis de Sanvitores*, translated by M. Higgins. Guam: Guam Recorder, April 1937.

Garcia, R. I., Mitchell, R. E., Bloom, J. and Szabo, G. 1977. The number of epidermal melanocytes, hair follicles and sweat ducts in skin of Solomon Islanders. *American Journal of Physical Anthropology* 47: 427–434.

Garn, S. M. 1956. Comparison of pinch-caliper and X-ray measurements of skin plus subcutaneous fat. *Science* 124: 178.

—— 1970. *The Earlier Gain and Later Loss of Cortical Bone in Nutritional Perspective.* Springfield, Ill.: Thomas.

Garn, S. M., Lewis, A. B. and Kerewesky, R. 1963. Third molar agenesis and size reduction of the remaining teeth. *Nature* 200: 488.

—— 1968. The magnitude and implications of the relationship between tooth size and bodysize. *Archives of Oral Biology* 13: 129–131.

Garn, S. M., Osbourne, R. H. and McCabe, K. D. 1979. The effect of prenatal factors on crown dimensions. *American Journal of Physical Anthropology* 51: 665–678.

Garn, S. M., Rohmann, C. G. and Guzman, M. A. 1966. Malnutrition and skeletal development in the pre-school child. In *Preschool Child*

Malnutrition, pp. 43–62. Washington DC: National Academy of Sciences – National Research Council.

Garn, S. M., Sandusky, S. T., Rosen, N. N. and Trowbridge, F. 1973. Economic impact on postnatal ossification. *American Journal of Physical Anthropology* 38: 1–4.

Garn, S. M., Sullivan, T. V., Decker, S. A. and Hawthorne, V. M. 1991. On the optimum number of metacarpals for Roentgenogrammetric measurement. *American Journal of Physical Anthropology* 85: 229–232.

Garrison, J. S. 1957. Cited by Behnke 1957.

Gaulin, S. J. C. and Boster, J. S. 1992. Human marriage systems and sexual dimorphism in stature. *American Journal of Physical Anthropology* 89: 467–475.

Genoves, S. 1967. Proportionality of the long bones and their relationship to stature in Mesoamericans. *American Journal of Physical Anthropology* 26: 67–77.

Gibbons, A. 1993. Pleistocene population explosions. *Science* 262: 27–28.

Gibson, T., Waterworth, R., Hatfield, P., Robinson, G. and Bremner, K. 1984. Hyperuricaemia, gout and kidney function in New Zealand Maori men. *British Journal of Rheumatology* 23: 276–82.

Gilbert, N. 1989. *Biometrical Interpretation.* Oxford: Oxford University Press.

Giles, E. and Friedlaender J. S. (eds). 1976. *The Measures of Man.* Cambridge, Mass.: Peabody Museum Press.

Gindhart, P. S. 1969. The frequency of appearance of transverse lines in the tibia in relation to childhood illnesses. *American Journal of Physical Anthropology* 31: 17–22.

Gingerich, P. D., 1979. The human mandible: lever, link, or both? *American Journal of Physical Anthropology* 51: 135–137.

Gingerich P. D., Smith, B. H. and Rosenberg, K. 1982. Allometric scaling in the dentition of primates and prediction of body weight from tooth size in fossils. *American Journal of Physical Anthropology* 58: 88–101.

Gladwin, T. 1970. *East is a Big Bird.* Cambridge, Mass.: Harvard University Press.

Glenister, T. W. 1976. An embryological view of cartilage. *Journal of Anatomy* 122:323–330.

Goodman, A. H. and Clark, G. A. 1981. Harris lines as indicators of stress in prehistoric Illinois populations. In *Biocultural Adaptation: Comprehensive Approaches to Skeletal Analysis,* ed. D.L. Martin and M.P. Bumsted, pp. 35–46. Amherst: University of Massachussets Department of Anthropology.

Goose, D. H. 1971. The inheritance of tooth size in British families. In *Dental Morphology and Evolution,* ed. A. A. Dahlberg, pp. 263–270. Chicago: University of Chicago Press.

Gopalan, C. and Belavady, B. 1961. Nutrition and lactation. *Federation Proceedings* 20 Suppl 7: 177–184.

Gordon, J. E. 1976. *The New Science of Strong Materials.* London: Penguin.

—— 1978. *Structures.* London: Penguin.

Gosden, C., Allen, J., Ambrose, W., Anson, D., Golson, J., Green, R., Kirch, P., Lilley, I., Specht, J. and Spriggs, M. 1989. Lapita sites of the Bismarck Archipelago. *Antiquity* 63: 561–586.

Gould, S. J. 1981. *The Mismeasure of Man.* New York: Norton.

Gould, S.J. and Lewontin, R. C. 1979. The spandrels of San Marco and the Panglossian paradigm: a critique of the adaptionist programme. *Proceedings of the Royal Society of London* B 205: 581–598.

Gower, J. C. 1972. Measures of taxonomic distance and their analysis. In *The Assessment of Population Affinities in Man*, ed. J. S. Weiner and J. Huizinga, pp. 1–24. Oxford: Clarendon Press.

Gracey, M. 1991. Nutrition and physical growth. In *Anthropometric Assessment of Nutritional Status*, ed. J. H. Himes, pp. 29–49. New York: Wiley-Liss.

Gray, J. P. and Wolfe, L. D. 1980. Height and sexual dimorphism of stature among human societies. *American Journal of Physical Anthropology* 53: 441–456.

Gray, R. H., Campbell, O. M., Apelo, R., Eslami, S. S., Zacur, H., Ramos, R. M., Gehret, J. C. and Labbok, M. H. 1990. Risk of ovulation during lactation. *Lancet* 335: 25–29.

Green, R. C. 1991. Near and Remote Oceania – disestablishing 'Melanesia' in culture history. In *Man and a Half: Essays in Honour of Ralph Bulmer*, ed. A. K. Pawley. pp. 491–502. Auckland: The Polynesian Society.

Green, R. C., Anson, D. and Specht, J. 1989. The SAC burial ground, Watom Island, Papua New Guinea. *Records of the Australian Museum* 41: 215–221.

Green, R. F. and Suchey, J. M. 1976 The use of inverse sine transformation in the analysis of non-metric cranial data. *American Journal of Physical Anthropology* 45: 61–69

Gregg, J. B., Steele, J. P. and Holtzhueter, A. M. 1965. Roentgenographic evaluation of temporal bones from South Dakota Indian burials. *American Journal of Physical Anthropology* 23: 51–61.

Grewal, M. S. 1962. The rate of divergence in the C57BL strain of mice. *Genetical Research* 3: 226–237.

Groube, L., Chappell, J., Muke, J., and Price, D. 1986. A 40,000 year-old occupation site at Huon Peninsula, Papua New Guinea. *Nature* 324: 453–455.

Groves, C. P. 1989. A regional approach to the problem of the origin of modern humans in Australasia. In *The Human Revolution*, ed. P. Mellars and C. Stringer, pp. 274–285. Edinburgh: University of Edinburgh Press.

Guglielmino-Matessi, C. R., Gluckman, P. and Cavalli-Sforza, L. L. 1977. Climate and the evolution of skull metrics in Man. *American Journal of Physical Anthropology* 50: 549–564.

Guppy, H. D. 1887. *The Solomon Islands and Their Natives*. London: Sonnescheim, Lowrey and Co.

Habgood, P. J. 1989. The origin of anatomically-modern humans in Australasia. In *The Human Revolution*, ed. P. Mellars and C. Stringer, pp. 245–273. Edinburgh: University of Edinburgh Press.

Hackett, C. J. 1976. *Diagnostic Criteria of Syphilis, Yaws and Treponarid Treponematoses and of some other Diseases in Dry Bones*. Berlin: Springer.

Haddon, A. C. 1937. *The Canoes of Melanesia, Queensland and New Guinea*. Bernice P. Bishop Museum Special Publication 28.

Haddon, A. C. and Hornell, J. 1936–38. *Canoes of Oceania*. Honolulu: Bernice P. Bishop Museum Press.

Hagelberg, E. and Clegg, J. B. 1993. Genetic polymorphisms in prehistoric Pacific islanders determined by analysis of ancient bone DNA. *Proceedings of the Royal Society of London* B 252: 163–170.

Hall, R. L. 1981. Misuse of statistics. 15. Analysis of dental traits to infer human migration. *New York Statistician* 33: 5–6.

Hamilton, M. E. 1982. Sexual dimorphism in skeletal samples. In *Sexual Dimorphism in Homo sapiens: a Question of Size*, ed. R. L. Hall, pp. 107–163. New York: Praeger.

Hammel, H. T. 1959. Thermal and metabolic responses of the Australian Aborigine exposed to moderate cold in summer. *Journal of Applied Physiology* 14: 605–615.

—— 1964. Terrestrial animals in cold: recent studies of primitive man. In *Handbook of Physiology, Section 4: Adaptation to the Environment*, ed. D. B. Dill, pp. 413–434. Washington: American Physiological Society.

Han, T.L. 1986. Hawaiian mortuary practices. In *Moe kau a ho'oilo: Hawaiian mortuary practices at Keopu, Kona, Hawaii*, ed. T. L. Han, S. L. Collins, S. D. Clark, and A. Garland, pp. 11–21. Honolulu: Bernice P. Bishop Museum.

Hanihara, K. 1969. Mongoloid dental complex in the permanent dentition. In *Proceedings of the 8th International Congress of Anthropological and Ethnological Sciences*, pp. 298–300.

—— 1992. Dental variation of the Polynesian populations. *Journal of the Anthropological Society of Nippon* 100: 291–302.

Hanken, J. 1983. Miniaturization and its effects on cranial morphology in plethodontid salamanders, Genus *Thorius (Amphibia, Plethodontidae)*: II. The fate of the brain and sense organs and their role in skull morphogenesis and evolution. *Journal of Morphology* 177: 255–268.

Hanson, D. B. 1987. Prehistoric mortuary practices and human biology. In *Archaeological Investigations on the North Coast of Rota, Mariana Islands: the Airport Road Project*, ed. B. M. Butler, pp. 355–413. Carbondale: Center for Archaeological Investigations.

—— 1993. Mortuary and skeletal analysis of human remains from Achugao, Saipan. In *Archaeological Investigations in the Achugao and Matansa Areas of Saipan, Mariana Islands*, ed. B. M. Butler, Chapter 16. Saipan: Division of Historic Preservation.

Harding, R. M. 1992. Polynesian migrations, genetic demography and patterns of polymorphisms in tandemly-repetitive DNA. *Annals of Human Biology* 19: 214.

Harding, T. G. 1967. *Voyagers of the Vitiaz Strait*. Seattle: University of Washington Press.

Harihari, S., Saitou, N., Hirai, M., Gojobori, T., Park, K. S., Misawa, S., Ellepota, S. B., Ishida, T. and Omoto, K. 1988. Mitochondrial DNA polymorphism among five Asian populations. *American Journal of Human Genetics* 43: 134–143.

Harihari, S., Hirai, M. and Suutou, Y. 1992. Frequency of a 9-bp deletion in the mitochondrial DNA among Asian populations. *Human Biology* 64: 161–166.

Harris, E. F. and Bailit, H. L. 1980. The metaconule: a morphologic and familial analysis of a molar cusp in humans. *American Journal of Physical Anthropology* 53: 349–358.

—— 1987. Odontometric comparisons among Solomon Islanders and other Oceanic peoples. In *The Solomon Islands Project*, ed. J. S. Friedlaender, pp. 215–264. Oxford: Clarendon.

—— 1988. A principal components analysis of human odontometrics. *American Journal of Physical Anthropology* 75: 87–99.

Harvey, R. G. 1974. An anthropometric survey of growth and physique of the populations of Karkar Island and Lufa subdistrict, New Guinea. *Philosophical Transactions of the Royal Society* B 268: 179–292.

—— 1985. Ecological factors in skin color variation among Papua New Guineans. *American Journal of Physical Anthropology* 66: 407–416.

Hasebe, K. 1928. The West Micronesians: a preliminary report on the physical anthropology of the Micronesians. *Publications of the Anatomical Institute of the Imperial University of Tohoku* 13: 197–205.

Hatfield, H. S. and Pugh, L. G. C. E. 1951. Thermal conductivity of human fat and muscle. *Nature* 168: 918–919.

Hauser, G. and De Stefano, G. F. 1989. *Epigenetic Variation in the Human Skull.* Stuttgart: E. Schweizerbart.

Hawley, T. G. and Jansen, A. A. J. 1971. Weight, height, body surface and overweights of Fijian adults from coastal areas. *New Zealand Medical Journal* 74: 18–21.

Hay, A. H. 1995. The Maori Femur. Unpublished doctoral thesis, University of Otago.

Hayward, M. G. and Keatinge, W. R. 1981. Roles of subcutaneous fat and thermoregulatory reflexes in determining ability to stabilise body temperature in water. *Journal of Physiology* 320: 229–251.

Healey, L. A., Caner, J. E. Z., Bassett, D. R. and Decker, J. L. 1966. Serum uric acid and obesity in Hawaiians. *Journal of the American Medical Association* 196: 364–365.

Heath, B. H. and Carter, J. E. L. 1971. Growth and somatotype patterns of Manus children, Territory of Papua New Guinea. *American Journal of Physical Anthropology* 35: 49–68.

Helm, P. and Helm, S. 1987. Uncertainties in designation of age at menarche in the nineteenth century: revised mean for Denmark 1835. *Annals of Human Biology* 14: 371–374.

Henderson, A. M. and Corruccini, R. S. 1976. Relationship between tooth and body size in American Blacks. *Journal of Dental Research* 55: 94–96.

Hendrikse, R. G. 1987. Malaria and child health. *Annals of Tropical Medicine and Parasitology* 81: 499–509.

Hertzberg, M., Mickleson, K. N. and Trent, R. J. 1988. Alpha-globin gene haplotypes in Polynesians: their relationships to population groups and gene rearrangements. *American Journal of Human Genetics* 43: 971–977.

Hertzberg, M., Mickleson, K. N., Serjeantson, S. W., Prior, J. F., Trent, R. J. 1989a. An Asian-specific 9-bp deletion of mitochondrial DNA is frequently found in Polynesians. *American Journal of Human Genetics* 44: 504–510.

Hertzberg, M., Jahromi, K., Ferguson, V., Dahl, H. H., Mercer, J., Mickleson, K. N. and Trent, R. J. 1989b. Phenylalanine hydroxylase gene haplotypes in Polynesians: evolutionary origins and absence of alleles associated with severe phenylketonuria. *American Journal of Human Genetics* 44: 382–387.

Heyerdahl, T. 1952a. *American Indians in the Pacific.* London: Allen and Unwin.

—— 1952b. *The Kon-Tiki Expedition.* London: Reprint Society.

Hezel, F. X. 1983. *The First Taint of Civilization. A History of the Caroline and Marshall Islands in Pre-colonial Days, 1521–1885*. Honolulu: University of Hawaii Press.

Hiernaux, J. and Froment, A. 1976. The correlations between anthropobiological and climatic variables in Sub-Saharan Africa: revised estimates. *Human Biology* 48: 757–767.

Higgs, D. R., Wainscoat, J. S., Flint, J., Hill, A. V. S., Thein, S. L., Nicholls, R. D., Teal, H., Ayyub, H., Peto, T. E. A., Falusi, Y., Jarman, A., Clegg, J. B. and Weatherall, D. J. 1986. Analysis of the human alpha-globin gene cluster reveals a highly informative genetic locus. *Proceedings of the National Academy of Sciences, USA* 83: 5165–5169.

Hill, A. V. S., Gentile, B., Bonnard, J. M., Roux, J., Weatherall, D. J. and Clegg, J. C. 1987. Polynesian origins and affinities: globin gene variants in eastern Polynesia. *American Journal of Human Genetics* 40: 453–463.

Hill, A. V. S., O'Shaughnessy, D. F. and Clegg, J. B. 1989. Haemoglobin and globin variants in the Pacific. In *The Colonization of the Pacific. A Genetic Trail*, ed. A. V. S. Hill and S. W. Serjeantson, pp. 246–285. Oxford: Clarendon Press.

Hill, A. V. S. and Serjeantson, S. W. (eds). 1989. *The Colonization of the Pacific. A Genetic Trail*. Oxford: Clarendon Press.

Hill, A. V. S., Bowden, D. K., Trent, R. J., Higgs, D. R., Oppenheimer, S. J., Thein, S. L., Mickleson, K. N. P., Weatherall, D. J. and Clegg, J. B. 1985. Melanesians and Polynesians share a unique alpha thalassaemia mutation. *American Journal of Human Genetics* 37: 571–580.

Himes, J. H. 1991. Introduction. In *Anthropometric Assessment of Nutritional Status*, ed. J. H. Himes, pp. 1–3. New York: Wiley-Liss.

—— 1991. Multivariate analyses and anthropometric dimensions and measures of total body composition. In *Anthropometric Assessment of Nutritional Status*, ed. J. H. Himes, pp. 141–150. New York: Wiley-Liss.

Himes, J. H., Martorell, R., Habicht, J. P., Yarbrough, C., Malina, R. M. and Klein, R. E. 1975. Patterns of cortical bone growth in moderately malnourished preschool children. *Human Biology* 47: 337–350.

Hiroko, G. 1926. An anthropometrical study of the Micronesians. *Proceedings of the 3rd Pan-Pacific Science Congress*, Tokyo.

Hogbin, H. I. 1957. The problem of depopulation in Melanesia as applied to Ontong Java, Solomon Islands. *Journal of the Polynesian Society* 39: 43–46.

Holt, S. B. 1968 *The Genetics of Dermal Ridges*. Springfield: Thomas.

Homoe, P. and Lynnerup, N. 1991. Pneumatization of the temporal bones in Greenlandic Inuit anthropological material. *Acta Oto-Laryngology* 111: 1109–1116.

Hood, X. and Elliot, R. B. 1975. A comparative study of the health of elite Maori and Caucasian children in Auckland. *New Zealand Medical Journal* 81: 242–246.

Horai, S. and Matsunaga, E. 1986. Mitochondrial DNA polymorphism in Japanese. II. Analysis with restriction enzymes of four or five base pair recognition. *Human Genetics* 72: 105–117.

Hornabrook, R. W. 1977a. Human adaptability in Papua New Guinea. In *Population structure and human variation*, ed. G. A. Harrison, pp. 285–312. Cambridge University Press.

—— 1977b. Human ecology and biomedical research: a critical review of the International Biological Programme in New Guinea. In *Subsistence and Survival. Rural Ecology in the Pacific*, ed. T. D. Bayliss-Smith and R. G. Feachem, pp. 23–61. New York: Academic Press.

Horvath, S. M. and Finney, B. R. 1976. Paddling experiments and the question of Polynesian voyaging. In *Pacific Navigation and Voyaging*, ed. B. R. Finney, pp. 47–54. Wellington: The Polynesian Society.

Houghton, P. 1974. The relationship of the pre-auricular groove of the ilium to pregnancy. *American Journal of Physical Anthropology* 41: 381–390

—— 1975a. A renal calculus from proto-historic New Zealand. *OSSA* 2: 11–14.

—— 1975b. The bony imprint of pregnancy. *Bulletin of the New York Academy of Medicine* 51: 655–661.

—— 1977. Trephination in Oceania. *Journal of the Polynesian Society* 86: 265–269.

—— 1978a. Polynesian mandibles. *Journal of Anatomy* 127: 251–260.

—— 1978b Dental evidence for dietary variation in prehistoric New Zealand. *Journal of the Polynesian Society* 87: 257–263.

—— 1980a. *The First New Zealanders*. Auckland: Hodder and Stoughton.

—— 1980b. *The People of Namu*. Dunedin: Department of Anatomy.

—— 1983. A nineteenth-century musket wound. *New Zealand Archaeological Association Newsletter* 26: 274–275.

—— 1989. Watom: the people. *Records of the Australian Museum* 41: 223–234.

—— 1990. The adaptive significance of Polynesian body form. *Annals of Human Biology* 17: 19–32.

—— 1991. Selective influences and morphological variation amongst Pacific *Homo sapiens*. *Journal of Human Evolution* 21: 49–59.

Houghton, P. and Kean, M. R. 1987. The Polynesian head: a biological model for *Homo sapiens*. *Journal of the Polynesian Society* 96: 223–242.

Houghton, P., Leach, B. F. and Sutton, D. G. 1975. The estimation of stature of prehistoric Polynesians in New Zealand. *Journal of the Polynesian Society* 84: 325–326.

Howell, N. 1979. *The Demography of the Dobe !Kung*. London: Academic Press.

—— 1982. Village composition implied by a paleodemographic life table: the Libben site. *American Journal of Physical Anthropology* 59: 263–270.

Howells, W. W. 1973a. *The Pacific Islanders*. New York: Scribner.

—— 1973b. *Cranial Variation in Man*. Papers of the Peabody Museum 67: 1–243.

—— 1978. Was the skull of the Moriori artificially deformed? *Archaeology and Physical Anthropology in Oceania* 13: 197–203.

—— 1984. Introduction. In *Multivariate Statistical Methods in Physical Anthropology*, ed. G. N. van Vark and W. W. Howells, pp. 1–11. Reidel: Dordrecht.

—— 1987. Introduction. In *The Solomon Islands Project*, ed. J. S. Friedlaender, pp. 3–13. Oxford: Clarendon.

—— 1989. *Skull shapes and the map*. Papers of the Peabody Museum of Archaeology and Ethnology Vol 79. Cambridge, Mass.: Harvard University Press.

Hrdlicka, A. 1935. Ear exostoses. *Smithsonian Miscellaneous Collections* 93: 1–100.

Hudson, E. H. 1963. Treponematosis and anthropology. *Annals of Internal Medicine* 58: 1037–1048.

Hummert, J. R. and Van Gerven, D. P. 1985. Observations of the formation and persistence of radiopaque transverse lines. *American Journal of Physical Anthropology* 66: 297–306.

Hunt, T. L. and Graves, M. W. 1990. Some methodological issues in Oceanic prehistory. *Asian Perspectives* 29: 107–115.

Hylander, W. L. 1975. The human mandible: lever or link. *American Journal of Physical Anthropology* 43: 227–242.

—— 1977. The adaptive significance of Eskimo craniofacial morphology. In *Orofacial Growth and Development*, ed. A. A. Dahlberg and T. Graber, pp. 129–169. The Hague, Mouton.

—— 1979. An experimental analysis of temporomandibular joint reaction force in macaques. *American Journal of Physical Anthropology* 51: 433–456.

Iampietro, P. F., Bass, D. E. and Buskirk, E. R. 1958. Heat exchanges of nude men in the cold; effects of humidity temperature and windspeed. *Journal of Applied Physiology* 12: 351–356.

Igarashi, Y. 1992. Pregnancy bony imprint on Japanese pelves and its relation to pregnancy experience. *Journal of the Anthropological Society of Nippon* 100: 311–329.

Igarashi, Y., Katayama, K. and Takayama, J. 1987. Human skeletal remains from Tuvalu with special reference to artificial cranial deformation. *Man and Culture in Oceania* 3: 105–124.

Ingalls, N. W. 1927. Studies on the femur. *American Journal of Physical Anthropology* 10: 297–321.

Into, M. 1986. Pigs in Micronesia: introduction or reintroduction by the Europeans? *Man and Culture in Oceania* 2: 1–25.

Irwin, G. 1985. The emergence of Mailu. *Terra Australis* 10.

—— 1992. *The Prehistoric Exploration and Colonisation of the Pacific*. Cambridge University Press.

Jamison, C. S., Jamison, P. I. and Meier, R. J. 1990. Dermatoglyphic and anthropometric relationships within the Inupiat Eskimo hand. *American Journal of Physical Anthropology* 83: 103–109.

Jantz, R. L. 1987. Anthropological dermatoglyphic research. *Annual Reviews in Anthropology* 16: 161–177.

Jelliffe, D. B. 1966. The assessment of the nutritional status of the community. WHO Monograph Series No. 53. Geneva: World Health Organisation.

Jenkins, C. 1988. Comment on Dennett and Connell, Acculturation and health in the Highlands of Papua New Guinea. *Current Anthropology* 29: 284.

Johansson, S. R. 1982. The demographic history of the native peoples of North America: a selective bibliography. *Yearbook of Physical Anthropology* 25: 133–152.

Johansson, S. R. and Horowitz, S. 1986. Estimating mortality in skeletal populations: Influence of the growth rate on the interpretation of levels and trends during the transition to agriculture. *American Journal of Physical Anthropology* 71: 233–250.

Johnston, F. E. and Zhen, O. 1991. Choosing appropriate reference data for the anthropometric assessment of nutritional status. In *Anthropometric Assessment of Nutritional Status*, ed. J. H. Himes, pp. 337–346. New York: Wiley-Liss

Jones, J. S. 1981. How different are human races? *Nature* 293: 188–190.

Jorde, L. B. 1985. Human genetic distance studies: present status and future prospects. *Annual Review of Anthropology* 14: 343–373.

—— 1986. Population structure studies and genetic variability in humans. In *Genetic Diversity and its Maintenance*, ed. D. F. Roberts and G. F. Stefano, pp. 199–230. Cambridge University Press.

Jungers, W. L. 1988. New estimates of body size in Australopithecines. In *Evolutionary History of the 'Robust' Australopithecines*, ed. F. E. Grine, pp. 115–126. New York: Aldine de Gruyter.

Kapandji, I. A. 1987. *The Physiology of the Joints, Volume 2, Lower Limb*, 5th ed. Edinburgh: Churchill Livingston.

Kariks, J. and Walsh, R. J. 1968. Some physical measurements and blood groups of the Bainings in New Britain. *Archaeology and Physical Anthropology in Oceania* 3: 129–142.

Katayama, K. 1988a. Geographic distribution of auditory exostoses in South Pacific human populations. *Man and Culture in Oceania* 4: 63–74.

—— 1988b. Human skeletal remains of late pre-European period from Mangaia, Cook Islands. In *People of the Cook Islands, Past and Present*, ed. K. Katayama and A. Tagaya, pp. 67–90. Rarotonga: Cook Islands Library and Museum Society Bulletin No. 5.

Katayama, K., Tagaya, A. and Houghton, P. 1988. Osteometric and somatometric analysis of Mangaians, Cook Islands. In *People of the Cook Islands, Past and Present*, ed. K. Katayama and A. Tagaya, pp. 91–113. Rarotonga: Cook Islands Library and Museum Society Bulletin No. 5.

Kaufmann, M. S. and Bothe, D. J. 1986. Windchill reconsidered. *Aeronautical, Space and Environmental Medicine* 57: 23–26.

Kean, M. R. and Houghton, P. 1982. The Polynesian head: growth and form. *Journal of Anatomy* 135: 423–435.

—— 1987. The role of function in the development of human craniofacial form – a perspective. *The Anatomical Record* 216: 107–110.

Keane, A. H. 1908. *The World's People*. London: Hutchinson.

Keatinge, W. R., 1969. *Survival in Cold Water*. Oxford: Blackwell.

Kelley, D. E., Moran, M. and Mandarino, L. J. 1993. Metabolic pathways of glucose in skeletal muscle of lean NIDDM patients. *Diabetes Care* 16: 1158–1166.

Kennedy, G. E. 1986. The relationship between auditory exostoses and cold water: a latitudinal analysis. *American Journal of Physical Anthropology* 71: 401–415.

Kerley, E. R. 1970. Estimation of skeletal age: after about age 30. In *Personal Identification in Mass Disasters*, ed. T. D. Stewart, pp. 57–70. Washington: Smithsonian Institution.

Kieser, J. A. and Groeneveld, H. T. 1988. Allometric relations of teeth and jaws in Man. *American Journal of Physical Anthropology* 77: 57–67.

Kimura, K. 1984. Studies on growth and development in Japan. *Yearbook of Physical Anthropology* 27: 179–214

King, J. 1967. Extracts from officers' journals. In *The Journals of Captain James Cook on his Voyages of Discovery. III (2) The Voyage of the Resolution and Discovery 1776–1780*, ed. J. C. Beaglehole, pp. 1361–1455. Cambridge University Press for Hakluyt Society.

Kirch, P. V. 1984. *The Evolution of the Polynesian Chiefdoms*. Cambridge: Cambridge University Press.

Kirch, P. V. and Ellison, J. 1994. Human colonization of remote Pacific Islands. *Antiquity* 68: 310–321.

Kirk, R. L. 1989. Population genetic studies in the Pacific: red cell antigen, serum protein, and enzyme systems. *In The Colonization of the Pacific. A Genetic Trail*, ed. A. V. S. Hill and S. W. Serjeantson, pp. 60–119. Oxford: Clarendon Press.

Konigsberg, L.W. 1992. Review of Green, R. C., Anson, D. and Specht, J. (eds). Some Lapita People. *Journal of the Polynesian Society* 101: 309–311.

Konigsberg, L.W. and Blangero, J. 1993. Multivariate quantitative genetic simulations in anthropology with an example from the South Pacific. *Human Biology* 65: 897–915.

Kotzebue, O. von. 1839. *A New Voyage Round the World in the Years 1823, 24, 25, and 26*. London: Colburn and Bentley.

Kowalski, C. J. 1972. A commentary on the use of multivariate statistical methods in anthropometric research. *American Journal of Physical Anthropology* 36: 119–132.

Kuhn, T. S. 1977. *The Essential Tension*. Chicago: University of Chicago Press.

Kurland, L. T. 1988. Amyotrophic lateral sclerosis and Parkinson's disease complex on Guam linked to an environmental neurotoxin. *Trends in Neuroscience* 11: 51–4.

Ladell, W. S. S. 1964. Terrestrial animals in humid heat: man. In *Handbook of Physiology*, Section 4: Adaptation to the Environment, pp. 625–659. Bethesda: American Physiological Society.

Lai, L. Y. C. and Bloom, J. 1982. Genetic variation in Bougainville and Solomon Islands populations. *American Journal of Physical Anthropology* 58: 369–382.

Lambert, S. M. 1941. *A Doctor in Paradise*. London: Dent.

Lampl, M., Johnston, F. E. and Malcolm, L. A. 1978. The effects of protein supplementation on the growth and skeletal maturation of New Guinea school children. *Annals of Human Biology* 5: 219–227.

Lancet editorial. 1989. Thrifty genotype rendered detrimental by progress? *Lancet* 7.10.89: 839–840.

Larnach, S. L. 1976. The parietal boss (or eminence): estimation or measurement. *Archaeology and Physical Anthropology in Oceania* 11: 70–74.

Larnach, S. L. and Macintosh, N. W. G. 1971. The mandible in eastern Australian Aborigines. *The Oceania Monographs* No. 17. Sydney: University of Sydney.

Laslett, P. 1985. Age at menarche in Europe since the eighteenth century. *Journal of Interdisciplinary History* 16: 221–236.

Laurence, B. R. 1991. Elephantiasis in early Polynesia. *Journal of the History of Medicine and Allied Sciences* 46: 277–290.

Lavelle, C. L. B. 1979. A study of craniofacial form. *Angle Orthodontist* 49: 65–72.

Lebot, V., Merlin, M. and Lindstrom, L. 1992. *Kava. The Pacific Drug.* New Haven: Yale University Press.

Lessa, W. A. and Lay, T. 1953. Somatology of Ulithi. *American Journal of Physical Anthropology* 11: 405–410.

Lesson, R. P. 1971. Voyage autour du monde sur la corvette *La Coquille.* In *Duperrey's visit to New Zealand in 1824,* ed. C. A. Sharp, pp. 51–108. Wellington: Alexander Turnbull Library.

Levison, M., Ward, R. G., and Webb, J. W. 1973. *The Settlement of Polynesia: a Computer Simulation.* Minneapolis: University of Minnesota Press.

Levy, D. and Houghton, P. In press. A computer simulation of survival at sea in the tropical Pacific. *Antiquity.*

Lewis, D. 1972. *We, the Navigators.* Wellington: Reed.

Lewontin, R. 1975. Foreword in Friedlaender, J. S. *Patterns of Human Variation.* Cambridge, Mass.: Harvard University Press.

Lie-Injo, L. E, Pawson, I. G. and Solai, A. 1985. High frequency of triplicated alpha-globin loci and absence or low frequency of alpha-thalassaemia in Polynesian Samoans. *Human Genetics* 70: 116–118.

Lind, A. R. 1980. Opening remarks on some of the physiological responses to isometric contractions and the mechanisms that control them. *Advances in Physiological Science* 18: 213–218.

Lister, A. M. 1989. Rapid dwarfing of red deer on Jersey in the last interglacial. *Nature* 342: 539–542.

Littlewood, R. A. 1972. *Physical Anthropology of the Eastern Highlands of New Guinea.* Seattle: University of Washington Press.

Loesch, D. Z. 1974. Genetical studies of sole and palmar dermatoglyphics. *Annals of Human Genetics* 37: 405–420.

Loesch, D. Z. and Lafranchi, M. 1990. Relationships of epidermal ridge patterns with body measurements and their possible evolutionary significance. *American Journal of Physical Anthropology* 82: 183–189.

Lourie, J. A. 1972. Anthropometry of Lau islanders, Fiji. *Human Biology in Oceania* 1: 273–277.

Lovejoy, C. O. and Heiple, K. G. 1972. Proximal femoral anatomy of Australopithecus. *Nature* 235: 175–176.

Lovejoy, C. O., Heiple, K. G. and Burstein A. H. 1973. The gait of Australopithecus. *American Journal of Physical Anthropology* 38: 757–780.

Lum, J. K., Rickards, O., Ching, C. and Cann, R. L. 1994. Polynesian mitochondrial DNAs reveal three deep maternal lineage clusters. *Human Biology* 66: 567–590.

Maclean, D. and Emslie-Smith, D. 1977. *Accidental Hypothermia.* Oxford: Blackwell.

Mahalanobis, P. C., Majundar, D. N. and Rao, C. R. 1949. Anthropometric surveys of the United Provinces 1941. *Sankya* 9: 89–324.

Mahotra, A. and Thopre, R. S. 1991. Experimental detection of rapid evolutionary response in natural lizard populations. *Nature* 353: 347–348.

Malcolm, L. A. 1970. *Growth and Development in New Guinea. A Study of the Bundi People of Madang District.* Institute of Human Biology

Monograph No. 1. Goroka: Papua New Guinea Institute of Medical Research.

Malina, R. M. 1978. Growth of muscle tissue and muscle mass. In *Human Growth*, ed. F. Faulkner and J. M. Tanner, pp. 273–294. London: Balliere Tindall.

Malina, R. M., Little, B. P., Shoup, R. F. and Buschang, P. H. 1987. Adaptive significance of small body size: strength and motor performance of school children in Mexico and Papua New Guinea. *American Journal of Physical Anthropology* 73: 489–500.

Maling, T., van Wissen, K., Toomath, R. and Siebers, R. In press. Blood pressure and metabolic markers of hyperinsulinaemia in normotensive Maori and Caucasian New Zealanders. *Asia Pacific Journal of Clinical Nutrition.*

Malinowski, B. 1922. *Argonauts of the Western Pacific.* London: George Routledge and Sons.

—— 1935. *Coral Gardens and their Magic.* London: George Allen and Unwin.

Mandell, G. L., Douglas, R. G. and Bennett, J. E. 1979. *Principles and Practice of Infectious Diseases.* New York: Wiley.

Margetts, E. L. 1967. Trepanation of the skull by the medicine-men of primitive cultures. In *Diseases in Antiquity*, ed. D. Brothwell, pp. 651–672. Springfield: Thomas.

Markham, A. H. 1970. *The Cruise of the Rosario.* Folkestone: Dawson.

Marsden, E. 1960. Radioactivity of soils, plants and bones. *Nature* 187: 192–195.

Marshall, D. S. and Snow, C. E. 1956. An evaluation of Polynesian craniology. *American Journal of Physical Anthropology* 14: 403–427.

Martin, A. D. 1991. Anthropometric assessment of bone mineral. In *Anthropometric assessment of nutritional status*, ed. J. H. Himes, pp. 185–196. New York: Wiley-Liss.

Martin, D. and Armelagos, G. J. 1979. Morphometrics and compact bone: an example from Sudanese Nubia. *American Journal of Physical Anthropology* 51: 571–578.

Martin, R. B. and Burr, D. B. 1989. *Structure, Function and Adaptation of Compact Bone.* New York: Raven Press.

Martin, R. and Saller, K. 1957. *Lehrbuch der Anthropologie.* Stuttgart: Gustav Fischer.

Martineau, L. and Jacobs, I. 1988. Muscle glycogen utilization during shivering thermogenesis in humans. *Journal. of Applied Physiology* 65: 2046–2050.

Matiegka, J. 1921 The testing of physical efficiency. *American Journal of Physical Anthropology* 4: 223–230.

Mayr, E. 1970. *Populations, Species, and Evolution.* Cambridge, Mass.: Harvard University Press.

McArthur, M. 1974. Pigs for the ancestors: a review article. *Oceania* 45: 87–123

McArthur, N. 1968. *Island Populations of the Pacific.* Canberra: Australian National University Press.

—— 1982. Isolated populations in enclaves on small islands. In *Melanesia: Beyond Diversity*, ed. R. J. May and H. Nelson, pp. 27–32. Canberra: Research School of Pacific Studies, Australian National University.

McArthur, N., Saunders, I. W. and Tweedie, R. L. 1976. Small population isolates: a microsimulation study. *Journal of the Polynesian Society* 85: 307–326.

McGregor, I. A., Rahman, A. K., Thompson, B., Billewicz, W. Z. and Thomson, A. M. 1968. The growth of young children in a Gambian village. *Transactions of the Royal Society for Tropical Medicine and Hygiene* 62: 341–352.

McHenry, H. M. 1968. Transverse lines in long bones of prehistoric California Indians. *American Journal of Physical Anthropology* 29: 1–18.

—— 1975. Fossil hominid body weight and brain size. *Nature* 254: 686–688.

—— 1988. New estimates of body weight in early hominids and their significance to encephalization and megadontia in 'robust' australopithecines. In *Evolutionary History of the 'Robust' Australopithecines.*, ed. F. E. Grine, pp. 133–148. New York: Aldine de Gruyter.

Mead, M. 1937. The Manus of the Admiralty Islands. In *Cooperation and Competition among Primitive Peoples*, ed. M. Mead, pp. 210–239. Boston: Beacon Press.

Mendana. 1901. *The Discovery of the Solomon Islands*. London: The Hakluyt Society.

Menkes, A., Mazel, S., Redmond, R. A., Koffler, K., Libanati, C. R., Gundberg, C. M., Zizic, T. M., Hagberg, J. M., Pratley, R. E. and Hurley, B. F. 1993. Strength training increases regional bone mineral density and bone remodeling in middle-aged and older men. *Journal of Applied Physiology* 74: 2478–2484.

Mensforth, R. P. 1990. Paleodemography of the Carlston Annis (Bt-5) Late Archaic skeletal population. *American Journal of Physical Anthropology* 82: 81–99.

Merbs, C. F. 1983. Patterns of activity-induced pathology in a Canadian Inuit population. *Archaeological Survey of Canada*, Paper No. 119. Ottawa: National Museum of Canada.

Mickleson, K. M., Dixon, M. W., Hill, P. J., Buck, R., Eales, M., Rutherford, J., Yakas, J. and Trent, R. J. 1985. Influence of alpha-thalassaemia on haematological parameters in Polynesian patients. *New Zealand Medical Journal* 98: 1036–1038.

Micozzi, M. S. 1988. Comment on acculturation and health in the Highlands of Papua New Guinea. *Current Anthropology* 29: 285.

Miklouho-Maclay, N. N. 1975. *New Guinea Diaries 1871–1883*. Madang: Kristen Press.

Miles, J. A. R. In press. *Infectious Diseases in the Pre-European Pacific*. Dunedin: University of Otago Press.

Mitchell, D., Nash, J., Ogan, E., Ross, H., Bayliss-Smith, T., Keesing, R. and Friedlaender, J. S. 1987. Profiles of the survey samples. In *The Solomon Islands Project*, ed. J. S. Friedlaender, pp. 28–60. Oxford: Clarendon.

Mitchell, J. W., Galvez, T. L., Hengle, J., Meyers, G. L. and Siebacker, K. L. 1970. Thermal response of human legs during cooling. *Journal of Applied Physiology* 29: 859–865.

Miyashita, T. and Takahashi, E. 1971. Stature and nose height of Japanese. *Human Biology* 43: 327–339.

Moller, I. J., Pindborg, J. J. and Effendi, I. 1977. The relation between betel chewing and dental caries. *Scandinavian Journal of Dental Research* 85: 64–70.

Møller-Christensen, V. 1967. Evidence of leprosy in earlier peoples. In *Diseases in Antiquity*, ed. D. Brothwell and A. T. Sandison, pp. 295–306. Springfield: Thomas.

Molnar, S. 1971. Human tooth wear, tooth function and cultural variability. *American Journal of Physical Anthropology* 34: 175–189.

—— 1983. *Human Variation*. Englewood Cliffs: Prentice-Hall.

Monkhouse, W. B. 1955. Journal. In *The Journals of Captain James Cook on his Voyages of Discovery. The Voyage of the Endeavour 1768–1771*, ed. J. C. Beaglehole, pp. 564–587. Cambridge University Press for Hakluyt Society.

Montague, M. F. A. 1960. *An Introduction to Physical Anthropology*. Springfield: Thomas.

Montesson, M. 1985. In *Early Eyewitness Accounts of Maori Life 2. Extracts from Journals Relating to the Visit to New Zealand in May–July 1772 of the French ships Mascarin and Marquis de Castries under Marion du Fresne*, transcription and translation by Isabel Ollivier, pp. 210–253. Wellington: Alexander Turnbull Library Endowment Trust with Indosuez New Zealand.

Montgomerie, J. Z. 1988. Leprosy in New Zealand. *Journal of the Polynesian Society* 97: 115–152.

Moore, W. J. 1981. *The Mammalian Skull*. Cambridge: Cambridge University Press.

Morant, G. M. 1936. A biometric study of the human mandible. *Biometrika* 28: 84–118.

Moss, M. L. 1961. Rotation of the otic capsule in bipedal rats. *American Journal of Physical Anthropology* 19: 301–307.

—— 1983. The functional matrix concept and its relationship to temporomandibular joint dysfunction and treatment. *Dental Clinics of North America* 27: 445–455.

Moyes, C. D., O'Hagen, L. C. and Armstrong, C. 1990. Anaemia in Maori infants – a persisting problem. *New Zealand Medical Journal* 103: 53.

Munilla, M. 1966. *The Journal of Fray Martin de Munilla*. Cambridge: The Hakluyt Society.

Murphy, A. M. C. and Wood, W. B. 1983. Lower limb proportions in the Australian Aborigine: a re-examination. *Pacific Science Association 15th Congress Programme*, Dunedin 1: 172.

Murrill, R. I. 1965. A study of cranial and postcranial material from Easter Island. In *Reports of the Norwegian Archaeological Expedition to Easter Island and the East Pacific*, ed. T. Heyerdahl and K. E. N. Ferdon, 2: 255–327. London: Allen and Unwin.

Neale, J. T. and Bailey, R. R. 1990. Chronic renal disease in Polynesians in New Zealand. *New Zealand Medical Journal* 13.6.90: 262.

Neave, M., Prior, I. A. M. and Toms, V. 1963. The prevalence of anaemia in two Maori rural communities. *New Zealand Medical Journal* 62: 20–28.

Neel, J. V. 1962. Diabetes mellitus: a thrifty genotype rendered detrimental by 'progress'. *American Journal of Human Genetics* 14: 353–361.

—— 1976. Towards a better understanding of the genetics of diabetes mellitus. In *The Genetics of Diabetes Mellitus*, ed. W. Creutzfeld, J. Kobberling and J. V. Neel, pp. 240–244. Berlin: Springer-Verlag.

Nei, M. and Roychoudhury, A. K. 1982. Genetic relationships and the evolution of human races. *Evolutionary Biology* 14: 1–59.

—— 1994. Evolutionary relationships of human populations on a global scale. *Molecular and Biological Evolution* 10: 927–943.

Neukirch, F., Chansin, R., Liarfid, R., Levallois, M. and Leproux, P. 1988. Spirometry and maximal expiratory flow-volume curve reference standards for Polynesian, European and Chinese teenagers. *Chest* 94: 792–798.

Newman, A. K. 1881. A study of the causes leading to the extinction of the Maori. *Transactions and Proceedings of the New Zealand Institute* 14: 459–477.

Newman, M. T. 1960. Adaptations in the physique of American aborigines to nutritional factors. *Human Biology* 32: 288–313.

—— 1970. Dermatoglyphics. In *Handbook of Middle American Indians* 9: 167–179. Austin: University of Texas Press.

Newman, R. W. and Munro, E. H. 1955. The relation of climate and body size in United States males. *American Journal of Physical Anthropology* 13: 1–17.

Nichol, C. R. 1989. Complex segregation analysis of dental morphological variants. *American Journal of Physical Anthropology* 78: 37–59.

Niswander, K. and Jackson, E. C. 1974. Physical characteristics of the gravida and their association with birth weight and perinatal death. *American Journal of Obstetrics and Gynecology* 119: 306–313.

Nurse, G. T. 1985. The pace of human selective adaptation to malaria. *Journal of Human Evolution* 14: 319–326.

—— 1988. Comment on acculturation and health in the Highlands of Papua New Guinea. *Current Anthropology* 29: 285–286.

O'Connell, J. F. 1972. *A Residence of Eleven Years in New Holland and the Caroline Islands.* Pacific History Series No. 4. Canberra: Australian National University Press.

O'Flynn, T. 1992. A Study of the Relationships between Form and Function in the Human Mandible. Unpublished BSc (hons) thesis: University of Otago.

Oliver, D. L. 1955. *A Solomon Islands Society.* Cambridge, Mass.: Harvard University Press.

—— 1988. *Oceania.* Honolulu: University of Hawaii Press.

Orbell, M. 1975. The religious significance of Maori migration traditions. *Journal of the Polynesian Society* 84: 341–347.

Ortner, D. J. 1975. Aging effects on osteon remodelling. *Calcified Tissue Research* 18: 27–36.

Ortner, D. J. and Putschar, W. G. J. 1981. Identification of pathological conditions in human skeletal remains. *Smithsonian Contributions to Anthropology* No. 28.

Ortner, D. J., Tuross, N. and Stix, A. J. 1992. New approaches to the study of disease in archeological New World populations. *Human Biology* 64: 337–360.

O'Shaughnessy, D. F., Hill, A. V. S., Bowden, D. K., Weatherall, D. J. and Clegg, J. B. 1990. Globin genes in Micronesia: origins and affinities of Pacific Island peoples. *American Journal of Human Genetics* 46: 144–155.

Owsley, D. W., Miles, A-M. and Gill, G. W. 1985. Carious lesions in permanent dentitions of protohistoric Easter Islanders. *Journal of the Polynesian Society* 94: 415–422.

Page, L. B., Rhoads, J. G., Friedlaender, J. S., Page, J. R. and Curtis, K. 1987. Diet and nutrition. In *The Solomon Islands Project,* ed. J. S. Friedlaender, pp. 65–88. Oxford: Clarendon.

Page, L. B., Friedlaender, J. and Moellering, R. C. 1977. Culture, human biology and disease in the Solomon Islands. In *Population Structure and Human Variation,* ed. G. A. Harrison, pp. 143–164. Cambridge: Cambridge University Press.

Palamino, H., Chakraboty, R. and Rothhammer, F. 1977. Dental morphology and population diversity. *Human Biology* 49: 61–70.

Palkovich, A. M. 1987. Endemic disease patterns in paleopathology: porotic hyperostosis. *American Journal of Physical Anthropology* 74: 527–537.

Park, E. A. 1964. The imprinting of nutritional disturbances on the growing bone. *Paediatrics* 33: 815–862.

Parker, G. H. and George, J. C. 1974. Effects of in vivo cold-exposure on intracellular glycogen reserves in the 'starling type' avian pectoralis. *Life Sciences* 15: 1415–1423.

Parkinson, R. 1907. *Dreissig Jahre in der Südsee.* Stuttgart: Strecker und Schroeder.

Parkinson, S. 1984. *A Journal of a Voyage to the South Seas in his Majesty's Ship Endeavour.* London: Caliban.

Parsons, P. A., 1991. Evolutionary rates: stress and species boundaries. *Annual Review of Ecology and Systematics* 22: 1–18.

—— 1994. Habitats, stress, and evolutionary rates. *Journal of Evolutionary Biology* 7: 387–397.

Pavlides, C. and Gosden, C. 1994. 35,000-year-old sites in the rainforests of West New Britain, Papua New Guinea. *Antiquity* 68: 604–610.

Pawley, A. K. and Green, R. C. 1973. Dating the dispersal of the Oceanic languages. *Oceanic Linguistics* 12: 1–67.

Pearson, K. 1899. Mathematical Contributions to the Theory of Evolution V: On the Reconstruction of the Stature of Prehistoric Races. *Philosophical Transactions of the Royal Society of London,* Series A. 192: 169–244.

—— 1901a. Editorial. *Biometrika* 1: 1–6.

—— 1901b. On the fundamental conceptions of biology. *Biometrika* 1: 320–344.

—— 1921. Was the skull of the Moriori artificially deformed? *Biometrika* 13: 338–346.

—— 1926. On the coefficient of racial likeness. *Biometrika* 18: 105–117.

Penrose, L. S. 1954. Distance, size and shape. *Annals of Eugenics* 18: 337–343

Perzigian, A. J. 1981. Allometric analysis of dental variation in a human population. *American Journal of Physical Anthropology* 54: 341–345.

Peterson, W. 1975. A demographer's view of prehistoric demography. *Current Anthropology* 16: 227–246.

Petrovsky, J. S. and Lind, A. R. 1975. Insulative power of body fat on deep muscle temperatures and isometric endurance. *Journal of Applied Physiology* 39: 639–642.

Pickersgill, R. 1961. Journal. In *The Journals of Captain James Cook on his Voyages of Discovery. The Voyage of the Resolution and Adventure 1772–1775,* ed. J. C. Beaglehole, 2: 767–775. Cambridge University Press for Hakluyt Society.

Picq, P. G., Plavcan, J. M. and Hylander, W. L. 1987. Nonlever action of the mandible: the return of the Hydra. *American Journal of Physical Anthropology* 74: 305–308.

Pietrusewsky, M. 1969. An osteological study of cranial and infracranial remains from Tonga. *Records of the Auckland Institute and Museum* 6: 287–402

—— 1976. Prehistoric human remains from Papua New Guinea and the Marquesas. *Asian and Pacific Archaeology Series* No. 7.

—— 1981. Metric and non-metric cranial variation in Australian Aboriginal populations compared with populations from the Pacific and Asia. *Occasional papers in Human Biology No. 3.* Canberra: Australian Institute of Aboriginal Studies.

—— 1983. Multivariate analysis of New Guinea and Melanesian skulls: a review. *Journal of Human Evolution* 12: 61–76.

—— 1989. A study of skeletal and dental remains from Watom Island and comparisons with other Lapita people. *Records of the Australian Museum* 41: 297–325.

Pietrusewsky, M. and Batista, C. 1980. *Human Skeletal and Dental Remains from Four Sites on Tinian and Saipan, Commonwealth of the Northern Mariana Islands.* Hawaii: Department of Anthropology, University of Hawaii-Manoa.

Pigafetta, A. 1906. *Magellan's Voyage around the World,* translated and edited by J. A. Robertson. Cleveland: Clark.

Pirie, P. 1971. The effect of treponematoses and gonorrhoea on the population of the Pacific Islands. *Human Biology in Oceania* 1: 187–206.

Pool, D. I. 1991. *Te Iwi Maori.* Auckland: Auckland University Press.

Pope, G. G. 1992. Craniofacial evidence for the origin of modern humans. *Yearbook of Physical Anthropology* 35: 243–298.

Portin, P. and Alvesalo, L. 1974. The inheritance of shovel shape in maxillary central incisors. *American Journal of Physical Anthropology* 41: 59–62 .

Powdermaker, H. 1933. *Life in Lesu.* London: Williams and Norgate.

Prentice, A. M. and Prentice, A. 1988. Energy costs of lactation. *Annual Review of Nutrition* 8: 63–79.

Prior, I. A. M. and Rose, B. S. 1966. Hyperuricaemia, gout and diabetic abnormality in Polynesian people. *Lancet* 1: 333–338.

Prior, I. A. M. and Stanhope, F. M. 1977. The Tokelau Island migrant study: fertility and associated factors before migration. *Journal of Biosocial Science* 9: 1–11.

Prior, I. A. M., Beaglehole, R., Davidson, F. and Salmond, C. E. 1978. The relationships of diabetes, blood lipids, and uric acid levels in Polynesians. *Advances in Metabolic Diseases* 9: 242–260.

Prior, I. A. M., Davidson, F., Salmond, C. E. and Czochanska, Z. 1981. Cholesterol, coconuts and diet on Polynesian atolls. *American Journal of Clinical Nutrition* 34: 1552–1561.

Prior, I. A. M., Hooper, A., Huntsman, J. W., Stanhope, J. M. and Salmond, C. G. 1977. The Tokelau Island migrant study. In *Population Structure and Human Variation,* ed. G. A. Harrison, pp. 165–186. Cambridge: Cambridge University Press.

Prior, I. A. M., Salmond, S. and Wessen, A. F. 1992. Chapter 12. In *Migration and Health in a Small Society: the Case of the Tokelau,* ed. A. F. Wessen, pp. 286–317. Oxford: Clarendon.

Prior, I. A. M., Welby, T. J., Ostbye, T., Salmond, C. E. and Stokes, V. M. 1987. Migration and gout: the Tokelau Island migrant study. *British Medical Journal* 293: 457–461.

Proctor, D. F., 1980. *Breathing, Speech and Song.* Vienna: Springer-Verlag.

Pugh, L. G. C. 1966a. Accidental hypothermia in walkers, climbers and campers. *British Medical Journal* 1: 123–129.

—— 1966b. Clothing insulation and accidental hypothermia in youth. *Nature* 209: 1281–1286.

—— 1967. Cold stress and muscular exercise, with special reference to accidental hypothermia. *British Medical Journal* 2: 333–337.

Pugh, L. G. C. and Edholm, O. G. 1955. The physiology of channel swimmers. *Lancet* 2: 761–768.

Quarry-Wood, W. 1920. The tibia of the Australian Aborigine. *Journal of Anatomy* 54: 232–257.

Quiros, P. F. de. 1904. *The Voyages of Pedro Fernandez de Quiros,* ed. C. Markham. London: Hakluyt.

Ramage, C. 1986. *El Nino. Scientific American* 254: 55–61.

Ranford, P. R. 1989. Genetic variants of the serum complement components. In *The Colonization of the Pacific. A Genetic Trail,* ed. A. V. S. Hill and S. W. Serjeantson, pp. 174–193. Oxford: Clarendon Press.

Rappaport, R. A. 1968. *Pigs for the Ancestors.* New Haven: Yale University Press.

Refshauge, W. F. and Walshe, R. J. 1981. Pitcairn Island: fertility and population growth 1790–1856. *Annals of Human Biology* 8: 303–312.

Reid, I. R., Cullen, S., Schooler, B. A., Livingston, N. E. and Evans, M. C. 1990. Calcitropic hormone levels in Polynesians: evidence against their role in interracial differences in bone mass. *Journal of Clinical Endocrinology and Metabolism* 70: 1452–1456.

Reid, I. R., Mackie, M. and Ibbertson, H. K. 1986. Bone mineral content in Polynesian and white New Zealand women. *British Medical Journal* 292: 1547–1548.

Rennie, D. W., Corvino, B. G., Howell, B. J., Hong, S. H., Kang, B. S. and Hong, S. K. 1962. Physical insulation of Korean diving women. *Journal of Applied Physiology* 17: 961–966.

Rhoads, J. G. 1983. Melanesian gene frequencies: a multivariate data-analytic approach. *Journal of Human Evolution* 12: 93–101

—— 1984. Improving the sensitivity, specificity, and appositeness of morphometric analyses, In *Multivariate Statistical Methods in Physical Anthropology,* ed. G. N. van Vark and W. W. Howells, pp. 247–259. Dordrecht: Reidel.

—— 1987. Longitudinal anthropometric changes in adults and adolescents. In *The Solomon Islands Project,* ed. J. S. Friedlaender, pp. 283–306. Oxford: Clarendon.

Rhoads, J. G. and Friedlaender, J. S. 1987. Blood polymorphism variation in the Solomon Islands. In *The Solomon Islands Project,* ed. J. S. Friedlaender, pp. 125–154. Clarendon: Oxford.

Riesenfeld, A. 1956. Shovel-shaped incisors and a few other dental features among the native peoples of the Pacific. *American Journal of Physical Anthropology* 14: 505–521.

Rightsmeier, J. T. and McGrath, J. W. 1986. Quantitative genetics of cranial non-metric traits in randombred mice: heritability and etiology. *American Journal of Physical Anthropology* 69: 51–58.

Riolo, M. L., Moyers, R. E., McNamara, J. A. and Hunter, W. S. 1974. *An Atlas of Craniofacial Growth.* Monograph No. 2: Craniofacial Growth Series, Center for Human Growth and Development. Ann Arbor: University of Michigan.

Rivers, J. P. W. and Payne, P. R. 1982. The comparison of energy supply and energy need: a critique of energy requirements. *Symposia of the Society for the Study of Human Biology* 22: 85–105.

Roberton, J. 1832. An enquiry into the natural history of menstrual function. *Edinburgh Medical and Surgical Journal* 38: 227–254.

Roberts, D. F. 1953. Body weight, race and climate. *American Journal of Physical Anthropology* 11: 533–558.

Roberts, D. F. 1978. *Climate and Human Variability.* Menlo Park: Cummings.

Rogers, A. R. and Harpending, H. C. 1983. Population structure and quantitative characters. *Genetics* 105: 985–1002.

Rogers, A. R. and Mukherjee, A. 1992. Quantitative genetics and sexual dimorphism in human body size. *Evolution* 46: 226–234.

Roggeveen, J. von. 1970. *The Journal of Jacob Roggeveen.* Oxford: Clarendon Press.

Rose, J. C., Condon, K. W. and Goodman, A. H. 1984. Diet and dentition: developmental disturbances. In *The Analysis of Prehistoric Diets,* ed. J. Mielke and R. Gilbert, pp. 281–306. New York: Academic Press.

Rosing, F. W. 1983. Stature estimation in Hindus. *Homo* 34: 168–171.

—— 1984. Discreta of the human skeleton: a critical review. *Journal of Human Evolution* 13: 319–323.

Ross, H. M. 1973. *Baegu.* Illinois Studies in Anthropology No. 8. Urbana: University of Illinois Press.

—— 1976. Bush fallow farming, diet, and nutrition: a Melanesian example of successful adaptation. In *The Measures of Man,* ed. E. Giles and J. Friedlaender, pp. 550–615. Cambridge, Mass.: Peabody Museum.

Rothhammer, F. and Silva, C. 1990. Craniometrical variation among South American prehistoric populations: climatic, altitudinal, chronological and geographic contributions. *American Journal of Physical Anthropology* 82: 9–17.

Roux, J. 1985. The Journal of Jean Roux. In *Early Eyewitness Accounts of Maori Life 2. Extracts from Journals relating to the visit to New Zealand in May–July 1772 of the French ships Mascarin and Marquis de Castries under the command of M.-J, Marion du Fresne,* transcription and translation by Isabel Ollivier, pp. 118–208. Wellington: Alexander Turnbull Library Endowment Trust with Indosuez New Zealand.

Roy, K. J. 1989. A Study of Early Marianas Islanders – the Skeletons under their Skins. Unpublished M.A. thesis, University of Otago.

Roy, K. J. and Tayles, N. 1989. *The People of Afetna.* Dunedin: Department of Anatomy.

Roydhouse, N. 1977. Running ears. *Patient Management* 6: 25–35.

Ruff, C. 1987. Sexual dimorphism in human lower limb bone structure: relationship to subsistence strategy and sexual division of labor. *Journal of Human Evolution* 16: 391–416.

—— 1988. Hindlimb articular surface allometry in *Homonoidea* and *Macaca,* with comparisons to diaphyseal scaling. *Journal of Human Evolution* 17: 687–714.

Ruff, C. and Hayes, W. 1983a. Cross-sectional geometry of the Pecos

Pueblo femora and tibiae – a biomechanical investigation. II. Sex, age and side differences. *American Journal of Physical Anthropology* 60: 383–400.

—— 1983b. Cross-sectional geometry of the Pecos Pueblo femora and tibiae – a biomechanical investigation. I. Method, and general pattern of variation. *American Journal of Physical Anthropology* 60: 359–381.

Saltin, B. and Gollnick, P. D. 1983. Skeletal muscle adaptability: significance for metabolism and performance. In *Handbook of Physiology* Section 10, ed. L. D. Peachey, pp. 555–631. Bethesda: American Physiological Society.

Salzano, F. M. 1985. The peopling of the Americas viewed from South America. In *Out of Asia*, ed. R. Kirk and E. Szathmary, pp. 19–29. Canberra: Journal of Pacific History.

Samejima, S. 1938. Anthropological research on the physical constitution of Micronesian tribes. *Kanezawa Medical University Hygiene Department Data*, 5th Compilation.

Sandford, M. K., Van Gerven, D. P. and Meglen, R. R. 1983. Elemental hair analysis: new evidence on the etiology of cribra orbitalia in Sudanese Nubia. *Human Biology* 55: 831–844.

Saville, P. D., Heaney, R. P. and Recker, R. R. 1976. Radiogrammetry at four bone sites in normal, middle-aged women. *Clinical Orthopaedics* 114: 307–315.

Schendel, S. A., Walker, G. and Kamisugi, A. I. 1980. Hawaiian craniofacial morphometrics. *American Journal of Physical Anthropology* 52: 491–500.

Schlaginhaufen, O. 1929. Zur anthropologie der Micronesischen Inselgruppe Kapingamarangi (Greenwich-Inseln). *Archiv der Julius Klaus-Stiftung für Vererbungsforschung, Sozialanthropologie und Rassenhygiene* 4: 219–287.

Schmitt, R. C. 1972. Garbled population estimates of Central Polynesia. In *Readings in Population*, ed. W. Petersen, pp. 71–76. New York: Macmillan.

Schofield, G., 1959. Metric and morphological features of the femur of the New Zealand Maori. *Journal of the Royal Anthropological Institute* 89: 89–105.

Scholander, P. F., Hammel, H. T., Hart, J. S., le Messurier, D. H. and Sheen, J. 1958. Cold adaptation in Australian Aborigines. *Journal of Applied Physiology* 13: 211–218.

Schreider, E. 1950. Geographical distribution of the body-weight/body-surface ratio. *Nature* 165: 286.

—— 1951. Anatomical factors of body-heat regulation. *Nature* 167: 823–824.

—— 1975. Morphological variations and climatic differences. *Journal of Human Evolution* 4: 529–539.

Schurr, T. G., Ballinger, S. W., Gan, Y-Y, Hodge, J. A., Merriwether, D. A., Lawrence, D. N., Knowler, W. C., Weiss, K. M. and Wallace, D. C. 1990. Amerindian mitochondrial DNAs have rare Asian mutations at high frequencies, suggesting they derived from four primary maternal lineages. *American Journal of Human Genetics* 46: 613–623.

Schwartz, J. H. and Brauer, J. L. 1990. The Ipiatuk dentition: implications for interpreting Sinodonty and Sundadonty. *American Journal of Physical Anthropology* 81: 293.

Scott, G. R. and Turner, C. G. 1988. Dental anthropology. *Annual Reviews in. Anthropology* 17: 99–124.

Scott, J. H., 1894. Contribution to the osteology of the Aborigines of New Zealand and of the Chatham Islands. *Transactions of the New Zealand Institute* 26: 1–64.

Seligmann, C. G. 1909. A classification of the natives of British New Guinea. *Journal of the Royal Anthropological Institute* 39: 246–275 and 314–333.

—— 1910. *The Melanesians of British New Guinea*. Cambridge: Cambridge University Press.

Serjeantson, S. W. 1985. Migration and admixture in the Pacific: insights provided by human leucocyte antigens. In *Out of Asia; Peopling the Americas and the Pacific*, ed. R. L. Kirk and E. Szathmary, pp. 133–146. Canberra: Journal of Pacific History

—— 1989. HLA genes and antigens. In *The Colonization of the Pacific. A Genetic Trail*, ed. A. V. S. Hill and S. W. Serjeantson, pp. 120–173. Oxford: Clarendon Press.

Serjeantson, S. W. and Hill, A. V. S. 1989. The colonization of the Pacific: the genetic evidence. In *The Colonization of the Pacific. A Genetic Trail*. ed. A. V. S. Hill and S. W. Serjeantson, pp. 286–294. Oxford: Clarendon Press.

Serjeantson, S. W., Kirk, R. L. and Booth, P. B. 1983. Linguistic and genetic differentiation in New Guinea. *Journal of Human Evolution* 12: 77–92.

Serjeantson, S. W., Ryan, D. P., Zimmet, P., Taylor, R., Cross, R., Charpin, M. and Le Gonidec, G. 1982. HLA antigens in four Pacific populations with insulin-dependent diabetes mellitus. *Annals of Human Biology* 9: 69–84.

Shabel'skii, A. P. 1990. Voyage to the Russian colonies in America. In *Russia and the South Pacific, 1696–1840, Volume 3, Melanesia and the Western Polynesian Fringe*, ed. G. Barratt pp. 192–194. Vancouver: University of British Columbia Press.

Shapiro, H. L. 1930. The physical characters of the Society Islanders. *Memoirs of the Bernice P. Bishop Museum* Vol 3: 4.

—— 1933. The physical characteristics of the Ontong Javanese. A contribution to the study of the non-Melanesian elements in Melanesia. *Anthropological Papers of the American Museum of Natural History* 33: 231–278.

—— 1943. Physical differentiation in Polynesia. *Papers of the Peabody Museum* 20: 3–8.

Shapiro, H. L. and Buck, P. H. 1936. The physical characters of the Cook Islanders. *Memoirs of the Bernice P. Bishop Museum* Vol 3: 1.

Sharp, A. 1963. *Ancient Voyagers in Polynesia*. Auckland: Paul.

—— 1968. *The Voyages of Abel Janzoon Tasman*. Oxford: Clarendon Press.

Shephard, R. I. 1991. *Body Composition in Biological Anthropology*. Cambridge: Cambridge University Press.

Sheridan, S. S., Mittler, D. M., Van Gerven, D. D. and Covert, H. H. 1991. Biomechanical association of dental and temporomandibular pathology in a medieval Nubian population. *American Journal of Physical Anthropology* 85: 201–205.

Shima, G. and Suzuki, M. 1967. Problems of race formation of the Maori and Moriori in terms of skulls. *Osaka City Medical Journal* 13: 9–54.

Short, R. V. 1976. The evolution of human reproduction. *Proceedings of the Royal Society of London* 195: 3–24.

Simmons, D. 1975. *The Great New Zealand Myth*. Wellington: Reed.

Simmons, R. T. 1962. Blood group genes in Polynesians and comparisons with other Pacific peoples. *Oceania* 32: 198–210.

Simpson, A. I. F. 1979. Assessment of Health of the Prehistoric Maori. Unpublished B.Med.Sc. thesis, University of Otago.

—— 1981. Regional occurrence of the fern-root plane in prehistoric New Zealanders. *New Zealand Journal of Archaeology* 3: 83–97.

Siple, P. A. 1968. Clothing and climate. In *Physiology of Heat Regulation*, ed. L. H. Newburgh, pp. 389–442. New York: Hafner.

Siple, P. A. and Passel, C. F. 1945. Measurements of dry atmospheric cooling in subfreezing temperatures. *Proceedings of the American Philosophical Society* 89: 177–199.

Sjovold, T. 1984. A report on the heritability of some cranial measurements and non-metric traits, in *Multivariate Statistical Methods in Physical Anthropology*, ed. G. N. Van Vark and W. W. Howells, pp. 223–24, Dordrecht: Reidel.

Skrobak-Kaczynski, I. and Lange-Anderson, K. 1974. Age dependent osteoporosis among men habituated to a high level of physical activity. *Acta Morphologica Neerlando-Scandinavica* 12: 283–92.

Smillie, A. C., Rodda, J. C. and Kawasaki K. 1986. Some aspects of hereditary defects of dental enamel including some observations on pigmented Polynesian enamel. *New Zealand Dental Journal* 82: 122–125.

Smith, P. 1977a. Selective pressure and dental evolution in hominids. *American Journal of Physical Anthropology* 47: 453–458.

—— 1977b. Regional variation in tooth size and pathology in fossil hominids. *American Journal of Physical Anthropology* 47: 459–466.

Snow, C. E. 1974. *Early Hawaiians*. Lexington: University Press of Kentucky.

Sofaer, J. A. 1970. Dental morphologic variation and the Hardy-Weinberg Law. *Journal of Dental Research* 49: 1505–1508.

Sofaer, J. A., Niswander, J. D. and MacLean, C. J. 1972. Population studies on South-West Indian tribes. V Tooth morphology as an indication of biological distance. *American Journal of Physical Anthropology* 37: 359–366.

Spate, O. H. K. 1979. *The Spanish Lake*. Canberra: Australian National University Press.

—— 1988. *Paradise Lost and Found. The Pacific Since Magellan*. London: Routledge.

Specht, J. 1968. Preliminary report on excavations on Watom Island. *Journal of the Polynesian Society* 77: 117–134.

Spriggs, M. and Anderson, A. 1993. Late colonization of East Polynesia. *Antiquity* 67: 200–217.

Spring, D. B., Lovejoy, C. O., Bender, G. N. and Duerr, M. 1989. The radiographic pre-auricular groove: its non-relationship to past parity. *American Journal of Physical Anthropology* 79: 247–252.

Stanhope, J. M. and Prior, I. A. M. 1975. Uric acid, joint morbidity and streptococcal antibodies in Maori and European teenagers. *Annals of Rheumatic Diseases* 34: 359–363.

Stanhope, J. M., Aitchison, W. R., Swindells, J. C. and Frankish, J. D. 1978. Ear disease in rural New Zealand school children. *New Zealand Medical Journal* 88: 5–8.

Stanhope, J. M., Prior, I. A. M. and Rees, R. O. 1976. Some health indicators in New Zealand adolescents: the Rotorua Lakes study 5. *New Zealand Medical Journal* 83: 271–272.

Staron, R. S., Hikida, R. S., Hagerman, F. C., Dudley, G. A. and Murray, T. F. 1984. Human muscle fibre type adapability to various workloads. *Histochemistry and Cytochemistry* 32: 146–152.

Stefansson, V. 1946. *Not by Bread Alone*. New York: Macmillan.

Steggerda, M. 1950. The anthropometry of South American Indians. In *Handbook of South American Indians*, ed. J. H. Steward, pp. 57–69. Smithsonian Institution Bureau of American Ethnology, Bulletin 143.

Steinbock, R. T. 1976. *Paleopathological Diagnosis and Interpretation*. Springfield: Thomas.

Steudel K. 1981. Body size estimators in primate skeletal material. *International Journal of Primatology* 2: 81–90.

Stevens, G. 1980. *New Zealand Adrift*. Wellington: Reid.

Stevenson, P. H. 1929. On racial differences in stature long bone regression formulae, with special reference to stature reconstruction formulae on the Chinese. *Biometrika* 21: 303–321.

Stewart, T. D. and Spoehr, A. 1952. Evidence on the paleopathology of yaws. *Bulletin of Historical Medicine* 26: 538–553.

Stini, W. A. 1969. Nutritional stress and growth: sex differences in adaptive response. *American Journal of Physical Anthropology* 31: 417–426

—— 1982. Sexual dimorphism and nutrient reserves. In *Sexual Dimorphism in Homo sapiens*, ed. R. M. Hall, pp. 391–419. New York: Praegar.

Stokes, J. P. and Fitzharris, S. J. 1991. *Relationship between Tooth Wear and Temporomandibular Joint Wear in Prehistoric New Zealand Skulls*. Dunedin: School of Dentistry.

Stoneking, M. 1993. DNA and recent human evolution. *Evolutionary Anthropology* 1: 60–73.

Stoneking, M. and Wilson, A. C. 1989. Mitochondrial DNA. In *The Colonization of the Pacific. A Genetic Trail*, ed. A. V. S. Hill and S. W. Serjeantson, pp. 215–245. Oxford: Clarendon Press.

Stott, S. and Gray, D. H. 1980. The incidence of femoral neck fractures in New Zealand. *New Zealand Medical Journal* 91: 6–9.

Strickland, S. A. and Ulijazek, S. J. 1990. Energetic cost of standard activities in Gurkha and British soldiers. *Annals of Human Biology* 17: 133–144.

Stuart-Macadam, P. 1985. Porotic hyperostosis: representative of a childhood condition. *American Journal of Physical Anthropology* 66: 391–398.

—— 1987a. A radiographic study of porotic hyperostosis. *American Journal of Physical Anthropology* 74: 511–520.

—— 1987b. Porotic hyperostosis: new evidence to support the anemia theory. *American Journal of Physical Anthropology* 74: 521–526.

Suchey, J. M., Wisely, D. V., Green, R. F. and Noguchi, T. T. 1979. Analysis of dorsal pitting in the os pubis in an extensive sample of modern American females. *American Journal of Physical Anthropology* 751: 517–540.

Sullivan, L. R. 1921. A contribution to Samoan somatology. *Memoirs of the Bernice P. Bishop Museum* 8 (2).

—— 1922. A contribution to Tongan somatology. *Memoirs of the Bernice P. Bishop Museum* 8 (4).

—— 1923. Marquesan somatology, with comparative notes on Samoa and Tonga. *Memoirs of the Bernice P. Bishop Museum* 3 (2).

Sundberg, J., 1977. The acoustics of the singing voice. *Scientific American* March: 82–91.

Surville, J. F. M. de. 1982. In *Early eyewitness acounts of Maori life 1. Extracts from Journals Relating to the Visit to New Zealand of the French ship St Jean Baptiste in December 1769 under the command of J. F. M. de Surville*, transcription and translation by Isabel Ollivier. Wellington: Alexander Turnbull Library Endowment Trust with Indosuez New Zealand.

Suzuki, T. 1985. Palaeopathological diagnosis of bone tuberculosis in the lumbosacral region – comparative study of an archaeological case with a modern autopsy case. *Journal of the Anthropological Society of Nippon* 993: 381–390.

—— 1987a. Paleopathological study on a case of osteosarcoma. *American Journal of Physical Anthropology* 74: 309–318.

—— 1987b. Cribra orbitalia in the early Hawaiians and Mariana Islanders. *Man and Culture in Oceania* 3: 95–104.

Swindler, D. R. 1962. *A Racial Study of the West Nakanai*. Philadephia: University Museum

Szathmary, E. J. E. 1993. Genetics of aboriginal North Americans. *Evolutionary Anthropology* 1: 202–220.

Tagaya, A. 1987. Interpopulation variation of sex differences: an analysis of the extremity long bone measurements of Japanese. *Journal of the Anthropological Society of Nippon* 95: 45–76.

Tanner, J. M. 1962. *Growth at Adolescence*. Oxford: Blackwell.

—— 1973. Trend towards earlier menarche in London, Oslo, Copenhagen, The Netherlands, and Hungary. *Nature* 243: 95–96.

Tasman, A. J. 1968. *The Voyages of Abel Janszoon Tasman*. Oxford: Clarendon Press.

Tayles, N. 1992. The People of Khok Phanom Di. Unpublished doctoral thesis, University of Otago.

Taylor, R. 1855. *Te Ika a Maui. New Zealand and its Inhabitants*. London: Wertheim and MacIntosh.

Taylor, R. 1989. Aetiology of non-insulin dependent diabetes. *British Medical Bulletin* 45: 73–91.

Taylor, R. M. S. 1963. Cause and effect of wear of teeth. *Acta Anatomica* 53: 97–157.

—— 1971. Dental report on archaeological material from Tonga. *Australian Dental Journal* 16: 175–181.

Templeton, A. R. and Rothman E. D. 1974. Evolution in heterogeneous environments. *American Naturalist* 108: 409–428.

Terrell, J. 1986. *Prehistory in the Pacific Islands*. Cambridge: Cambridge University Press.

Thomas, W. L. 1963. The variety of physical environments among Pacific islands. In *Man's Place in the Island Ecosystem*, ed. F. R. Fosberg, pp. 7–38. Honolulu: Bishop Museum Press.

Thompson, D. W. 1942. *On Growth and Form*. Cambridge: Cambridge University Press.

Thompson, L. M. 1932. Archaeology of the Marianas Islands. *Bernice P. Bishop Museum Bulletin* 100.

—— 1940. Southern Lau, Fiji: an Ethnography. *Bernice P. Bishop Museum Bulletin* 162.

Thomson A. S. 1854. Contribution to the natural history of the New Zealand race of men; being observations on their stature, weight, size of chest, and physical strength. *Journal of the Statistical Society* 17: 27–33.

—— 1859. *The Story of New Zealand: Past and Present, Savage and Civilized.* London: Murray.

Thomson, E. Y. 1915–17. A study of the crania of the Moriori, or Aborigines of the Chatham Islands. *Biometrika* 11: 82–135.

Toner, M. M. and McArdle, W. D. 1988. Physiological adjustments of Man to the cold. In *Human Performance Physiology and Environmental Medicine at Terrestrial Extremes,* ed. K. B. Pandolf, M. N. Sawka and R. R. Gonzalez, pp. 361–399. Indianapolis: Benchmark Press.

Tonkin, S. L. 1960. Anaemia in Maori infants. *New Zealand Medical Journal* 59: 329–333.

—— 1970. Height, weight and haemoglobin study of adolescent Maoris and Europeans. *New Zealand Medical Journal* 72: 323–327.

—— 1974. Polynesian child health: effects on education. In *Polynesian and Pakeha in New Zealand education. II: Ethnic differences and the school,* ed. D. Bray and C. Hill, pp. 16–32. Auckland: Heinemann.

—— 1975. Preliminary report on Tokelau children under five years of age in New Zealand examined in 1972. In *Migration and Related Social Health Problems in New Zealand and the Pacific,* ed. J. M. Stanhope and J. S. Dodge. Wellington:

Torroni, A., Schurr, T. G., Yang, G., Szathmary, E. J. E., Williams, R. C., Schanfield, M. S., Troup, G. A., Knowler, W. C., Lawrence, D. N., Weiss, K. M. and Wallace, D. C. 1992. Native American mitochondrial DNA analysis indicates that the Amerind and NaDene populations were founded by two independent migrations. *Genetics* 130: 153–162.

Townsend, G. C. and Brown, T. 1978. Heritability of permanent tooth size. *American Journal of Physical Anthropology* 49: 497–504.

Trent, R. J., Mickleson, K. N. P., Wilkinson, T., Yakas, J., Dixon, M. W., Hill, P. J. and Kronenberg, H. 1986. Globin genes in Polynesians have many rearrangements including a recently described γγγγ*/*. *American Journal of Human Genetics.* 39: 350–360.

Trent, R. J., Mickleson, K. N. P., Yakas, J. and Hertzberg, M. 1988. Population genetics of the globin genes in Polynesians. *Hemoglobin* 12: 533–7.

Trinkaus, E. 1975. Squatting among the Neandertals: a problem in the behavioural interpretation of skeletal morphology. *Journal of Archaeological Science* 2: 327–351.

—— 1981. Neandertal limb proportions and cold adaptation. In *Aspects of Human Evolution,* ed. C. B. Stringer, pp. 187–224. London: Taylor and Francis.

Trotter, M, A. 1954. A preliminary study of estimation of weight of the skeleton. *American Journal of Physical Anthropology* 12: 537–552.

Trotter, M. A. and Gleser, G. C. 1952. Estimation of stature from long bones of American whites and negroes. *American Journal of Physical Anthropology* 10: 463–514.

Turner, C. G. 1985. The dental search for Native American origins. In *Out of Asia*, ed. R. Kirk and E. Szathmary, pp. 31–78. Canberra: Journal of Pacific History.

—— 1989a. Dentition of Watom Island, Bismarck Archipelago, Melanesia. *Records of the Australian Museum* 41: 293: 296.

—— 1989b. Teeth and prehistory in Asia. *Scientific American* 260: 70–77.

—— 1990. Major features in Sundadonty and Sinodonty, including suggestions about East Asian microevolution, population history, and late Pleistocene relationships with Australian Aborigines. *American Journal of Physical Anthropology* 82: 295–317.

Turner, C. G. and Swindler, D. R. 1978. The dentition of the New Britain West Nakanai Melanesians. *American Journal of Physical Anthropology* 49: 361–371.

Ullrich, H. 1975. Estimation of fertility by means of pregnancy and childbirth alterations at the pubis, the ilium, and the sacrum. *OSSA* 2: 23–29.

Underwood, J. 1969. Human skeletal remains from Sand Dune Site (H1), South Point (Ka Lae), Hawaii: *Pacific Anthropological Records* Number 9.

Van de Water, N. S., Ridgeway, D. and Ockelford, P. A. 1991. Restriction fragment length polymorphisms associated with the factor VIII and factor IX genes in Polynesians. *Journal of Medical Genetics* 28: 171–176.

Van Dijk, N. 1993. The Evolution of the Polynesian Phenotype: an analysis of skeletal remains from Tangatapu, Tonga. Unpublished M.A. thesis, University of Auckland.

Van Gerven, D., and Armelagos, G. 1983. 'Farewell to paleodemography'? Rumours of its death have been greatly exaggerated. *Journal of Human Evolution* 12: 353–360.

Van Vark, G. N. and Howells, W. W. eds. 1984. *Multivariate Statistics in Physical Anthropology*. Dordrecht: Reidel

Van Wieringen, J. C. 1972. *Secular Changes of Growth, 1964–1966. Height and Weight Surveys in the Netherlands in Historical Perspective*. Leiden: University of Leiden.

—— 1978. Secular growth changes. In *Human Growth*, ed. F. Falkner and J. M. Tanner, Vol 2: 445–473. New York: Plenum.

Vanggaard, L. 1975. Physiological reactions to wet-cold. *Aviation Space and Environmental Medicine* 46: 33–36.

Visser, E. P. 1995. The People of Sigatoka. Unpublished Doctoral thesis, University of Otago.

Von Bonin, G. 1931. Craniology of the Easter Islanders. *Biometrika* 23: 249–270

—— 1936. On the craniology of Oceania: crania from New Britain. *Biometrika* 28: 123–148

Wade, A. J., Marbut, M. M. and Round, J. M. 1990. Muscle fibre type and the aetiology of obesity. *Lancet* 1: 805–808.

Wagner, K. 1937. *The Craniology of the Oceanic Races*. Oslo: I Kommisjon hos Jacob Dybwad.

Wainscoat, J. S., Hill, A. V. S., Boyce, A. L., Flint, J., Hernandez, M., Thein, S. L. and Old, J. M., Lynch, J. R., Falusi, A. G., Weatherall, D. J. and Clegg, J. B. 1986. Evolutionary relationships of human populations from an analysis of nuclear DNA polymorphisms. *Nature* 319: 491–493.

Walensky, N. A. 1965. A study of anterior femoral curvature in Man. *Anatomical Record* 151: 559–570.

Wales, W. 1961. Journal of William Wales. In *The Journals of Captain James Cook on his Voyages of Discovery. The Voyage of the Resolution and Adventure 1772–1775*, ed. J. C. Beaglehole, pp. 776–869. Cambridge: Hakluyt Society.

Wallace, D. C. and Torroni, A. 1992. American Indian prehistory as written in the mitochondrial DNA: a review. *Human Biology* 64: 403–416.

Wallis, H. 1964. The Patagonian giants, In *Byron's Journal of His Circumnavigation*, ed. R. E. Gallagher, pp. 185–196. Cambridge: Hakluyt Society.

Walters, T. J. and Constable, S. H. 1993. Intermittent cold exposure causes a muscle-specific shift in the fiber type composition in rats. *Journal of Applied Physiology* 75: 264–267.

Wang, Y., Weng, J. and Hu, B. 1979. Estimation of stature of long bones from Chinese male adults in Southwest District. *Acta Anatomica Sinica* 10: 16.

Waterlow, J. C. 1986. Metabolic adaptation to low intakes of energy and protein. *Annual Review of Nutrition* 6: 495–526.

Waters, A. P., Higgins, D. G. and McCuthan, T. F. 1991. *Plasmodium falciparum* appears to have arisen as a result of lateral transfer between avian and human hosts. *Proceedings of the National Academy of Science* 88: 3140–3144.

Weidenreich, F. 1936. The mandibles of *Sinanthropus pekinensis*. A comparative study. *Palaeontologica Sinica*, Series D, 7: 1–132.

Weiner, J. S. 1954. Noseshape and climate. *American Journal of Physical Anthropology* 12: 615–618.

Weiss, K. M. 1973. Demographic models for anthropology. *American Antiquity* 38: 1–88.

Wells, L. H. 1969. Stature in earlier races of mankind. In *Science and Archaeology*, ed. D. Brothwell and E. Higgs, pp. 453–467. London: Thames and Hudson.

Wessen, A. F. (ed.) 1992. *Migration and Health in a Small Society. The Case of Tokelau*. Oxford: Clarendon Press.

White, P., Allen, J., and Specht, J. 1988. Peopling the Pacific: the Lapita homeland project. *Australian Natural History* 22: 410–416.

Williams, H. U. 1935. Pathology of yaws. *Archives of Pathology* 20: 596–630.

Williams, H. W. 1972. *A Dictionary of the Maori Language*. 7th ed. Wellington: Government Printer.

Williams, T. 1858. *Fiji and the Fijians*. London: Alexander Heylin.

Williams-Blangero, S. and Blangero, J. 1989. Anthropometric variation and the genetic structure of the Jirels of Nepal. *Human Biology* 61: 1–12.

Williamson, I. and Sabath, M. D. 1984. Small population instability and island settlement patterns. *Human Ecology* 12: 21–34.

Wilmore, J. H. and Behnke, A. R. 1968. Predictability of lean body weight through anthropometric assessment in college men. *Journal of Applied Physiology* 25: 349–355.

—— 1969. An anthropometric estimation of body density and lean body weight in young men. *Journal of Applied Physiology* 27: 15–31.

Wilson, E. A. 1950. The basal metabolic rates of South American Indians. In *Handbook of South American Indians*, ed. J. H. Steward, pp. 97–104. Washington: Smithsonian Institution Bureau of American Ethnology, Bulletin 143.

Wilson, P. W. and Mathis, M. S. 1930. Epidemiology and pathology of yaws, based on study of 1423 consecutive cases. *Journal of the American Medical Association* 94: 1289–1292.

Wilson, S. R., 1984. Towards an understanding of data in physical anthropology. In *Multivariate Statistical Methods in Physical Anthropology*, ed. G. N. Van Vark and W. W. Howells, pp. 261–282. Dordrecht: Reidel.

Wiriyaromp, W. 1984. The Human Skeletal Remains from Ba Na Di. Unpublished M.A. thesis, University of Otago.

Wolpoff, M. H. 1989. Multiregional evolution: the fossil alternative to Eden. In *The Human Revolution*, ed. P. Mellars and C. Stringer, pp. 62–108. Edinburgh: University of Edinburgh Press.

Wolpoff, M. H., X. Z. Wu and A. G. Thorne, 1984. Modern *Homo sapiens* origins: a general theory of hominid evolution involving the fossil evidence from East Asia. In *The Origins of Modern Humans*, F. H. Smith and F. Spencer, pp. 411–483. New York: Alan R. Liss.

Wood, C. S. and Gans, L. P. 1984. Some hematological findings in children of Western Samoa. *Journal of Tropical Pediatrics* 30: 104–110.

Woodfield, D. G., Simpson, L. A., Seber, G. A. F. and McInerney, P. J. 1987. Blood groups and other genetic markers in New Zealand Europeans and Maoris. *Annals of Human Biology* 14: 29–37.

Wright, S. 1951. The genetical structure of populations. *Annals of Eugenics* 15: 323–354.

Wrischnik, L. A., Higuchi, R. G., Stoneking, M., Erlich, H. A., Arnheim, N. and Wilson, A. C. 1987. Length mutations in human mitochondrial DNA: direct sequencing of enzymatically amplified DNA. *Nucleic Acids Research* 15: 529–542.

Wurm, S. A. 1983. Linguistic prehistory in the New Guinea area. *Journal of Human Evolution* 12: 37–60.

Young C. M., Martin E. K., Tensuan, R. and Blondin, J. 1962. Predicting specific gravity and body fatness in young women. *Journal of the American Dietetic Association* 40: 102–107.

Yuen, S. 1984. Height and weight standards for school children in Tahiti. *Journal of Tropical Pediatrics* 30: 122–126.

Zaino, D. E. and Zaino, E. C. 1975. Cribra orbitalia in the Aborigines of Hawaii and Australia. *American Journal of Physical Anthropology* 42: 91–4.

INDEX